BUILDING ROBOTS WITH JAVA BRAINS

Brian Bagnall

VARIANT PRESS

Designed by Hayden Sundmark
Original artwork by Calvin Innes
Edited by Sylvia Philipps
Cover by Hayden Sundmark
Printed in Canada

Library and Archives Canada Cataloguing in Publication

Bagnall, Brian, 1972-,
author

Maximum Lego Mindstorms EV3 : building robots with Java brains /
Brian Bagnall.

Includes index.

ISBN 978-0-9868322-9-1 (pbk.)

1. Lego Mindstorms toys. 2. Robots--Programming.

3. Robots--Design and construction. 4. Java (Computer program language). I. Title.

TJ211.15.B33 2014 629.8'92 C2014-905280-4

VARIANT PRESS
3404 Parkin Avenue
Winnipeg, Manitoba
R3R 2G1

Table of Contents

Acknowledgements

Thanks to Lawrie Griffiths, Professor Roger Glassey, Andy Shaw, Aswin Bouwmeester and Sven Köhler for their incredible work on leJOS. Thanks to Kevin Clague and Philippe "Philo" Hurbain for their assistance with the rendered instructions. And a big thanks to LEGO for coming through with another great product!

Brian Bagnall

Foreward

Inspiration is vital for any robot builder. As robotics is a medium that takes from all subjects, books such as Brian's are incredibly important. He has taken the wonderfully simple tools of LEGO® MINDSTORMS™ EV3™ and transformed them into another set of amazing robots as well as providing tools for the novice to the seasoned user.

Brian had been an early inspiration for me. His six wheeled, rocker bogie suspension rover was one of the first things I ever built with the NXT sets. The wonderful mechanism was a marvel; driving through obstacles was so fun; and inspiring young minds at the local Community Space Center was incredible. I wasn't ready to program with Java at the time, but even the NXT G programming made that beautiful machine move gracefully. Later, I would try to use this mechanism in other robots, and though none of those other ones became anything, the inspiration was there. Building Robots with Java Brains has a lasting place on my bookshelf and it never collects dust.

Having been part of the EV3 development team, it was no easy task to create models and content. Everything came from nothing, prototypes were just dust before they were 3D printed into white, powdery shells that we could put into models. Eventually they would have PCBs that stuck out of the sides, with barely working screens. Imagine trying to build a robot that could balance itself, when you could not even put all the parts together, and also without a full working program. Each week of development produced new challenges with something working and something else not, but slowly it came together. Seeing Gyroboy balance for the first time, seeing the first 3D videos, clicking through the first tutorials and finally getting those molded bricks and holding them in your hands were amazingly sweet moments.

It's with great pleasure to see the bricks become so much more than they are. To see the simple collection of bricks become fuel for the imagination of authors like Brian is its own reward. The tools in this book can be the continuation of a journey you will make in creating your own robots. Learn them, master them, and begin to build something absolutely amazing of your own design. Bruce Lee taught me to, "adapt what is useful, add what is essentially your own." And I say, make something beautiful, make it work. Build some amazing robots with Brian's hand guiding you, it will make your investment in the EV3 that much more worth it.

Lee Magpili
Designer LEGO Education
LEGO Mindstorms EV3 Core Set 45544

Preface

Here we are, over 15 years after LEGO released the groundbreaking Robotics Invention System. All this time later and LEGO still holds the crown of the world's most popular DIY robotics kit.

This is the third generation of Mindstorms books I've written and I'm doing things differently this time around. Rather than try to be one of the first LEGO Mindstorms books to market, I decided to wait until the Java programming tools had enough time to mature. Also, if you own the EV3 kit and a computer, you will be able to complete every single project in this book without any additional parts!

So why should someone use the EV3 kit to create robots instead of hacking them together from homemade parts? For starters, it's much easier to get everything you need in one refined package. But on a deeper level, a uniform kit with the same parts allows you to share your robot designs with others and use their designs, as you will be able to do with the robots in this book. Because you have the exact same kit as others, it's possible to create a set of instructions that guarantees you will have success replicating a robot that someone else has spent time building. Only you can build in half-hour what took the designer weeks or months to build.

The compliment of this is code sharing. Code sharing allows you to instantly have access to code someone else has developed (with the case of the leJOS API, this includes years of refinement). You don't have to reinvent the wheel every time you want to do something new. Take A* search. The leJOS API has a set of classes that help the robot search through an area for the shortest route. It can also take a long time to create an application to visualize data on the screen as the robot is moving. leJOS includes applications, such as Map Command, that run on your PC and let you see the robot moving through the environment.

It's amazing that a box of disparate parts can contain so much possibility, but it does. A computer is not a robot, although it is a key component of robots. Motors do not make something a robot either. Nor do sensors. And we know for sure that a box of plastic parts is not enough to make a robot. Even together, those parts still don't guarantee you have a robot. What it really requires is the ability to move around in the environment and interact with objects. This book will take a stab at all of the concepts you need to know to make your dream robot. Let's get started!

Brian Bagnall
August 1, 2014

CHAPTER 1

Third Evolution

TOPICS IN THIS CHAPTER

- A Brief History of MINDSTORMS
- Unboxing the EV3 Kit
- Using the LEGO software
- The Future of Mindstorms

The EV3 kit is the third iteration—or evolution as LEGO describes it—in the Mindstorms universe that began back in 1998. In fact, EV3 stands for EVolution 3. With this iteration, LEGO has literally blown apart any limitations there were with the previous Mindstorms kits.

You can add as much memory as you want via a microSD slot. You can plug USB devices into a dedicated host USB port. The brick has a full Linux operating system! And finally, there are now four motor ports.

In this chapter, we will review the background of the EV3 kit, its most prominent features, as well as making a few big decisions, such as what kind of batteries to get. There are also a few low-cost additions to the EV3 brick that are absolute essentials: a micro-SD card, a WiFi USB adapter, and maybe even a Bluetooth adapter. Before going out to buy those items, read this chapter to make sure you purchase the right ones.

A Brief History of MINDSTORMS

Before getting started, lets look back on where it all began. In 1987, the MIT Media Laboratory began developing a device they called the Programmable Brick, under sponsorship from the LEGO Group. The main designer was Fred G. Martin, and between 1987 and 1998 he turned out several different versions of the Programmable Brick. The final version is a big red brick with four output ports for motors (an ideal number) and six input ports for sensors (see Figure 1-1).

Figure 1-1 The MIT Programmable Brick

LEGO's Robotics Invention System soon followed. The heart of the RIS Kit was a yellow RCX brick that bears a resemblance to Spongebob Squarepants (see Figure 1-2). It was modest compared to today's technology, using an 8-bit processor at 16 MHz and a scant 32 kilobytes of memory—less memory than the Commodore 64 had in 1982. Even with these limitations, MINDSTORMS users built incredible machines.

Figure 1-2 The RCX brick.

More than a million people said, "This is what I have been waiting for" and purchased an RIS kit. The MINDSTORMS community became the largest robotics community on earth, easily able to share their robot designs and code with others. (Perhaps these users have helped fuel the current robotics boom that has led to Google acquiring over a dozen robotics companies in 2013.)

The big surprise for LEGO was the hacker community. Hackers soon unlocked the RCX brick and began using alternate programming languages such as C and Java. Many attribute the success of the RIS kit to the availability of these programming languages.

Between 1998 and 2006 there was a long wait for a true sequel to the RIS kit. Unknown to MINDSTORMS users, LEGO began work on a sequel in early 2004. The project was led by a brilliant Dane, Søren Lund, who eventually created a robot brick that used standards such as Bluetooth and I^2C (more on these later). When development began, Lund hung a sign in the development lab: "We will do for robotics what iPod did for music".

On August 2, 2006 LEGO rolled out the NXT 1.0 kit, followed by an upgraded NXT 2.0 kit in July 2009. The NXT Intelligent brick contained an Atmel® 32-bit ARM processor running at 48 MHz. This processor has direct access to 64 KB of RAM, and 256 KB of flash memory.

Fast forward another seven years from the release of the LEGO NXT, and LEGO surprised the world with the EV3 kit on September 1, 2013. Although there are superficial similarities with the NXT, the EV3 kit is deceptively different. In fact, there are probably more substantial changes between NXT and EV3 as there were between RCX and NXT.

Unboxing the EV3 Kit

So much in the computing landscape has changed since LEGO released the NXT in 2006. Tablets and smartphones are now everywhere. Battery technology has improved dramatically. Broadband wireless Internet connectivity is ubiquitous. With the rise of digital downloads, it's a rarity to need a DVD anymore to install software.

Indeed, LEGO has assumed that all of their customers are now online. The EV3 kit no longer comes with a DVD ROM. Users are required to download the software online—all 591 megabytes. This move saves LEGO on production costs, which they can theoretically use to improve other parts of the EV3.

Let's examine what you will find inside the kit.

EV3 Intelligent Brick

Superficially, the EV3 brick looks very similar to the NXT brick. Gray and white are still the main colors, but this time there is no signature orange button. Instead, the only color is provided by the prominent red LEGO logo. The LEGO brand has certainly increased in prominence since 2006, and this is perhaps reflected in the boldness of the logo, which was previously quite subdued on the NXT brick.

Physically the brick contains the same number of LEGO pin connector holes as the NXT brick, but the side hole layout is reversed compared to last time (see Figure 1-3). Overall the dimensions are identical to the NXT (7.2 cm by 11.2 cm), except for where the LCD screen itself protrudes slightly out of the rectangular box. Without batteries, the EV3 weighs in at 178 grams (18 grams heavier than the NXT).

The EV3 brick is built very durable, but out of all the parts in the kit this is the one you should treat gently due to the complex parts inside and the overall cost of the brick. The buttons are made of hard plastic this time, rather than the rubbery feel of the NXT brick. There are six buttons on the face of the brick—two more than the NXT had. Additionally, there is a signature glow around the buttons that changes color depending on what the brick is doing. I find this effect similar to the Xbox 360's ring of LED's, which magnificently adds character to the device.

The new brick also contains several differences with the USB ports. The NXT brick contained one USB standard-B port, which is the typical port used for USB printers. This time the EV3 has two USB ports. The standard-B has been replaced with a micro-B port, which is more appropriate for a device of this size. This is a slave port, meaning the EV3 brick can hook up to your PC with this port. It also includes a full sized standard-A USB host port! This new addition allows USB devices to connect to the EV3 brick. The most common device that will be used here is likely to be a WiFi adapter.

Figure 1-3: The EV3 Intelligent Brick.

Memory

The previous two generations of Mindstorms were measured in mere kilobytes. As consumer devices with terabytes of data are now freely available, LEGO has joined the age of gigabytes. However, this amount of memory does not come right out of the box. In fact, the EV3 "only" has 80 MB of memory (64 MB of RAM and 16 MB of flash). This is already a giant step above the 320 KB of memory in the NXT.

Where the EV3 really launches into memory orbit if due to the inclusion of a micro-SD card slot. SD cards containing gigabytes of memory are available at low prices—

often under $5 for 4 GB of memory, which will be more than enough for all your EV3 robot needs. If you want, you can insert up to 32 GB of memory into the slot.

This new injection of memory will allow projects and system software that was previously impractical in Mindstorms robotics. For example, it now has enough memory to store large, detailed maps of its environment. Advanced artificial intelligence programs of any size will now comfortably fit in the EV3 memory, such as chess playing games or other complex and data heavy AI routines. And finally, as we will examine in the next section, the EV3 sports a much better operating system (OS) than previous generations.

The LEGO NXT brick contained an ARM7 processor running at 48 MHz. This time, the EV3 brick contains an ARM9 processor running at 300 MHz! Specifically, it is a Texas Instruments Sitara AM1808 processor with an ARM9 core. While 300 MHz may not sound very fast compared to phones and tablets operating above 1 GHz, it is still much faster than the previous mindstorms generation.

Operating System

Previous versions of Mindstorms bricks did not really have what anyone would define as an operating system. Instead, it was called firmware and was a basic system for presenting a menu and launching applications. The EV3 however contains a full blown Linux OS. For those who have been living on Mars for the past 20 years, Linux is an open

BIG DECISIONS

Choosing a Micro SD Card

Your first big decision is what kind of micro SD card to use in your EV3. If you already have a micro-SD card, great! Your decision is made. If you need to buy one, you have some considerations. Size can range anywhere from 2 GB up to 32 GB—and there's really no point in going over 2 GB. The total leJOS install is much less than 100 MB, leaving the rest for your Java code and data. I can't imagine anyone requiring over a gigabyte for robotics projects. There is also a speed rating for cards. Class 2 is the slowest, at 2 MB/s, whereas class 10 is the fastest at 10 MB/s. You can find cards on eBay, Amazon, Kijiji, or your local electronics store. I recommend buying brand name cards only and avoiding no-name SD cards available on eBay from China. Don't forget to get an standard SD card adapter so you can plug it into your PC.

NOTE: *SDXC cards are not compatible with leJOS.*

source operating system (an "OS OS") that is every bit as powerful as Microsoft Windows and Apple's OS X. However, it is in third place in terms of installations on computers, largely due to a real or perceived lack of user friendliness.

Both the official LEGO software and leJOS run under a version of Linux compiled specifically for the Sitara AM1808 processor. This version of Linux is stripped down compared to what you might find on a desktop version of Linux, but was we will discover later in the book it is still very complete.

The LEGO software takes only 30 seconds to boot up. This is slower than it took for the NXT to boot up. It also takes a long time to turn off because, now that we are using a full blown operating system, it needs to close programs and system processes properly.

Batteries

One of the key early decisions you will make with your new EV3 kit is what kind of batteries to use. Unless you already have a good set of rechargeable batteries on hand, this is an important decision for a number of reasons. First, it is one of the most costly items you can buy for your EV3. And second, it can affect how long your EV3 runs for and how powerful the motors will turn.

Fun With Linux!

If you want to play around with Linux, you can skip ahead to chapter 28. This chapter will show you how to log into Linux from your PC and enter some commands.

The EV3 brick uses six 1.5V AA batteries to provide a total of 9 volts. However, if you use your EV3 a lot, disposable AA batteries will end up costing you a fortune. There are two off-the-shelf options for rechargeable batteries.

WARNING: *Avoid purchasing cheap generic rechargeable batteries from eBay or other online sources! The prices might look great, but you will find they don't hold much of a charge and will not last long. In practice, you will be lucky if they can keep your EV3 robot running for longer than 15 minutes.*

The first option is the rechargeable lithium ion battery from LEGO (see Figure 1-4). This battery provides at least 7.4 volts (closer to 8.2 volts after recharging). It fits into the regular battery case, but it also increases the depth of the EV3 brick slightly.

LEGO also sells an AC adapter for charging the lithium ion battery which, conveniently, can be done while it is still inside the EV3 brick. If you are a hardware hacker, you could conceivably devise a robot that drives up to a recharging station when the batteries are low!

Lithium batteries provide power even while they are being charged, meaning your robot can also feed directly from household current. People who want to create robots that operate 24 hours a day, seven days a week (such as an Internet controlled robot) will find the lithium battery an even more useful accessory.

Figure 1-4 LEGO Lithium Ion battery.

Another option is to use six rechargeable AA batteries. There are at least 20 different combinations of chemicals that can be used in rechargeable batteries, but the two most popular are: Ni-MH (Nickel Metal Hydride) and Ni-Cd (Nickel Cadmium). Both work well, but the Ni-MH battery supplies 1.2 volts while the Ni-Cd battery supplies 1.25 volts. They don't store as much charge as lithium, however (see Table 1-1).

WARNING: *Rechargeable batteries provide 7.2 to 7.5 volts to the NXT, which means your motors will not operate as fast or powerfully as they would with 9 volts. Other than that, rechargeable batteries work well.*

	Recharge?	mAh	Volts	Duration
Alkaline	No	1700-3000	9.0	Longest
Ni-Cd	Yes	600-1000	7.5	Lowest
Ni-MH	Yes	1300-2900	7.2	40% more than Ni-Cd
Lithium	Yes	2050	7.4	2 x Ni-Cd

Table 1-1: Comparing battery options

The term milliampere hour, or mAh, is used to describe the total charge in a battery. Basically it tells you how much fuel is in the tank. A higher value means the batteries can supply power for longer. However, batteries have a discharge rate that occurs even while the battery is sitting on the shelf. This discharge rate is quite high for most rechargeables (around 30% per month). In other words, regular rechargeables have a small hole in the tank. If only there was a battery that plugged that leak!

There is a special type of battery released in November 2005 that is called low self-discharge NiMH. These batteries can be charged up and retain at least 80% of the charge one year later. And, they can be recharged 1800 times without wearing out. In other words, they will last decades.

These special batteries are marketed under the name Eneloop by Panasonic (previously by Sanyo). A typical 2nd or 3rd generation Eneloop battery will provide 2000 mAh. If you want more, there are Eneloop XX AA batteries which contain 2500 mAh, but they are only good for about 500 recharges, and they cost about 50% more.

NOTE: *As of this writing, 4th generation Panasonic Eneloop batteries are on the market (model number BK-3MCC). These batteries allow 2100 repeat charges over lifetime, but in my opinion they are too costly at present for the marginal improvement in performance.*

Just for Fun!

Did you know you can charge your batteries using solar power only? There are AA battery chargers available online (eBay, Amazon, Aliexpress.com) that cost around $10. These charges take around 20 hours to charge 4 AA batteries. Be sure to get the 1W version, not the 0.5W version which takes twice as long to charge.

Eneloops also suffer no memory effect, meaning you can partially drain the battery and recharge it at any time with no ill effects to your batteries. You can also partly charge them and start using them anytime if you are in a hurry.

Speakers

The EV3 contains a sound amplifier chip that can play sampled sound. You can even make your own recordings and upload them to the EV3 brick—and here is where the EV3 has an advantage over the NXT. With the NXT, sound samples were generally poor quality, due to memory limitations. The low sample

BIG DECISIONS

Choosing Batteries

Your best battery options are probably the Eneloops or the Lithium Ion battery from LEGO. Both have advantages and disadvantages when compared to each other. Let's look at the similarities first!

Both battery options have a high number of estimated recharge cycles in them before wearing out (500 for Li-Ion and 1800 for Eneloops). Both store a lot of charge, both have a low discharge rate and neither suffers from the memory effect.

There's a bit of an allure associated with getting the official LEGO battery pack for the EV3. From an aesthetic design angle, this is the only battery out there specifically made for the EV3. But there are other, more practical reasons too.

The LEGO EV3 battery is significantly lighter than six eneloops. The Li-Ion battery weighs in at 108g, compared to 168g total using 6 eneloops (including the 13g plastic battery cover). This means your robots will move a little faster and use a little less energy by using the official LEGO battery.

There are other advantages too. When it comes to recharging batteries, you might need to disassemble your robot in order to access the battery cover to pull out the six AA batteries. In contrast, the LEGO battery can remain in the EV3 while it is recharging due to access to the charge port.

Finally, in a classroom setting, you are less likely to lose the EV3 battery since it is larger and does not have to be removed for recharging.

Now let's look at advantages to using Eneloop batteries. The most obvious difference is price. It only costs around $45 for a pack of 12 AA's, 4 AAA's, a charger and carrying case. That's essentially enough for two sets of EV3 batteries! In contrast, it costs $64.99 for a LEGO Li-Ion battery plus $29.99 for the charger pack.

Physically, the LEGO Li-Ion battery pack is larger, sticking out from the bottom of the EV3 brick. The smaller footprint of AA batteries is a minor plus. You can also use AA batteries in multiple devices, not just the EV3 brick. And the lifespan of Eneloops is much longer than LEGO's battery (1800 cycles vs. 500). Plus, Li-Ion batteries age even when they are not in use. A typical Li-Ion battery lasts around three years (although my NXT battery lasted about double that and is still going—perhaps due to using and storing it in a cool, air conditioned environment), while Eneloops last decades under all conditions.

These are the main factors when evaluating batteries. What you choose might depend on a variety of factors (including economical and design factors). The final decision is up to you!

rate of sound files resulted in low fidelity sound. Now you have enough on-board memory to upload MP3 quality sound.

Do things now sound better? Absolutely! When you first boot up leJOS it plays a piano sound. The clarity of this sample is noticeably better than it was with the NXT.

Input and Output Ports

As previously mentioned, the EV3 has four sensor ports and four motor ports. There are three kinds of sensors that can connect to the sensor ports: analog, I^{2}C, and UART. Let's explore what these are.

Analog sensors are sensors that provide values as raw voltage. For example, the touch sensor provides a voltage when the button is down and no voltage when it is up. There is no complicated data bus to transmit data with analog sensors.

Digital sensors transmit data values. As such, they need a data bus to transmit data. Both UART and I^{2}C are digital data busses. The main difference between them is speed.

UART stands for Universal Asynchronous Receiver/Transmitter. It has been around since 1971 when Western Digital came up with the standard. The most important thing to know about UART right now is that it is capable of speeds up to 460.8 Kbit/s, which compares favorably to the 9.6Kbit/s available from I^{2}C.

All EV3 sensors (except touch) are UART based sensors. As such, they are incompatible with the NXT brick, which only supports I^{2}C.

Inter-Integrated Circuit or I^{2}C (pronounced I-squared-C) is a legacy protocol supported by the EV3 in order to remain compatible with NXT sensors. Philips invented the standard in the early 1980s and since then it has seen use in cell phones and other small devices. For example, monitors have used various standards through the years, including VGA, DVI and HDMI. All three of these standards have an I^{2}C bus to allow the monitor to tell the computer what type of monitor it is.

Although there are four physical ports, the ^{12}C ports are capable of using far more than four sensors at once. As long as the sensor uses Auto Detecting Parallel Architecture (ADPA), you can connect additional sensors using an expander.

The EV3 contains the same physical port (RJ12) as the NXT, which means all NXT motors and sensors can be used on the EV3. Conversely, EV3 motors can be used on the NXT, but not EV3 sensors.

Cables

EV3 cables are virtually identical to the NXT, other than length and perhaps a more waxy/rubbery feel. These cables contain six wires, giving them a rigid quality. The EV3 kit contains seven cables (one for each motor/sensor included in the kit) with three different lengths (see Table 1-2). If any of these prove too short, there are other lengths available online.

Short x4	Medium x2	Long x1
25cm (10 inches)	35cm (14.5 inches)	50cm (18.5 inches)

Table 1-2: Comparing cable lengths.

LEGO Mindstorms uses a connector known as RJ12, which looks much like a phone connector (see Figure 1-7). Since the connectors for sensors and motors are identical, you might think you can hook up and motor to a sensor port. This does not pan out, unfortunately, because the wire signals are different.

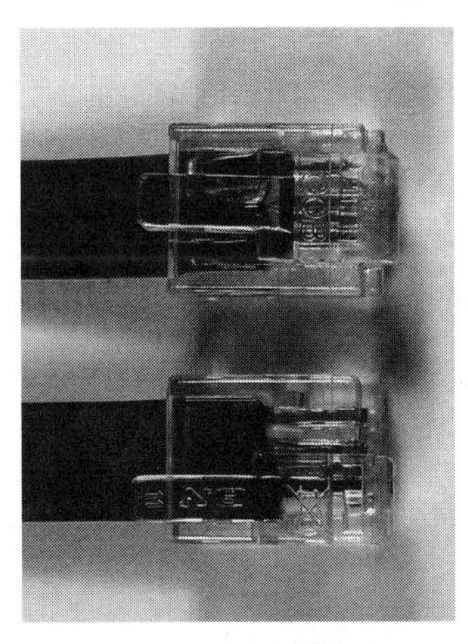

Figure 1-5 phone connector (top) compared to a RJ12 connector.

Fun Fact!

The old RCX connectors plug in just like LEGO bricks. They were novel, but if you rotated the connector 90 degrees the motor rotated in the opposite direction, which was often confusing. There is no chance of improperly plugging in a cable with RJ12 connectors.

LCD Display

Another area where the improvement doesn't look impressive at first glance is the LCD screen. The NXT has a resolution of 100 x 64 pixels over an area of 40.6 x 26.0 mm. Compare that to the EV3, which has a resolution of 178 x 128 over 43 x 28 mm. Ho-hum you might say, until you realize that's a jump from 6,400 pixels to a whopping 22,784 pixels on the EV3! There are almost four times the number of pixels on display!

Where the LCD display is not improved is with color. The display is black and white only, with no intermediate grey. It requires 17 ms to draw a new screen. It can refresh the display almost 60 times per second (60 Hz) and easily displays animations such as the introductory animation when the EV3 is first powered on.

Once again, LEGO's menu system is brilliant. You can access any number of functions using the five navigation buttons (see Figure 1-3). The functions contained in the menu system range from Bluetooth settings to playing sound files. LEGO has done a great job making a lot of content easily accessible.

Data Transfer

In order to program your robots, you need a method to transfer the code from your PC to the EV3 brick. The EV3 provides two standard methods to transfer code: USB and Bluetooth. There is also a third option that is the most convenient, but you need a WiFi adapter for the EV3 brick. All three are fast and reliable.

USB Slave Port

USB can be used to upload code and data to the EV3, and it is the only method for transmitting data that is available to you without purchasing an additional device (Bluetooth adapter or WiFi adapter). The USB port can transmit data at 12 Mbits per second. This solution is familiar to most computer users and easy to use. LEGO even supplies a nice long USB cable with micro-B USB end.

Chapter 2, installing leJOS and Eclipse, will go into more detail about using the USB slave port for uploading code to your EV3 brick.

Try it!

With the LEGO EV3, you don't even need a computer to program a robot. Instead you can use the menu system to write a very simple program with up to 16 commands. For this example we will create a program for the Everstorm that reverses and turns when you tap the touch sensor on his shoulder.

1.1 Connect motors to ports B and C.

1.2 Connect a touch sensor to port 1.

1.3 From the main menu hit right twice, then hit down three times to select Brick Program and hit the square enter button.

1.4 We'll insert the first command in between the play and loop icons by hitting the up button. Select the icon showing two motors (hit right once then enter).

1.5 Hit right once, then hit up to insert another command. This time select the touch icon with an hourglass (up twice, left twice).

1.6 Hit right once, then hit up to insert another command. Select the icon showing two motors (hit right once then enter). We will change this value. This time hit enter, then down twice to select the backup and turn icon. Hit enter.

1.7 Hit right once, then hit up to insert another command. Select the stopwatch icon (up four times, right once). Hit enter.

1.8 We'll change the program to loop forever until escape is pressed. Hit right twice then enter. Press up 6 times until the infinity icon is displayed. Hit enter.

1.9 Now you can run the program by hitting left until you see the play icon, then hit enter. Press the touch sensor to reverse direction, or press escape to end.

You can even save the program with a filename by selecting the disk icon. That's it!

USB Host Port and WiFi

The other big USB addition is a USB host port. We won't get into any details about using it here, but there is one important topic to discuss: buying a USB WiFi adapter. I recommend ordering one immediately. You won't regret it.

BIG DECISIONS

WiFi Adapter

There are only two WiFi adapters that are guaranteed to work with both the LEGO software and the leJOS software (as of this writing). The Netgear WNA1100 is the adapter officially sold by LEGO for the EV3 (see Figure 1-6). It is rather large, but for about a dollar you can purchase a 90 degree USB adapter to make the WiFi adapter face a different direction, which might help with some LEGO builds (see Figure 1-7).

WARNING: *There is a smaller Netgear WNA1100m, but it does not work! Do not purchase this one.*

If you want a smaller WiFi adapter, try the Edimax EW-7811UN (see Figure 1-6). This one only protrudes out of the side of the EV3 brick by about 6 mm, making it much easier to build around.

Both adapters mentioned above will cost you under $20 from sites like eBay. Sometimes you can find used ones under $5 if you get lucky.

Figure 1-6: Choosing a Netgear or Edimax adapter.

Figure 1-7: Redirecting a WiFi adapter.

Bluetooth

Once again we come to another feature that looks identical to the NXT if you look at a specs sheet, but is in fact radically improved for the EV3. First, a bit of background on Bluetooth.

Most people are somewhat familiar with Bluetooth headsets, which are used with mobile phones. This is merely one use of Bluetooth, and there are many more. But what is Bluetooth? Briefly, it is a wireless connection, like a USB port without cables. It is ideal for computer peripherals like keyboards and mice.

What isn't Bluetooth? Well, for starters it isn't the same as WiFi (otherwise known as IEEE 802.11), which is used for networks. You wouldn't want your wireless keyboard and mouse to be controlled through your network due to delay issues, which is where Bluetooth comes in.

Bluetooth (otherwise known as IEEE 802.15.1) operates on nearly the same frequency band as Wi-Fi (2.45 GHz vs. 2.4 GHz). However, you will probably not experience any interference problems even if you use both on the same computer, since they automatically change frequencies.

One of the biggest immediate improvements for this generation of Mindstorms is that the EV3 uses Bluetooth 4.0 rather than the Bluetooth 2.0 of the NXT. Bluetooth 4.0 transmits data at 24Mbit per second (2.86 MB/sec)—8 times faster than it was on the NXT.

The EV3 is also capable of creating it's own wireless network, known as a Personal Access Network (PAN). This makes communications with the EV3 much easier if you decide to use Bluetooth to upload code to your brick or control robots wirelessly. More on this later.

Bluetooth supports a broad range of devices. Each of these devices is supported by one or more profiles. For example, the HID profile (Human Interface Devices) is used to support keyboards, mice, and game controllers. The A2DP profile is used for audio transmission. In total, there are 34 Bluetooth profiles as of this writing.

And here is where the EV3 has so much more potential in this area than the NXT had. Because the EV3 is essentially a Linux PC with full Bluetooth, it can therefore connect to the same sort of Bluetooth devices that a PC can. For example, the EV3 can connect to HID devices like game controllers (including those by Nintendo and Sony Playstation), keyboards, audio devices, and so on.

Keep in mind I'm saying it has the potential to connect to them. It all depends on how many Bluetooth profiles end up being supported by leJOS. As of this writing, this is something we are exploring with the leJOS project, so look for exciting developments in this area in the future.

Bluetooth also allows robots to interact with other devices, such as your PC. This can be handy if you want to display data from your robot onto something larger than the EV3's LCD. Basically the Bluetooth adapter will open up incredible possibilities for what you can do with your robot.

BIG DECISIONS

Bluetooth Adapter

To use Bluetooth, you need a Bluetooth adapter (sometimes called a Bluetooth dongle). The EV3 kit does not include an adapter since Bluetooth isn't required to upload programs to the EV3. Do you really need one of these? The short answer is yes, go right now and purchase a Bluetooth adapter (see Figure 1-8). You can find Bluetooth adapters in electronics stores, from LEGO, and on eBay.

Be sure to get a Bluetooth 4.0 adapter for optimal performance. As with most eBay items, the cheapest Bluetooth adapters are from China. As of this writing, they can sell for under $5 US with free shipping.

NOTE: *You can also purchase Bluetooth versions higher than 4.0, such as 4.1. This is because the Bluetooth protocol is always backward compatible with previous versions of Bluetooth. If you already own an existing Bluetooth 2.0 dongle, this will also work with the EV3. Keep in mind that it will not transmit data as quickly as a 4.0 or higher dongle.*

Figure 1-8: Bluetooth adapter.

Chapter 13 will go into more detail on how to use Bluetooth. For now, let's look at a quick comparison between Bluetooth and WiFi.

WiFi Advantages:

- Faster data transfer rate (with 802.11n)
- Larger area of coverage
- Automatically connects to EV3

Bluetooth Advantages:

- Potentially less lag with 4.0
- No adapter protruding from brick
- Can connect to other unique Bluetooth devices (GPS, game controller, etc…)

LEGO Building Parts

LEGO has totally revamped the parts included with the EV3 kit. There are 601 parts in all, which is a little more than the 577 parts in the NXT 1.0 kit and a little less than the 612 parts in the NXT 2.0 kit.

LEGO reached a nadir of including incredible LEGO parts in the old RIS kit (with the RCX brick). The old RIS kit contained 18 wheels of various types, but this time there are fewer wheel options. This is a significant hindrance if you want to build a specialized robot with six wheels, such as rocker-bogie suspension. We'll look more closely at the parts in your EV3 kit in Chapter 5.

Using the LEGO Software

LEGO calls its software the LEGO Mindstorms EV3 Home Edition. This time it is not included on a CD, but rather as a free download from the official LEGO website. The development environment is called LabVIEW, made by a company called National Instruments. The language itself is called "G". Programmers use a graphical interface to develop code. The language is almost identical to the previous NXT generation.

As with the original software, users create programs by dragging building blocks into an open area from a reservoir of predefined blocks (see Figure 1-9). Each block is essentially a method, and the user selects different parameters for the method using radio buttons, sliders and drop down menus.

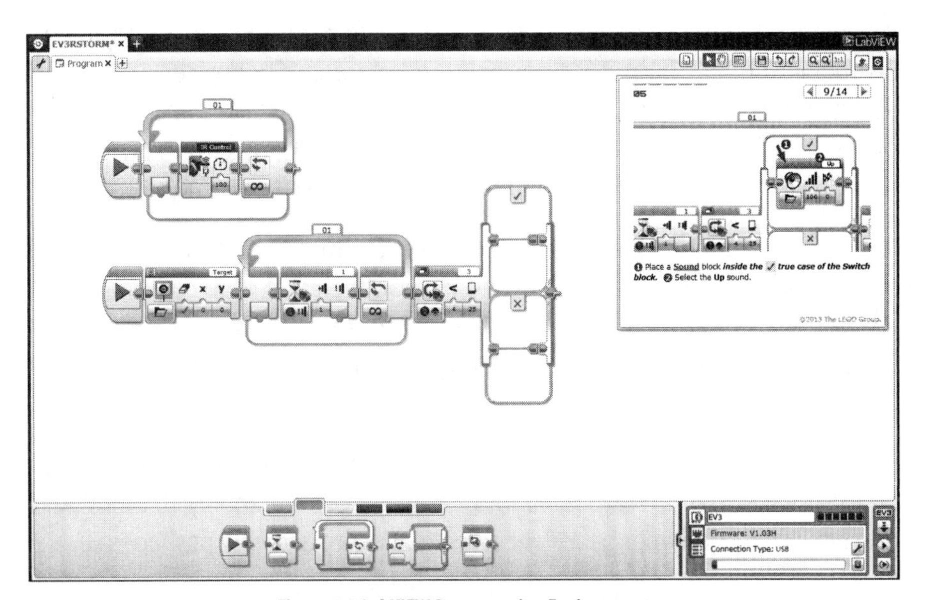

Figure 1-9 LabVIEW Programming Environment

The programming environment is more full-featured than might be expected. You can create custom blocks (like methods) that contain lots of code for a specific task. This makes it easier to fit a lot of code into the limited graphical area. The software also allows import of new blocks as new devices become available for the EV3 brick.

Although the graphical paradigm is easy to get into, especially for new programmers, it can be hard to manage code beyond a certain size and complexity. If you are used to Java or some other language, it isn't always obvious what a program is doing or how the blocks function.

The Future of Mindstorms

With LEGO hitting the mark in virtually every category this time around, where can they possibly go from here? They definitely nailed the memory and processing speed. Perhaps a lower price would make the product more attractive, especially for younger users who don't have $350 to throw around. A color touch screen would also be nice, but this comes at a price. This section examines some avenues that Mindstorms can take.

It's notable that the EV3 brick duplicates a lot of features that are already found on mobile phones and other portable Android and iOS devices. I'm talking about memory, a fast microprocessor, Bluetooth, WiFi, speakers and a small display screen. In fact, mobile devices even have GPS units, gyros and tilt sensors! These built in sensors make it an appealing device to control robots.

In order to make this work, we would need to cut the cords to motors and sensors. How? By adding a battery and Bluetooth chip to each sensor and motor. This would make it a little more of a pain to manage battery charging, but it would sure make the robots behave better without having to worry about cable interference.

Batteries can store more and more charge every year. There are even batteries on the horizon that use carbon nanotubes to increase storage by 10 times! And thin Lithium Ion polymer batteries can be made into a variety of shapes.

Finally, LEGO would need to create an Android and iOS application to control robots from the phone. If LEGO ever goes down this road, they could potentially decrease the cost of their kits while increasing sales, and increasing the potential of LEGO robots even further. Here's to more innovation!

CHAPTER 2

Installing leJOS

TOPICS IN THIS CHAPTER

- Installation Overview
- The Installer
- Writing the SD Card
- Installing the Eclipse Development Environment
- Installing the leJOS Eclipse Plugin

In this chapter, we will install Java, leJOS, and a graphical development environment. If you feel yourself tensing up, don't worry. Installing these components is ultra-easy.

Java programming is possible with a text editor and a command line. However, it's easier to click on buttons to make things happen rather than typing commands and optional parameters. Also, standard text editors included with Windows and Macintosh systems do not have many features to help you enter code. They won't tell you when you've misspelled the name of a class or forgotten a bracket.

An IDE, or *Integrated Development Environment*, is a tool that allows you to enter, compile, and upload code to your EV3 using simple buttons. It also monitors code syntax, coloring your code so you can more easily identify keywords and variables. This section will suggest a free, open source IDE for your leJOS needs.

One of the best open source IDEs is Eclipse by IBM (see Figure 2-1). It's free, powerful, and easy to use. It makes sense to use an advanced IDE like Eclipse with the EV3 since your code can grow quite large.

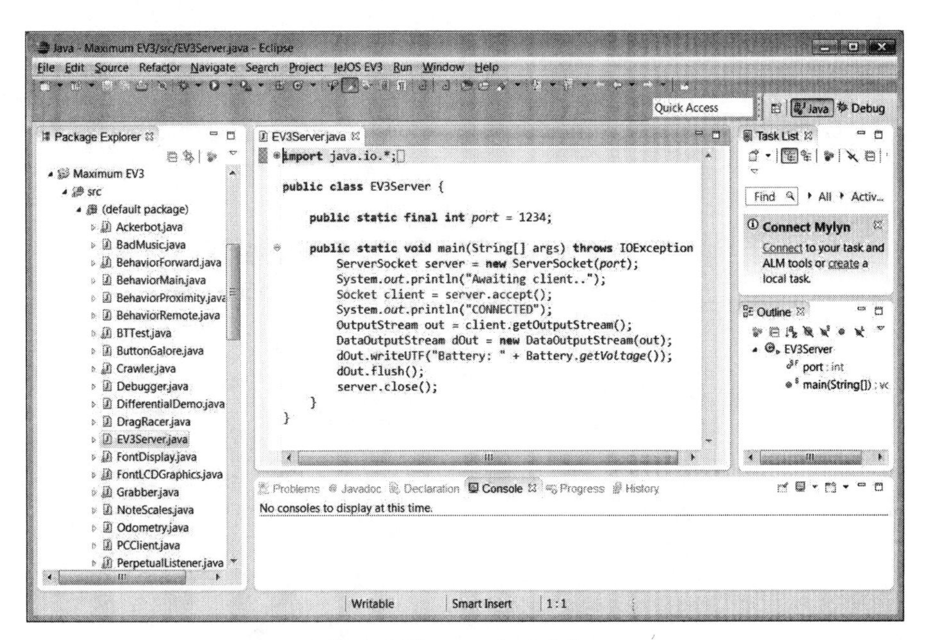

Figure 2-1 Programming in Eclipse.

In this chapter we will install the following (in order):

1. leJOS (along with Java) to your PC
2. leJOS to your EV3
3. Eclipse
4. a leJOS plug-in for Eclipse

Whew! That sounds like a lot, but again, it is very easy to install each of these packages and does not take a lot of time. In step 2, we'll copy leJOS to an SD card and insert it into the EV3 brick. Once Eclipse is installed in step 3, along with the leJOS plugin in step 4, we can try out a simple "Hello world" program. Let's get started.

The Installer

Due to the many steps required to install leJOS (including burning an image to a micro-SD card) one of the developers of leJOS created an installer program.

1. Download the installer from Sourceforge. Choose the highest version number in the list (the one at the top):

   `https://sourceforge.net/projects/lejos/files/lejos-EV3/`

2. When you run the setup you will see a Welcome screen (see Figure 2-2). Click Next.

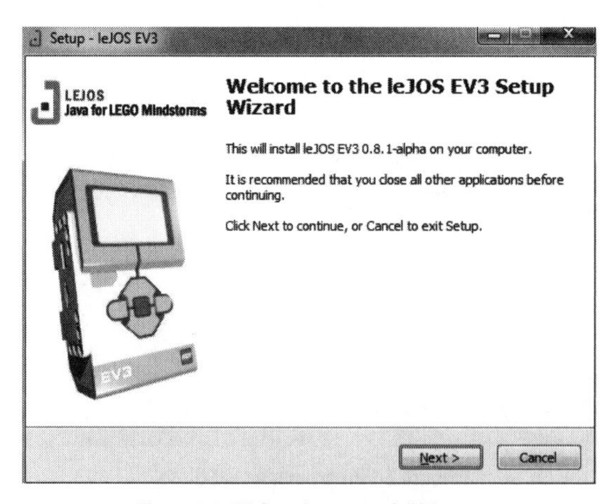

Figure 2-2: Welcoming a new leJOS user

3. The next screen allows you to choose a Java Development Kit (see Figure 2-3). You can either choose an existing one, or download a version by clicking Download JDK. Java 8 (JDK 1.8) is recommended. Once you have chosen the JDK, click Next.

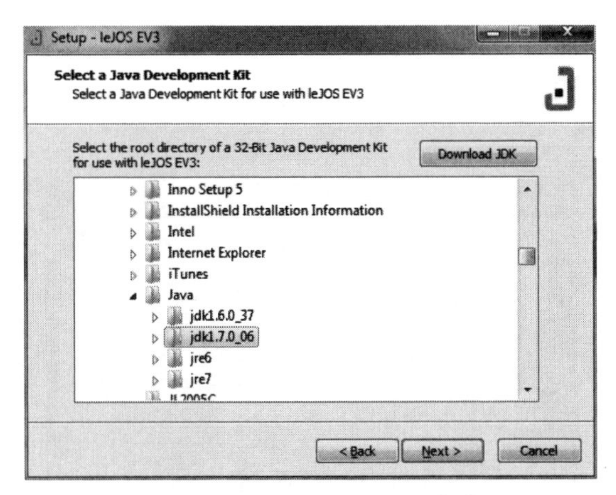

Figure 2-3: Choosing your version of Java

4. Choose an install folder. The default folder is recommended. Click Next.

5. The next screen allows you to choose components to install (see Figure 2-4). It is best to select all, then click Next.

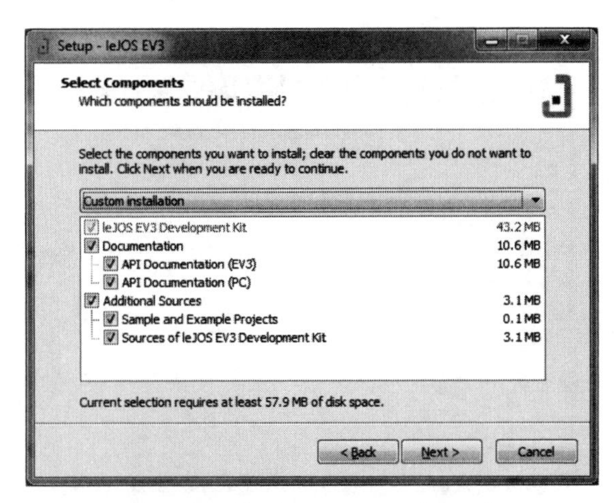

Figure 2-4: Selecting components to install

6. At this stage you can select the folders to install the components to. Again, it is best to select the defaults. Click Next.

7. The next screen allows you to select the start menu folder. It is up to you if you want this to show up in the start menu. Click Next.

8. This screen summarizes the install (see Figure 2-5). Click Install and everything will be installed. If a previous installation was detected, it will want to delete it first. Click OK.

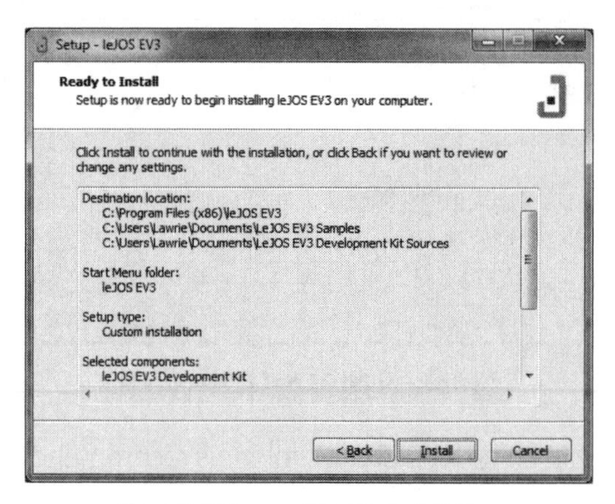

Figure 2-5: Summarizing the installation

9. Once completed, it is best to launch the SD Card creator (see Figure 2-6). Click Finish and the SD Card creator will launch.

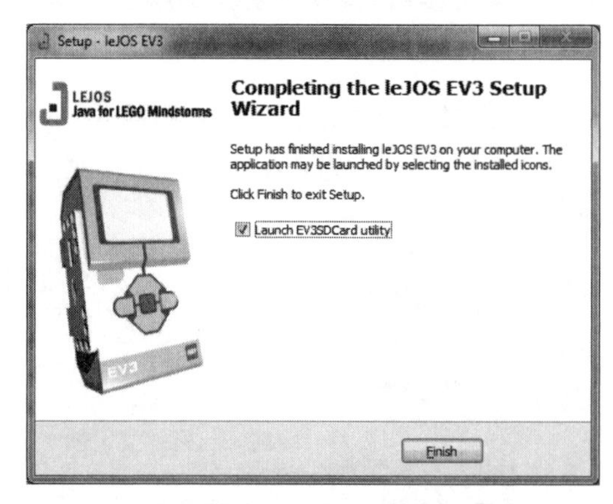

Figure 2-6: Getting ready to launch the SD card creator

Writing the SD Card

For this stage, you will need a micro-SD card, along with an SD card adapter so you can plug it into your PC. The micro-SD card must be at least 2 GB in size, but no more than 32 GB. NOTE: SDXC cards are not supported.

1. Insert the SD card with adapter into the SD card slot in your computer.
2. With the EV3 Card creator, click refresh (see Figure 2-7). You should see your card appear in the selection list at the top. Select the card.

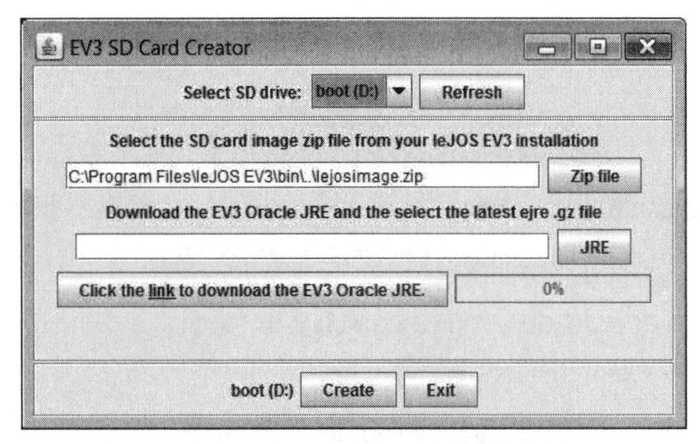

Figure 2-7: Writing an image to the SD card

3. Choose a leJOS image to burn. Likely this is already filled in for you, but if not you can browse to the directory that you installed leJOS to above.
4. You also need to select the Oracle Java Runtime Environment (JRE) which will run on your EV3 brick. Click the link to download it.
5. You will be taken to the Oracle website (see Figure 2-8). The file you want starts with "ejre" and ends with ".gz". Choose version 7 and download it. You will need to create an account here if you do not already have one.

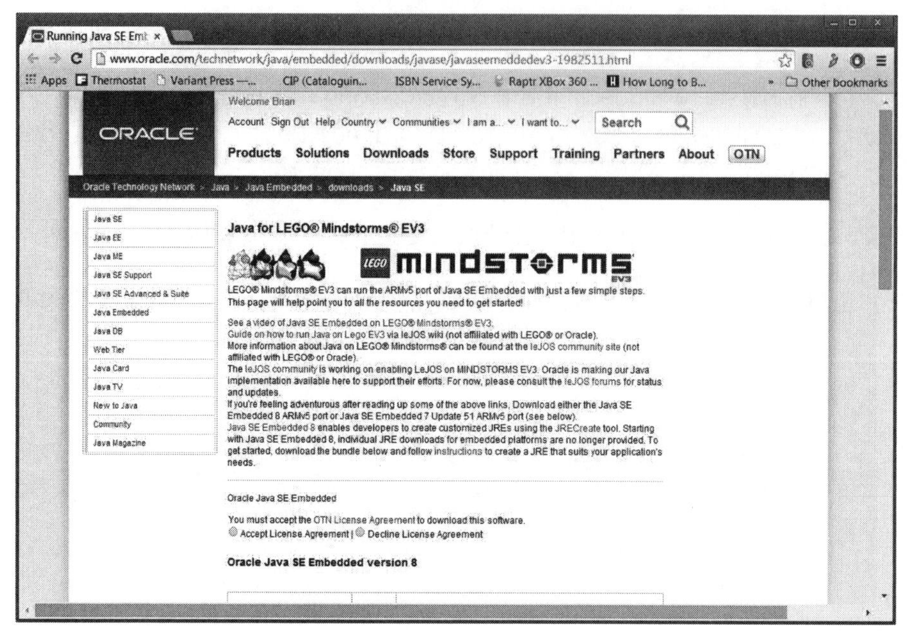

Figure 2-8: Downloading the JRE for EV3.

6. Once it is downloaded, you can go back to the SD Card creator. Browse to the JRE you just downloaded.

7. Click Create and the program will begin writing the image to your card. This will take a few minutes.

8. When done, safely eject the SD card from your computer by right clicking the "Safely Remove Hardware" tools icon. Then physically eject the card.

9. Remove the micro-SD card from the adaptor, then insert it into the (powered off) EV3 brick with the label facing up.

10. Turn on the EV3 and it will begin installing. This will take about 8 minutes.

Almost there! Time to install an IDE.

Setting Up Eclipse

Now that you have Java installed on both your PC and the EV3, we can install Eclipse. The Eclipse Foundation releases a new version every year and they name the different versions after the moons of Jupiter. This book uses Eclipse Luna, released in 2014. The version after Luna is Eclipse Mars, with a release date of June 2015. The loading screen of Eclipse tells you what version you are using.

1. Download Eclipse from: www.eclipse.org in the download section. There are several different distributions of Eclipse for different types of users. The one we want is the Eclipse IDE for Java Developers. Make sure to download the 32-bit version even *if you are using a 64-bit computer*, because our leJOS plug-in will not work with the 64-bit version.

2. The download is a zip file. Decompress the files into a directory. This will be the permanent location for Eclipse.

3. That's it. Eclipse uses no setup and doesn't store registry settings or copy native library classes to other directories. This applies to all versions of Eclipse, including Linux.

4. To run Eclipse, double click the executable file in the Eclipse directory (or create a shortcut to this). To uninstall Eclipse, merely delete the Eclipse directory from your computer.

5. The first time you run Eclipse it will ask for a workspace location. Unless you have a cloud drive you want to save files to, it is probably best to go with the default folder.

Eclipse will also guide you to some optional help files and tutorials. If you want to go right to Eclipse programming, click Go to Workbench or close the Welcome tab.

It's a good idea to have Eclipse automatically search for updates, in case there are software patches or new features. Click on Window > Preferences. Double click Install/Update in list, and highlight Automatic Updates (see Figure 2-9). Place a checkmark next to Automatically find new updates and notify me and click OK. This will update not only Eclipse but your plugins as well.

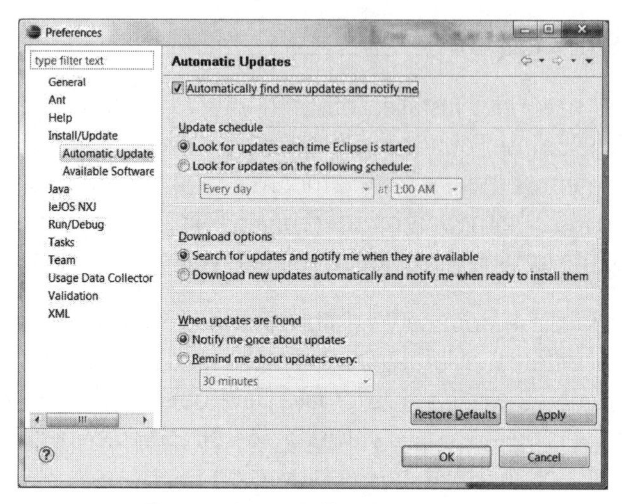

Figure 2-9 Automatically updating Eclipse

Using Eclipse with leJOS

Now that you have Eclipse installed it's time to install the leJOS Plugin. Note: You can also search for leJOS EV3 in the Eclipse Marketplace and install it from there.

1. In Eclipse, select Help > Install New Software… (see Figure 2-10).

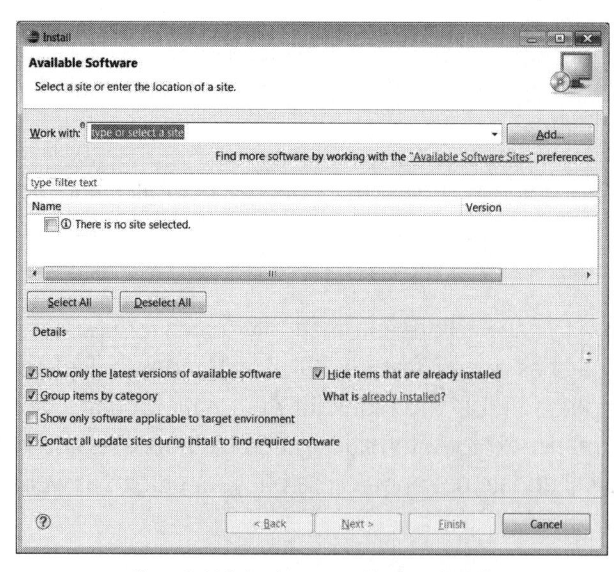

Figure 2-10 Selecting new software to install

2. You will receive a dialog to input a URL. Click the Add... button and you will see another dialog box (see Figure 2-11). Enter the name "leJOS EV3" and for the location enter the following:

```
http://lejos.sourceforge.net/tools/eclipse/plugin/ev3/
```

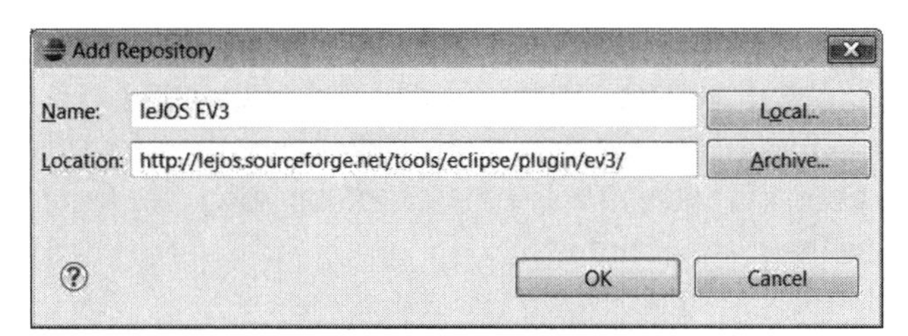

Figure 2-11 Adding a repository

3. Click OK. You should see a new item in the main box. Place a check-mark in the box next to the new item and click Next (see Figure 2-12).

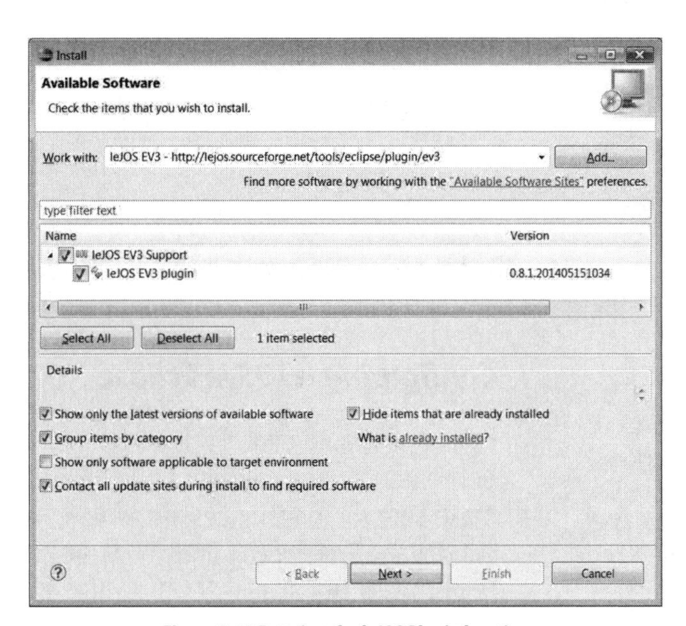

Figure 2-12 Entering the leJOS Plugin location

4. Read and accept the license agreement, and click Next. The plugin will automatically install.

5. When complete, it will ask to restart Eclipse. Once it has restarted, you will see some subtle changes with Eclipse. The plug-in will add new leJOS menu items to a variety of places within Eclipse. Let's explore the new options.

Help Files

In case you ever need a helping hand, there is a section in the help files specifically for leJOS. These files are continually updated as the plug-in is updated, so they are the final word on what you need to do to use the latest plug-in. Select Help > Help Contents and you will see a section for leJOS (see Figure 2-13). Feel free to browse them at any time.

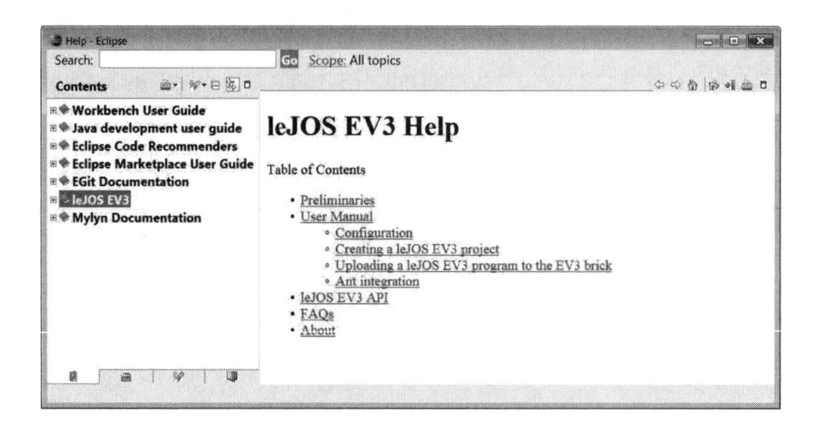

Figure 2-13 Browsing the help files for leJOS

> ***Tip!***
>
> You can find the Java docs in your leJOS install directory. These are extremely useful to reference if you are going to spend anytime with leJOS. I recommend making a shortcut to it so you can access it at any time. It is located in the docs directory under ev3 (index.html is the file to open).

Configuring leJOS in Eclipse

Eclipse will automatically look for the EV3_HOME environment variable to locate leJOS. Let's check to make sure the preferences are to your liking. Select Window > Preferences and then select leJOS EV3 from the list (see Figure 2-14). If the leJOS directory is incorrect, either type in the location or browse to it. Make sure you browse to the main directory and not one if its subdirectories.

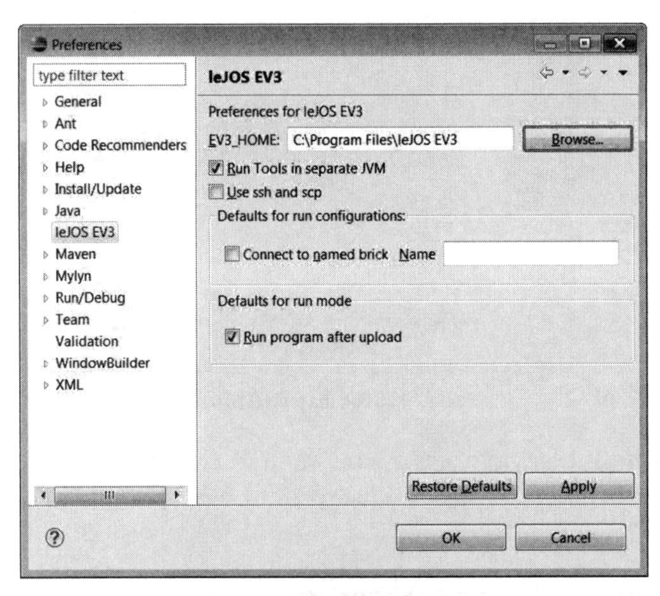

Figure 2-14 Selecting preferences

You can also select other options according to your own tastes, but I recommend setting them as shown above. You can also enter the name of your brick if there are other bricks in the vicinity by specifying an IP address. If you leave this field blank, leJOS will connect to the first brick it finds via WiFi or Bluetooth.

Project Wizards

Now let's create a place to enter some code. Eclipse keeps individual Java projects in their own project directories. For example, if you are creating a large, multi-class project dealing with mapping, you would create your own project with its own directory to store the class and data files. Let's try creating a project which we will use to store the code in this book.

1. Select File > New > Project...

2. In the next window, double-click LeJOS EV3 to expand the folder options (see Figure 2-15). We want to create a leJOS EV3 Project. Select leJOS EV3 Project and click Next.

3. For the Project Name, enter Maximum EV3. Then click Finish.

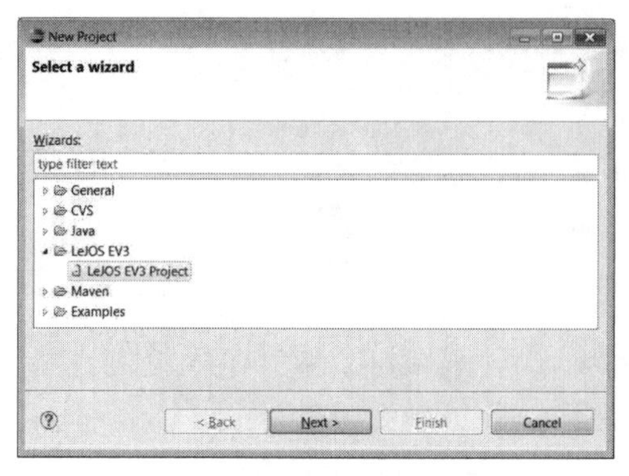

Figure 2-15 Using the project wizard

You can create as many projects as you want. You might want another project for your own programs, so go ahead and create another right now if you like.

Creating and uploading a program

Let's try entering a simple Hello World program using Eclipse and then uploading it using the leJOS plug-in.

1. To add a new class file, select File > New > Class (see Figure 2-16). Enter HelloWorld in the Name field. Eclipse also offers other options, such as automatically adding a main() method. Check this if you want Eclipse to do some of the typing for you.

Figure 2-16 Creating a new Java class file

2. Click Finish when you are done and you should see a new class file with some starter code.

3. Enter the HelloWorld code below into the file.

4. Click the save button, turn on your EV3 brick, then click the green Run button in the Eclipse toolbar. A popup window will appear the first time you click this for a class file (see Figure 2-17). Select leJOS EV3 Program and click OK. The program will begin uploading.

Figure 2-17 Selecting how to run a class

5. Your EV3 will beep when it is uploaded and automatically run, assuming you are using the default settings (see Figure 2-18). If you do not want the program to run automatically every time it is uploaded, select Window > Preferences > leJOS and uncheck Run Program after Upload.

```
import lejos.hardware.*;
public class HelloWorld {
  public static void main (String[] args) {
    System.out.println("Hello World");
    Button.waitForAnyPress();
  }
}
```

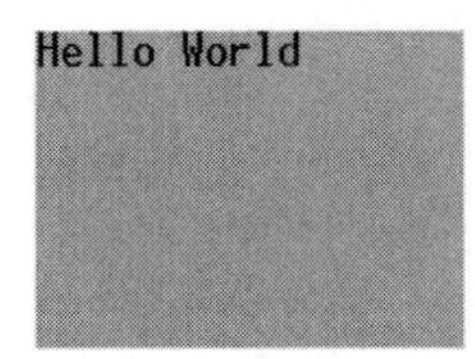

Figure 2-18 Viewing the HelloWorld program

Run Configuration

You can also configure the profile for each class you are working on. Instead of clicking the green button, click the arrow next to it and select Run Configurations... Select the configuration for HelloWorld (see Figure 2-19). This allows you to choose specific settings for particular classes you are working on. For example, if you do not want a project to run automatically you can change the setting here. You can also choose the name of the EV3 brick to upload to or other settings.

Tip!

Eclipse can be enhanced using a variety of plugins. For example, it can be difficult and time consuming to create a graphical user interface (GUI) using raw Java code. You can download a plugin that allows you to graphically design a GUI for your program which then automatically generates the code. Most plugins are free for non-commercial use. Eclipse plugins can be found at: www.eclipseplugin-central.com

Plugins are installed through Eclipse (Help > Software Updates > Find and Install...) however you need to enter the remote location of the plugin supplier. The remote location is typically located on the web page for the project.

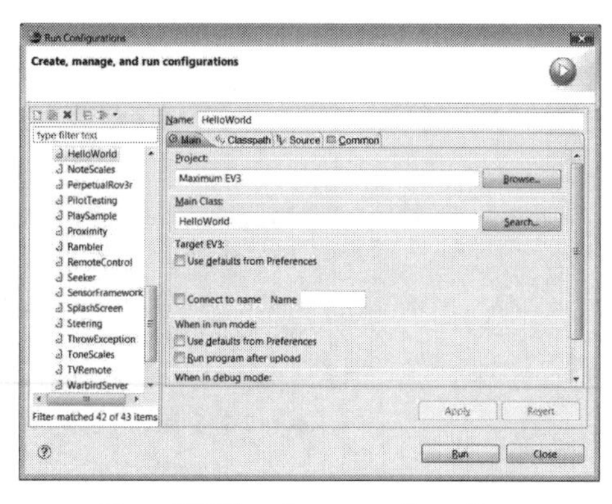

Figure 2-19 Choosing run configurations

Eclipse Summary

So what is so good about Eclipse? Try making an error in your code. Notice that right away it tells you where there is a problem? It lists all the errors or warnings, shows you exactly where they are, and even tells you what is wrong and how to remedy the problem. When you type an object name and press the period key it even pops up all the methods you can access.

Although Eclipse is simple, it has many tools to make your coding experience easier. Briefly, click on the Source menu item and look at some of the tools. Cleanup lets you import code and clean it up to your preferred style, or clean up your own code. If your indentations are wrong, highlight your code and select Correct Indentation. Instantly it fixes everything. There are also functions for generating try-catch blocks and getter/setter methods.

Sometimes you need to overhaul your code in major ways. Check out the Refactor menu item, which you can use when you want to rename a class. If you want to rename a variable, right click it and select Refactor > Rename. You can even overhaul the architecture of your code, such as making an interface from a class. These methods go through your entire code making changes.

CHAPTER 3

The First

TOPICS IN THIS CHAPTER

- Building the first robot
- Uploading and running code
- Using the Drag Racer

In the previous chapter we entered a simple Hello World code example. Now let's try a more substantial robot project that covers all the bases: building a robot, uploading code, and operating the robot.

The First Robot

Our first project is to familiarize you with building and programming a robot, namely a drag racer. Both the robot construction and the Java code are minimal, but there are a few tricky parts to get through. Let's try building the robot first.

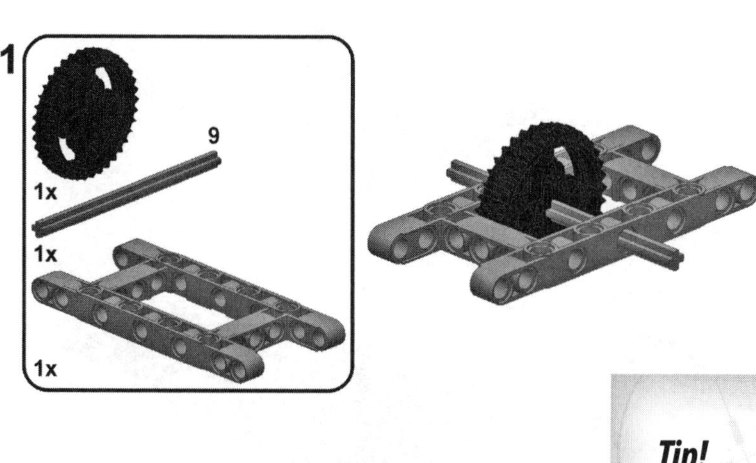

Tip!

The following LEGO plans are easier to understand if you first gather up the parts for the *current* step (shown in the upper left corner) and then assemble the step. If you pick up parts one at a time it will take longer and you might lose track of whether you've used the correct number of parts for the step.

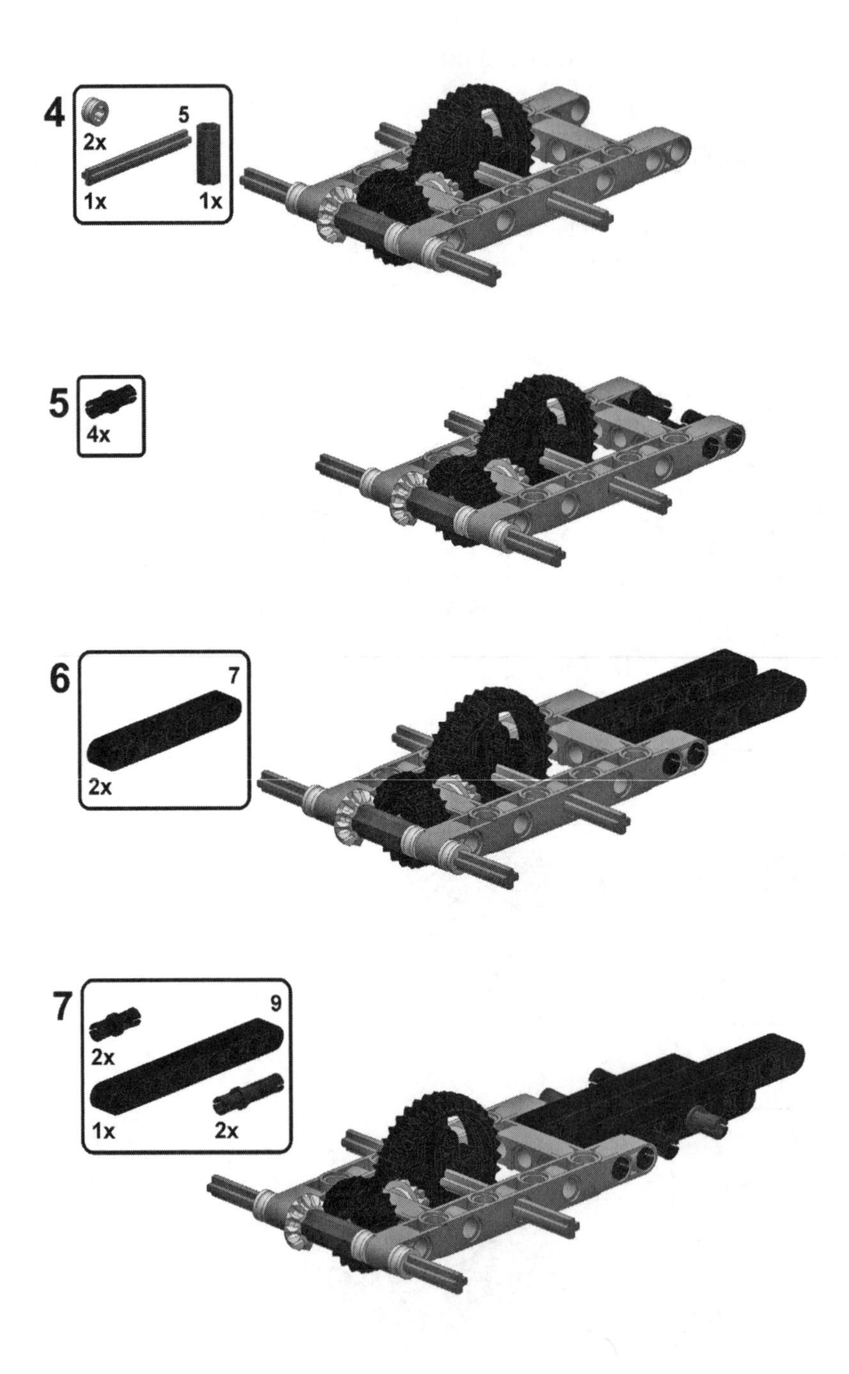
4
2x
5
1x
1x
5
4x
6
7
2x
7
9
2x
1x
2x

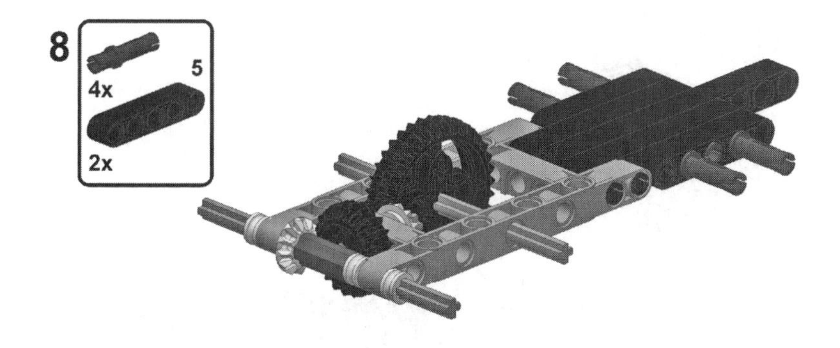
8
4x
5
2x

9
2x
2x

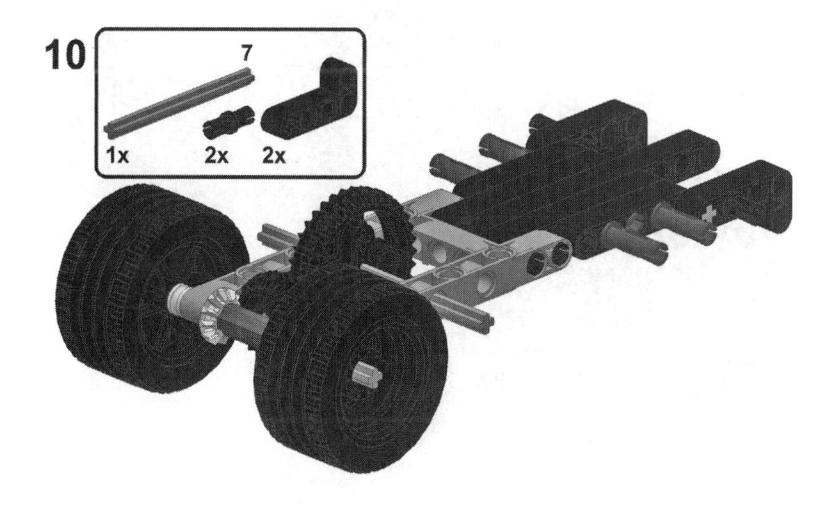
10
7
1x
2x
2x

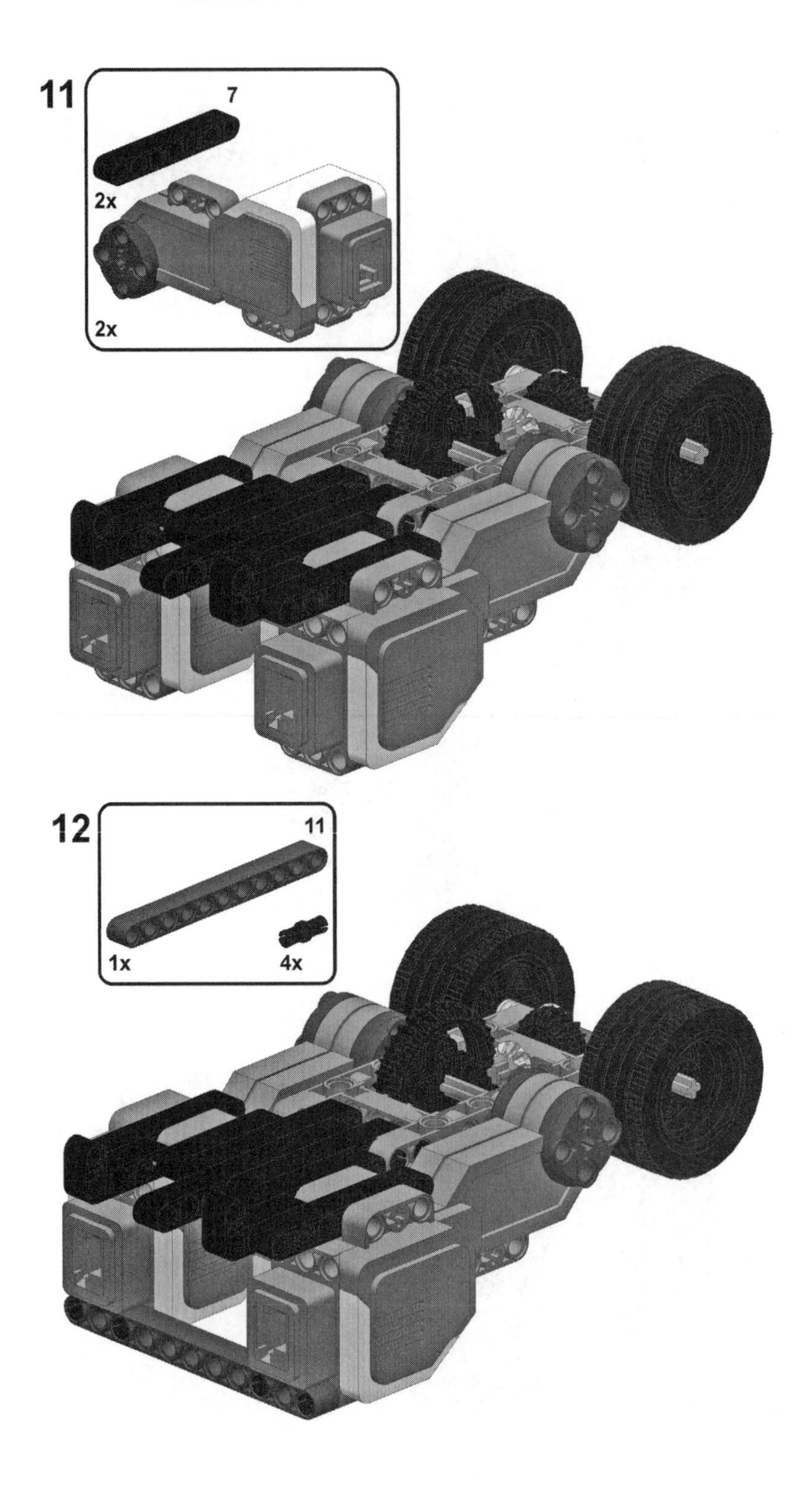
11
7
2x
2x
12
11
1x
4x

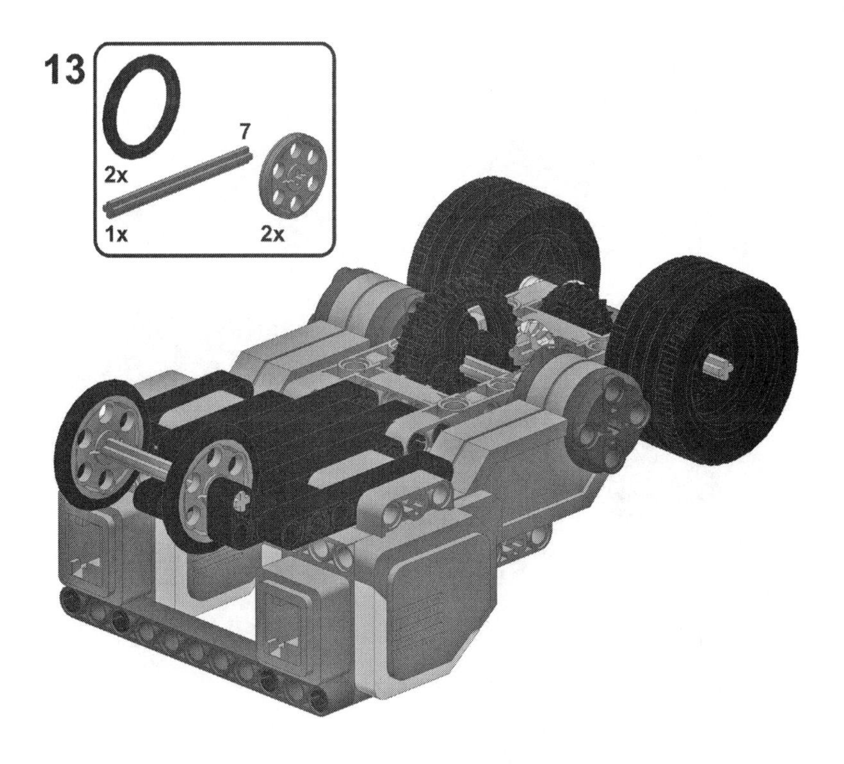
13
2x
7
1x
2x

14
4x
5
2x

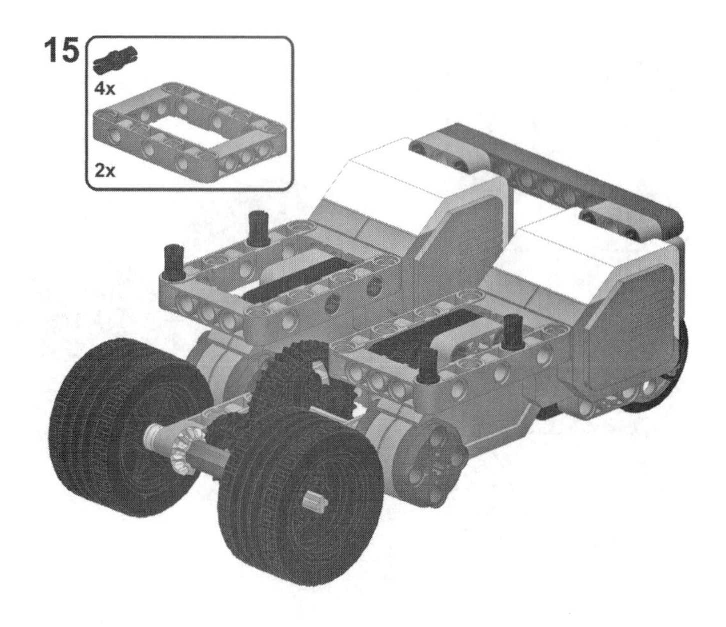
15
4x
2x

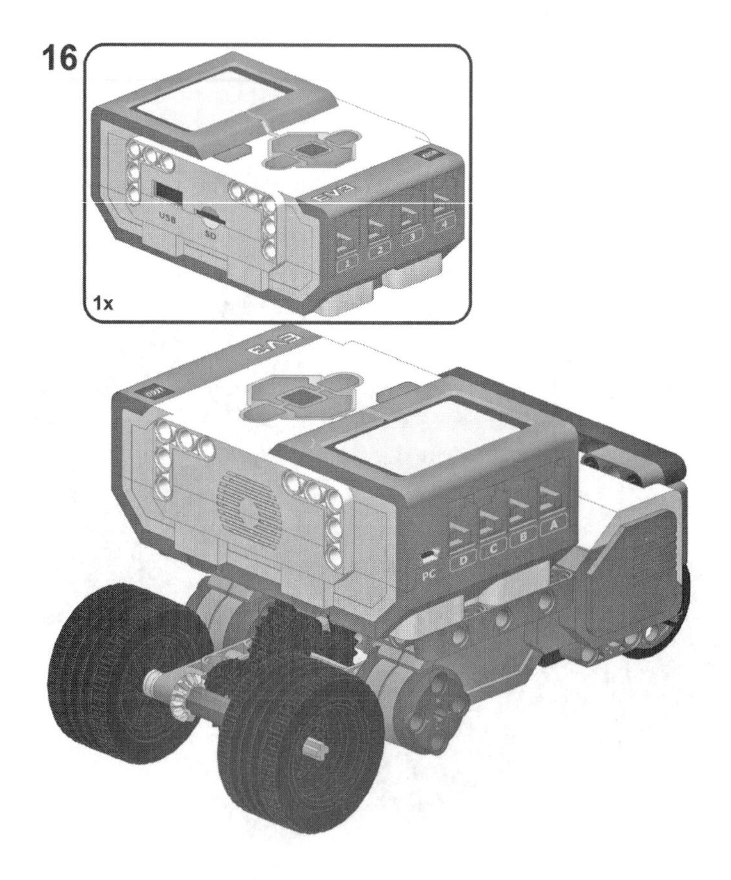
16
USB
SD
EV3
1
2
3
4
1x
EV3
PC
D
C
B
A

Cables

After each build we need to make sure the sensors and motors are attached to the correct ports. Find one medium and one short cable and attach them to ports B and C, then attach the other ends to the motors (it doesn't matter which for this particular robot).

Entering Code

The code for the drag racer is fairly straightforward. It simply rotates both motors at the same time for a short period of time, then stops.

To enter the code in Eclipse, create a new class in your project by selecting File > New > Class (see Figure 3-1). Type the class name as DragRacer and click Finish.

Figure 3-1 Starting a new class in Eclipse.

NOTE: *I strongly recommend downloading all the code in this book straight from the book's website. This code is continually updated in case leJOS changes and it saves you from typing the entire code yourself. The website is at: www.variantpress.com*

```
import lejos.hardware.*;
import lejos.hardware.motor.*;
import lejos.hardware.port.*;
import lejos.utility.Delay;

public class DragRacer {
  public static void main(String[] args) {
    Button.waitForAnyPress();
    // Play warning tones:
    for(int i=0;i<3;i++) {
      Sound.playTone(500, 800);
      Delay.msDelay(200);
    }
    Sound.playTone(1000, 800);

    // Initialize motors:
    UnregulatedMotor b = new
UnregulatedMotor(MotorPort.B);
    UnregulatedMotor c = new
UnregulatedMotor(MotorPort.C);
    b.setPower(100);
    c.setPower(100);

    // Start rotating motors:
    b.forward();
    c.forward();
    Delay.msDelay(1500);

    // Stop motors and close:
    b.flt();
    c.flt();
    b.close();
    c.close();
  }
}
```

This code starts by announcing the imminent departure of the drag-racer using a series of beeps. It then rotates the motors briefly for 1500 ms (1.5 seconds) and then ends. Notice the code is commented with friendly notes to make it easier to understand.

WARNING: *The code will begin executing once it is uploaded. As a safety precaution it requires you to press a button first to start the countdown. Made sure it does not crash into a wall and damage your EV3 brick. As usual, neither the publisher nor the author is responsible for any damage that might happen to you or your property.*

Compiling and Uploading in Eclipse

We will review the procedure for uploading once more (briefly this time). The remainder of the book will assume you are familiar with the uploading procedure. Before starting, make sure you have at least 12 feet (around 4 meters) of space for the drag racer, otherwise it might go crashing into a wall.

Enter the DragRacer code above into Eclipse and save it.

Turn on your EV3 brick and plug in the USB cable. Or, if you are using WiFi, just turn on your EV3 brick.

Try compiling DragRacer by pressing the green Play button. The program will compile and then begin uploading to the EV3 brick. It will then begin running unless you changed the settings of the Eclipse plugin to not run automatically.

> ***Tip!***
>
> To stop an EV3 program that is currently running, press Enter and Down at the same time. This will stop all programs and is useful if your program has been caught in a loop. However, you should still make it clear how to exit your own program without using this button chord.

Using the Drag Racer

As mentioned above, the drag racer needs about 12 feet of space to travel. Sit it down and after the beeps it will take off and stop. Fast, isn't it? The speed attained by this particular robot is possible through gear multiplier, which is covered more in chapter 5.

In a future chapter we will build onto the Drag Racer chassis to create a fast steerable robot. Don't disassemble the Drag Racer yet if you wish to try this out!

CHAPTER 4

Motors, Sensors, and Accessories

TOPICS IN THIS CHAPTER

- Motors
- Sensors
- Kits and Accessories
- Surveying the Competition

Motors and sensors are the heart of the LEGO EV3 kit. There are three different sensors in the EV3 kit, plus a few more are available from the LEGO website. You can also look to third party developers for some EV3 sensors, but as of this writing there are far more NXT sensors (which are EV3 compatible) than EV3 sensors.

LEGO generously includes three motors in the EV3 kit: two large and one medium.

Proprioception is the ability to keep track of internal conditions of body parts. When you move your arm, you know its position even with your eyes closed. Robots can also have proprioception by sensing the position of the motor axle using encoders.

All three EV3 motors use encoders to keep track of axle rotation. The motors can turn in a direction for thousands of rotations and come back to the exact starting position at any time. This feature opens up incredible possibilities for robot creation, especially with navigation and robot arms.

EV3 Large Motor

The large EV3 motors each weigh 82 grams, slightly heavier than the 81 gram NXT motors. There are very few differences between the NXT and EV3 motors, other than the LEGO hole alignment (which is discussed in Chapter 6). Also, the EV3 motor can produce faster rotation and higher torque than the NXT motor (more on that later).

The motor contains gears that reduce the speed of the motor by 48 to 1 and thus increase power. This gives the motors an unusual shape, making them look like R2-D2 legs (see Figure 4-1). Sometimes this shape can make it more difficult to integrate motors into your design, but often it can help form the structure of a robot. Most motors have an axle, but the EV3 motor axle is inverted into an axle hole. This makes it easier to choose the appropriate axle size.

Figure 4-1 Examining the Large EV3 Motor

In a head to head pushing match, the EV3 motor will outperform the NXT motor, yielding 17.3 N cm of torque. It outputs almost three times more torque than the smaller EV3 medium motor, but curiously uses less power (see table 4-1). However, the gear train is not as precise as the smaller motor, and it contains around 3 degrees of backlash.

On the plus side, large EV3 motors are cheap! In fact, they are cheaper than most sensors (only the touch sensor is cheaper). Online, they can be purchased for around $25 each. Given that there are 4 motor ports available, this is a worthwhile purchase.

Website!

A detailed analysis of different LEGO motors can be found on Philippe Hurbain's website:

philohome.com/motors/motorcomp.htm

	NXT	EV3 Large	EV3 Medium
Torque	16.7 N cm	17.3 N cm	6.64 N cm
Current	60 mA	60 mA	80 mA
Rotation Speed	170 RPM	175 RPM	260 RPM
Weight	81 grams	82 grams	39 grams

Table 4-1: Motor characteristics under no load.

EV3 Medium Motor

The medium EV3 motor is obviously smaller and lighter than its larger sibling, but it has many advantages other than being more compact. It spins at a noticeably faster rate, and is more accurate in its rotations. Let's examine these concepts in more detail.

The medium motor contains *epicyclic gearing* which is more commonly called a *planetary gear train* (see Figure 4-2). A planetary gear train contains a sun gear (1) three or four planet gears (2) and an annular gear (3) around the outside.

The LEGO motor has only three planet gears. The sun gear connects to the axle of the electric motor inside the housing, while the annular gear is attached to the axle hole (the hole you insert an axle into). It also contains two complete planetary gear trains stacked on each other, making it a two-stage planetary gear train (with one annular gear connected to one sun gear).

Not only is the planetary gear train compact, but it results in almost no backlash! Because of this, the medium EV3 motor is useful for tasks requiring higher precision.

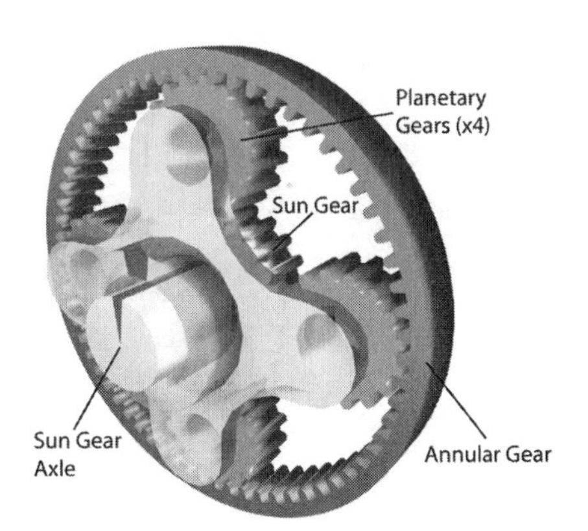

Figure 4-2: A planetary gear train.

The DC motor in the medium motor spins faster than the large EV3 motor, requiring more electric current when running (see Table 4-1). Note that these figures are when the motors are under no load—the axle is not connected to anything. These characteristics will change when the motor axle is connected to something requiring more force to turn.

Sensors

The sensors in the EV3 kit are identical in size and shape, except for the infrared sensor which looks like a robot head (the infrared sensor is wider at the front, but the rear is identical to the other sensors). All of the sensors use the same means of attaching to the robot (two holes for friction pins plus one axle hole), making the sensors interchangeable.

> ***Website!***
>
> More information can be found on the Wikipedia page for planetary gear trains.
>
> http://en.wikipedia.org/wiki/Epicyclic_gearing

Touch Sensor

The touch sensor is the most basic sensor in the EV3 kit (see Figure 4-3). It has a simple switch activated by the red button on the front. Because it is a simple on-off switch, there is no need for a complicated data protocol. Therefore it is an analog sensor—the only one of its type included in the EV3 kit.

The button has an axle hole, allowing you to connect parts directly to the sensor switch. There is a single touch sensor in the EV3 kit.

Figure 4-3 EV3 Touch Sensor

EV3 Color Sensor

The EV3 kit contains a single color sensor (see Figure 4-4). This sensor can detect different colors of light, and the built in lamp can emit either red, blue, or green light from a multicolored LED. There are four modes available:

1. Color ID: Identifies colors by name. Each number constant represents a different color.

2. Red: With the red LED lit, identifies light levels.

3. RGB: Identifies the proportion of red, green, and blue in any object placed in front of the sensor.

4. Ambient: Identifies ambient light in front of the sensor. (Similar to 2 but without the red LED).

By pointing the sensor down, the robot can follow a black line. Sometimes the sensor is used to prevent a robot from driving off the edge of a table, since the sensor values decrease significantly when an object (such as a floor) is farther away (far objects do not reflect as much light as near objects). The light sensor can also distinguish dark objects from light objects, since dark objects reflect less light.

Operation of the color sensor is much more complicated than the touch sensor in terms of data, therefore it uses UART. This makes it incompatible with the NXT, which does not support UART.

Figure 4-4: Color sensor

Try it!

The leJOS menu software has a tool for reading sensors. We can use this to detect IR light from a TV remote control or even the EV3 IR Remote Control.

1. Connect the color sensor to any sensor port.
2. From the main menu on the EV3 brick, select Tools > Test Sensors.
3. Select the proper serial port and then select EV3 Color.
4. For the Sensor mode, select Ambient.
5. The changing floating point number indicates the reading from the light sensor. If you point the sensor at a light source, such as a window, the value will increase.
6. Try holding a TV IR remote about five inches from the light sensor and press some buttons. The values on the EV3 will increase momentarily for each press, like pointing a flashlight at the sensor.

Infrared Sensor

The infrared sensor looks like it has a pair of red eyes, similar to the EV3 ultrasonic sensor. However, these eyes are merely cosmetic paint. If you hold the infrared sensor up to a light source so you can see through the blue tinted plastic, you will notice there is nothing behind the red eyes except air (see Figure 4-5). The actual sensor electronics are located in the center of the head. In fact, the infrared sensor could have taken the same box shape as other EV3 sensors, but then it wouldn't have looked like a robot head.

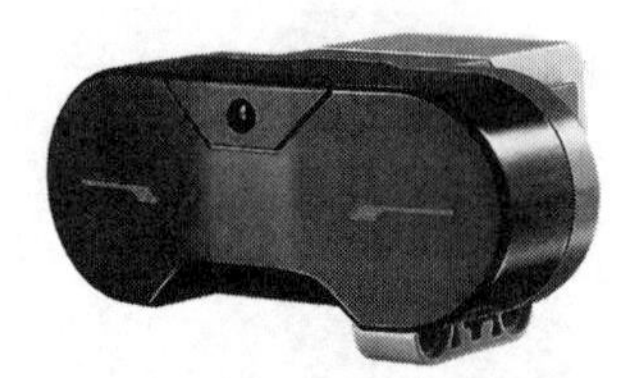

Figure 4-5: Viewing the EV3 IR Sensor

A lot of people think distance is measured by the strength of the reflected IR light back, but this is not true. The infrared sensor works on the same principles as the long-running line of Sharp GP sensors, such as the GP2Y0A02YK. Light is emitted from the infrared LED visible at the top of the front of the sensor. If the light strikes an object, it is reflected back and strikes a linear CCD array (charge-coupled device) located between the two red eyes. CCD's are found in cameras, but this one is a little less elaborate.

A precision lens reflects light onto different parts of the CCD based on the incident angle the light enters the lens at (Figure 4-6). For example, if the sensor is close to an object, the light comes into the sensor at a very wide angle. But if the light is reflected from a distant object, the angle is much more narrow. By detecting the angle, the sensor uses simple triangulation to calculate the distance.

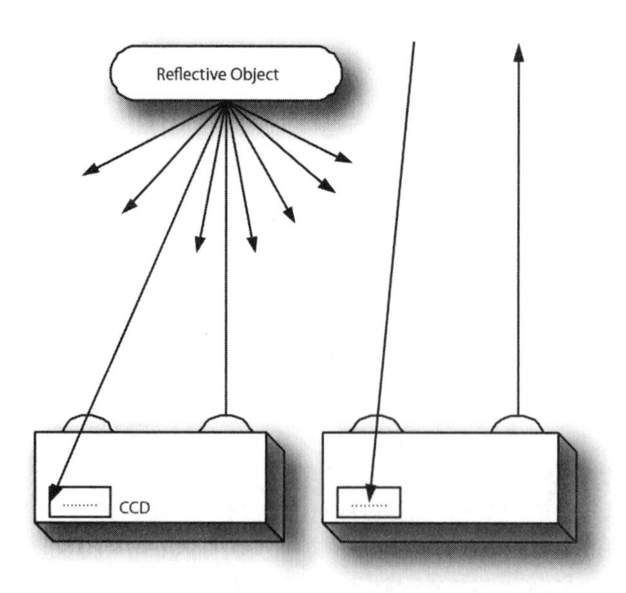

Figure 4-6: Reflecting IR light back to the CCD.

This sensor is also a UART sensor. Chapter 17 will exclusively cover this sensor, along with the IR remote beacon.

Try it!

Let's use the leJOS menu to read distance values from the IR sensor.

1. Connect the IR sensor to any sensor port.
2. From the main menu on the EV3 brick, select Tools > Test Sensors.
3. Select the proper serial port and then select EV3 IR.
4. For the sensor mode, select Distance.
5. Stand in front of a wall with the IR sensor pointed at it. The changing number on the LCD indicates the distance, in centimeters.
6. Try moving the sensor slowly right up to the wall, and notice you get quite good readings even a few centimeters away. Now go the other direction and notice readings are good until about 100 cm.

Other LEGO EV3 Sensors

There are a large number of NXT sensors which are compatible with the EV3. LEGO also sells a few other UART sensors specifically for EV3, which are included in the education kit (more on that below) or they can be purchased separately. Let's examine these sensors!

EV3 Gyro Sensor

In order for a robot to perform dynamic balancing, a gyroscope sensor is required (gyro for short). A gyro measures the angular velocity along an axis. For example, a spinning wheel of an automobile might have an angular velocity of 100 degrees per second as it is backing out of the driveway. Degrees per second is the unit of measurement used for gyro sensors.

The technology for digital gyroscopes has advanced enormously to the point that they are tiny and cheap. The gyroscopes used in mobile phones are housed on a microchip; however there is still a mechanical component to them. They use a special technology called MEMS, which stands for Micro-Electro-Mechanical Systems. Figure 4-7 shows a magnification of this mechanism.

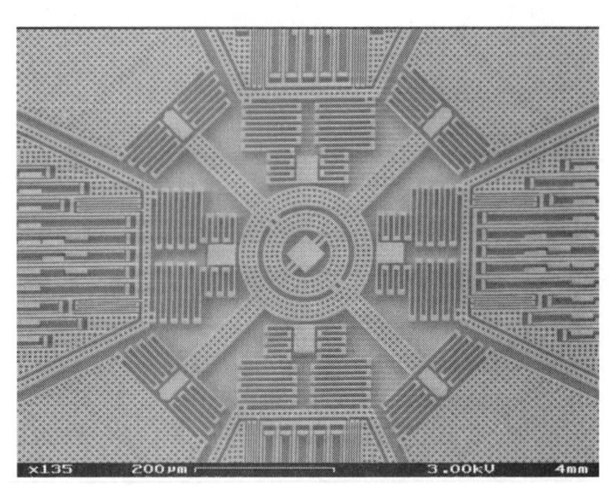

Figure 4-7: Magnifying the MEMS gyro mechanism.

The LEGO gyro sensor is about the same size and shape of most other sensors (see Figure 4-8). It contains arrows on top of the gyro sensor to indicate the axis of rotation. There are three modes which determine which measurements you can read from this sensor: rate of change, angle, and both.

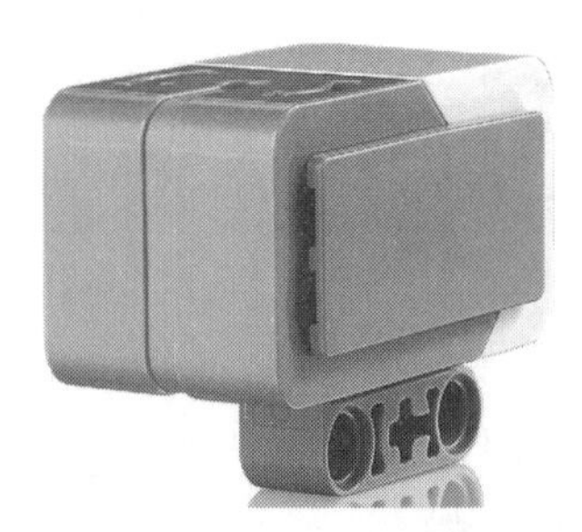

Figure 4-8: Examining the LEGO Gyro.

When in angle mode, it has an accuracy of +/- 3 degrees. Meanwhile, gyro mode can detect changes in rotation as high as 440 degrees/second. Because this sensor uses high speed UART, it can take as many as 1000 samples per second.

The EV3 gyro sensor costs only $29.99 as of this writing, which is down quite a lot from the NXT gyro sensor, which previously cost more than $50.

Ultrasonic EV3 Sensor

The EV3 ultrasonic sensor has a few differences compared to the IR sensor (see Figure 4-9). It can detect objects from farther distances, but it is not as accurate at closer distances. And it is not compatible with the IR remote control. As of this writing, it costs $29.99. Let's examine it in detail.

Figure 4-9: Measuring distances with the ultrasonic sensor

Even though the ultrasonic sensor looks like a pair of eyes, it actually has more in common with the sound sensor than a camera. The ultrasonic sensor sends out a sound signal that is inaudible to humans (like a bat), then measures how long it takes for the reflection to return. Since it knows the speed of sound, it can calculate the distance the signal traveled.

The ultrasonic sensor is also a UART based sensor. It measures distances to solid objects in centimeters or inches. The sensor is capable of measuring distances up to 255 centimeters, though returns are inconsistent at these distances because the return ping becomes weaker. The sensor is accurate from 6 to 180 centimeters, with objects beyond 180 centimeters not reliably located. It has an accuracy of plus or minus three centimeters, though the accuracy is better for close objects.

The ultrasonic sensor produces a sonar cone, which means it detects objects in front of it within a cone shape. The cone opens at an angle of about 30 degrees (see Figure 4-10). This means that at a distance of 180 centimeters the cone is about 90 centimeters in diameter. The cone shape is ideal for most robots, since it is better to scan a large area in front of the robot for possible collisions.

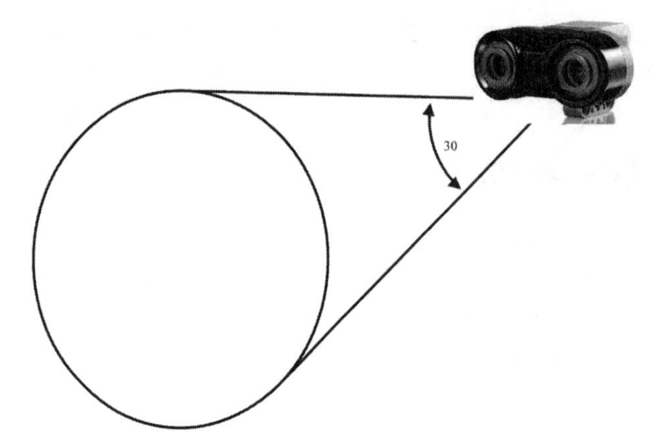

Figure 4-10 The 30 degree ultrasonic sensor cone.

Try it!

Theoretically a soft object returns less of a sonar ping than a solid object. Let's see how this works in practice.

1. Find a pillow and a solid object, like a book.
2. Place the objects next to each other at equal distances from the LEGO EV3.
3. Plug the ultrasonic sensor into the EV3.
4. From the leJOS main menu, select Tools > Test Sensors.
5. Select the proper serial port and then select EV3 Ultrasonic.
6. For the Sensor mode, select distance.
7. Watch carefully for differences between the pillow and the book. In my tests, it seems the surface of an object does not affect the distance reading from the ultrasonic sensor, however the solid object is detected at longer distances.

Kits and Accessories

The EV3 kit provides the basics to create a large number of interesting robots, but you can take your projects even further by expanding the number of parts at your disposal.

Education EV3 Core Set – 5003400 (45544 in Europe)

LEGO Education sells a kit tailored to the educational market. It contains a rechargeable battery and adapter, two large motors, one medium motor, two touch sensors, a color sensor, an ultrasonic sensor, and a gyro. In total, there are 541 parts, including additional tires, a castor wheel (which they call a ball wheel), a differential gear, and most of the parts in the consumer kit. It costs approximately $339.95.

Education EV3 Expansion Set - 45560

The education resource set contains a huge assortment of 853 parts. Perhaps most notable are the large turntable gears, gear rails, and perpendicular joiners, which are missing from the retail set. Most of the other parts are already included in the retail kit, but there are more of them. It costs $99.95 as of this writing.

Tetrix

LEGO Education North America sells a set of advanced metal parts called Tetrix that are compatible with LEGO Technic parts. These are high precision, high quality aluminum parts that are meant for semi-permanent construction. Parts include gears, servo motors, DC motors, beams, wheels, tubes and axles (see Figure 4-11).

The parts are attached with screws that are tightened by a hex key, and there are even tools to cut parts to custom lengths. Parts may be purchased individually or in larger kits. Most of the prices are reasonable, although some parts are very costly, such as gears ($20 or more).

Figure 4-11 LEGO Tetrix parts

Tetrix is designed for control by an EV3 or NXT brick. The Tetrix servo controller, which plugs into the EV3 brick, controls up to six servo motors. However, unlike the motors in the EV3 kit, the servo motors require a separate battery pack. There is also a separate DC motor controller available.

Website!

To find out more, search for "Tetrix" on the LEGO Education website:

www.legoeducation.com

Or visit:

parts.ftcrobots.com/store/

An overview of the Tetrix Parts system is available on Youtube:

www.youtube.com/watch?v=YGn8kfktwJo

www.youtube.com/watch?v=ilSuuchQC8E

The main kit, called the *Tetrix Robotics Base Set*, is available from Lego Education for $399 (product ID W739108). It includes 556 parts and comes with two DC motors and two servo motors (but no motor controllers).

Surveying the Competition

We'll conclude this chapter by looking at some of the other robot kits on the market. Although LEGO dominates the field of do-it-yourself robotics, there are other competitors. This section will examine the major competition to LEGO.

VEX IQ kits

VEX is LEGO MINDSTORMS strongest competitor. It was originally produced and invented by Radio Shack and later sold to Revell, makers of plastic models.

VEX released the VEX IQ line of robot kits to compete with the NXT. The main programmable unit, called the Robot Brain, is reminiscent of the NXT or EV3, with up to 12 sensors or motors (called Smart Ports). It also has a USB port for programming, and a radio port for wireless programming from a PC.

You can use the popular RobotC language, which is also compatible with EV3 and NXT. The Microcontroller is a Texas Instruments Tiva ARM Cortex-M4 Processor. It contains 256K flash and 32K RAM, putting it closer to the NXT than EV3 in terms of specs.

Unlike the friendly plastic LEGO parts, the VEX kit contains mostly metal pieces with plastic for gears and wheels (see Figure 4-12). Parts attach using regular nuts and bolts. It comes with three analog motors and a servo motor. There are even kits available for pneumatic control.

Figure 4-12: Controlling a VEX robot arm

VEX is more complex but less refined than MINDSTORMS. The FIRST competition (For Inspiration and Recognition of Science and Technology) indicates how MINDSTORMS compares to VEX. The competition has three categories: the LEGO competition, the VEX competition, and the original FIRST Robotics competition. The LEGO competition is the easiest, VEX is moderate, and the non-kit competition is the hardest.

The Super-Kit from Vex contains 4 motors and 7 sensors, plus 800 parts for only $299. That's a pretty good deal and plenty competitive with the LEGO Mindstorms kits.

Website!

For more information visit:

www.vexrobotics.com/vexiq/products/super-kit.html

Raspberry Pi with BrickPi

Dexter Industries has a history of producing excellent sensors for LEGO Mindstorms. They held a successful Kickstarter in 2013 which raised over $100,000 to produce the BrickPi. This unit allows LEGO NXT and EV3 sensors to be controlled by the Raspberry Pi.

What is the Raspberry Pi? Well, if you haven't heard of it yet, where have you been? This little computer only costs $25 to $35, yet has an abundance of outputs, including HDMI, two USB ports and many input/output pins. It even runs Linux, just like the EV3! By plugging the BrickPi into the Raspberry Pi (see Figure 4-13) you can also plug in LEGO Mindstorms sensors and motors.

Website!

The BrickPi costs $80 from Dexter Industries:

www.dexterindustries.com/BrickPi/

Figure 4-13: Viewing the Raspberry Pi with BrickPi

CHAPTER 5

LEGO Parts

TOPICS IN THIS CHAPTER

- Beams
- Liftarms
- Pins
- Axles
- Axle Accessories
- Tires and Wheels
- Gears
- Cables
- Other Parts

EV3 parts are incredibly versatile and, when combined with an active imagination, can be used to build thousands of different machines. It is hard to appreciate the variety of machines you can create with this kit. For example, just within the radio controlled car category, this book contains several different models. Looking further out over the Internet you truly see a huge variety of cars. Some cars have different features from others, such as high speeds and sharp turning radius. Most will have radically different appearances. The diversity of machines, even within one minor category, is astounding.

This book alone describes plans many diverse projects. Yet, despite the how dissimilar these machines may be, the basic parts used to create them are the same.

LEGO Parts Overview

Most parts in the EV3 kit are LEGO TECHNIC™ parts. The product line was first introduced in 1977 as the *Expert Builder* series and was renamed as the TECHNIC series in 1986. These parts differ from the standard LEGO bricks in that they can be used to build complex moving machines.

All the parts in your EV3 kit are of high quality. The accuracy of LEGO's injection molding process is seemingly flawless, with no defects or vestigial plastic. On top of that, the

BIG DECISIONS

A Storage Container

The cardboard LEGO box is not really meant for holding parts on a permanent basis. If you plan to build robots for any length of time, a plastic storage container is a definite asset (see Figure 5-1). But before diving in, it can pay to think a bit about your requirements. Most people will want something portable, easy to carry (a handle), occupy minimal space on a shelf, with removable trays or drawers (so you can lay them out when building), lots of compartments in each tray/drawer, and preferably customizable compartments. It would also be nice if the plastic storage container had room for an assembled robot.

Figure 5-1: Organizing LEGO parts in a storage container.

I recommend looking for something made out of plastic, such as a tool box with removable drawers. A fishing tackle-box or a crafts organizer will also fit the bill. Search for "tool box" on Amazon to get an idea of what is out there.

quality of the plastic is impressive. Each part is almost indestructible, and won't easily scratch, dent, bend or break. I've been especially hard on the gears and, surprisingly, they did not shatter or wear down.

Website!

All LEGO parts and sets have a part number. You can look up the EV3 kit by its number (31313 for the consumer EV3 kit and 45544 for the Education Core Set) on the *Brickset* web site and see all the numbered parts. You can also look up a particular part to determine which LEGO set it belongs to, as well as sources to purchase individual parts. Visit www.brickset.com.

Just for Fun!

The Good, the Bad and the Ugly

The EV3 kit gets some things better than previous Mindstorms kits, while some things are decidedly not better.

GOOD: Lots of gears. More small tires.

BAD: Where is my 4 unit axle? There are no yellow frictionless axle pins. There are no motorcycle wheels.

UGLY: Perpendicular pins are absent. The price is higher.

Not including electrical parts such as cables and sensors, the EV3 kit contains 586 pieces and 90 unique building parts. (The NXT 2.0 kit contained 612 pieces and 83 unique parts, while the older RIS kit contained 141 unique parts.) Perhaps the greatest addition to the EV3 kit in terms of plastic parts are the frames, which allow for rapid prototyping and strong skeletons for your robots.

Most of the parts in the LEGO kit were chosen based on the five demonstration models included in the LEGO software. EV3RSTORM, the humanoid robot showcased on the box cover, had the most influence on the parts selection. If you build EV3STORM you will notice it uses the exact number of many parts in the kit.

The biggest challenge for first time LEGO builders is to figure out what all the odd-shaped parts are capable of building. Let's look through the parts contained in your kit and see what they can do.

NOTE: *Part names are taken from the Brickset website inventory. The number following some parts refers to the length in LEGO units. Sometimes the name itself has a number followed with the letter "L," which also refers to LEGO units. The next number is the part number, and the number in parenthesis refers to the number of parts in the consumer EV3 kit.*

letter name length — part number total color

A: **Beam 2 with hole and axle hole** 60483 (10 black)

Beams

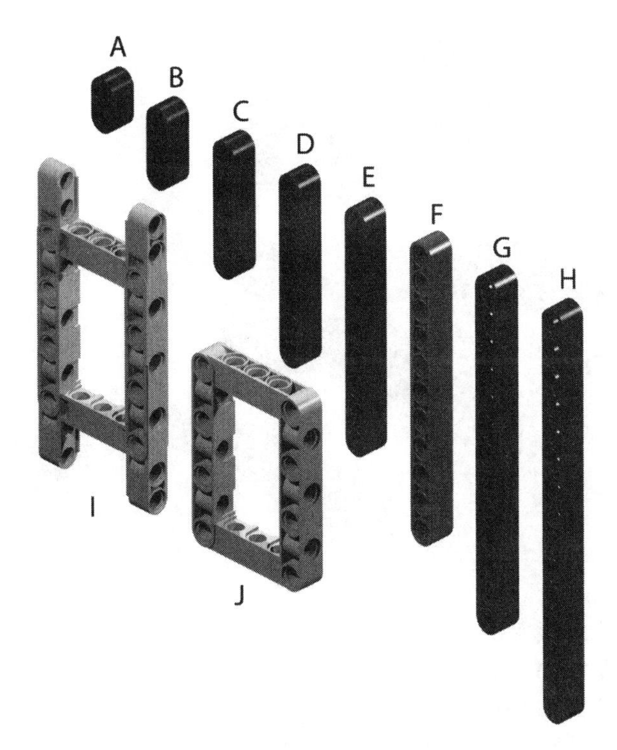

A: **Beam 2 with hole and axle hole** 60483 *(10 black)*
B: **Beam 3** 32523 *(12 black)*
C: **Beam 5** 32316 *(10 black)*
D: **Beam 7** 32524 *(6 black)*
E: **Beam 9** 40490 *(8 black)*
F: **Beam 11** 32525 *(4 red)*
G: **Beam 13** 41239 *(4 black)*
H: **Beam 15** 32278 *(4 black)*
I: **Beam Frame 11x5** 64178 *(2 gray)*
J: **Beam Frame 7x5** 64179 *(2 gray)*

Beams (sometimes called girders) often form the chassis of robotic creations. They have no studs, only axle holes. When beams are bent they are called liftarms (see section below). All lengths are an odd number of LEGO units (except the two unit beam), so most of your creations will have holes along the center axis.

There are also two types of frames included in the EV3 kit which make life a lot easier. Since they are solid pieces, they also make for sturdier robots.

Liftarms

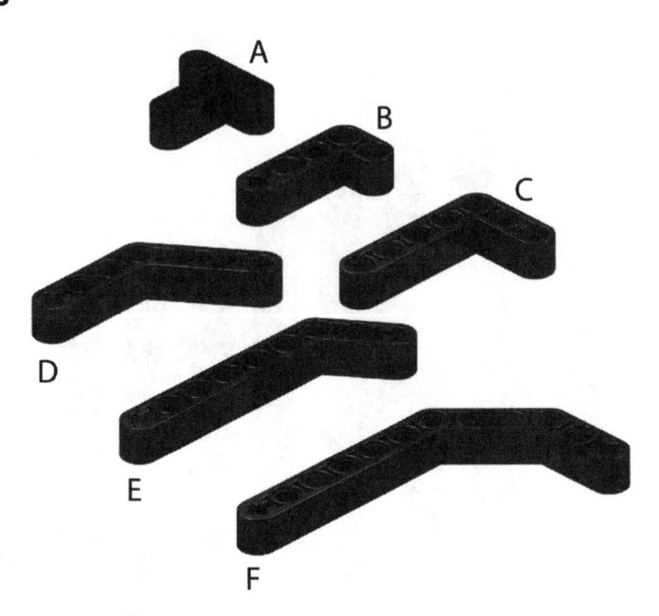

A: **T-Beam 3 x 3** 60484 *(4 black)*
B: **Liftarm 5 Bent 90** 32140 *(8 black)*
C: **Liftarm 7 Bent 90** 32526 *(6 black)*
D: **Liftarm 1x7 Bent 53.5** 32348 *(4 black)*
E: **Liftarm 9 Bent 53.5** 32271 *(12 black)*
F: **Liftarm 11.5 Bent 45 Double** 32009 *(12 black)*

Liftarms are useful for altering the angle of your construction and attaching motors to the EV3 brick. The EV3 kit contains no studs (other than on a few token bricks), which means you will need to find clever ways to stack beams on top of other beams. Liftarms are ideal for constructing boxes with beams.

Pins

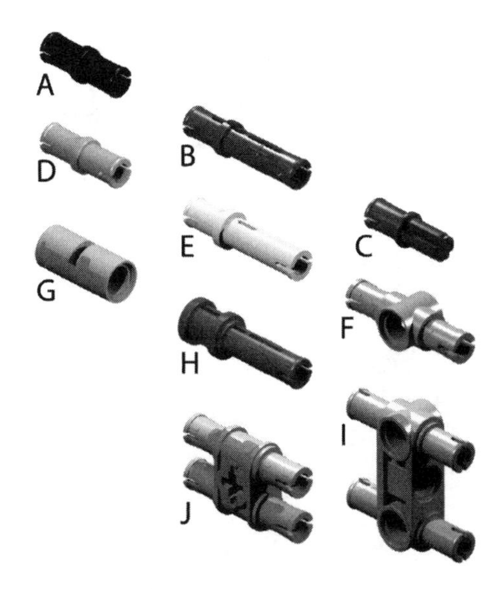

A: **Pin with Friction** 4459 *(95 black)*
B: **Pin Long with Friction** 6558 *(38 blue)*
C: **Axle Pin with Friction** 43093 *(28 blue)*
D: **Pin** 3673 *(4 gray)*
E: **Pin Long** 32556 *(4 tan)*
F: **Pin Long with Pin Hole** *(6 gray)*
G: **Pin Joiner Round with Slot** 62462 *(2 gray)*
H: **Pin Long with Stop Bush** 32054 *(10 red)*
I: **Cross Block with 4 Pins** 48989 *(12 gray)*
J: **Pin 3L Double** 32136 *(4 gray)*

TECHNIC pins are the glue that holds your robot together. As a result, you will probably end up using a lot of them. These pins are inserted in the holes of beams to connect to other beams. There are more black friction pins in the EV3 kit than any other part, which is a clear indication that you will use a lot of these in everything you build. Blue axle pins with friction are useful when you need to join a beam to an X-shaped axle hole.

The gray non-friction pins are used when you need a joint between two beams that moves freely. If you need more than four non-friction pins, you will have to rely on the black friction pins (which don't move as freely), or the long tan non-friction pins.

Sorely missed from previous kits are the non-friction axle pins, which were useful for attaching idler gears or wheels to beams. Instead, you will have to make due with blue axle-pins, which do not turn as freely.

Gone also are the 90-degree connectors, which were very useful for connecting beams at perpendicular angles. The addition of square frames perhaps cuts down on the need for these, but I still find myself wishing they were present.

Axles

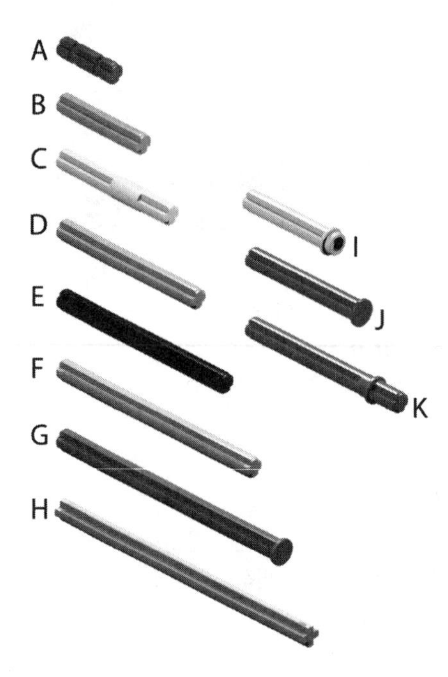

A: **Axle 2 Notched** 32062 *(12 red)*
B: **Axle 3** 4519 *(22 gray)*
C: **Axle 4 with middle cylindrical stop** 99008 *(3 tan)*
D: **Axle 5** 32073 *(9 gray)*
E: **Axle 6** 3706 *(9 black)*
F: **Axle 7** 44294 *(2 gray)*
G: **Axle 8 with stop** 55013 *(6 dark gray)*
H: **Axle 9** 60485 *(1 gray)*
I: **Axle 3 with Stud** 6587 *(4 yellow)*
J: **Axle 4 with stop** 87083 *(4 dark gray)*
K: **Axle 5.5 with stop** 32209 *(2 dark gray)*

Axles are used for rotating parts, but occasionally can be used to connect a series of beams together. They can also be used to create legs, arms, or other specialized items. The color of the axle has no bearing on the function; they all exert the same friction.

It would have been nice to have some regular 4-unit axles. A few times I've been forced to use two 2-unit axles when a 4-unit axle would have been better.

Axle Accessories

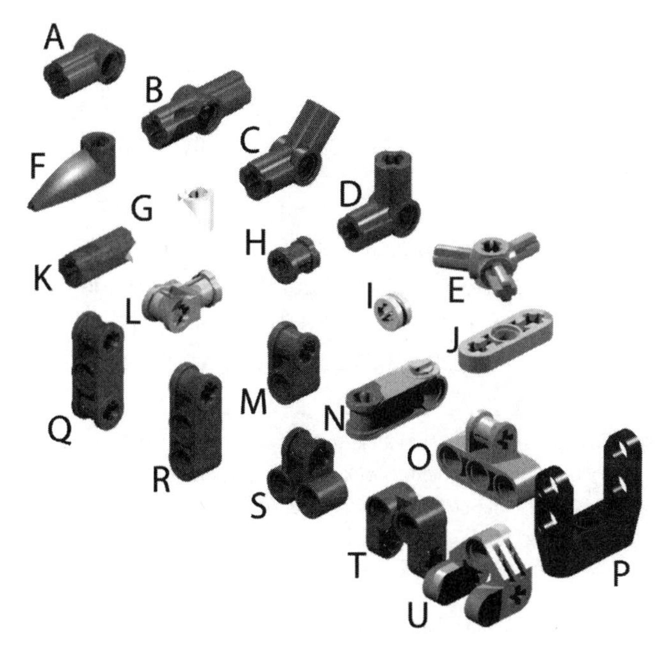

A: **Angle Connector #1** 32013 *(4 red)*
B: **Angle Connector #2** 32034 *(6 red)*
C: **135 Degree Angle Connector #4** 32192 *(4 red)*
D: **90 Degree Angle Connector #6** 32014 *(1 red)*
E. **Bush with 3 Axles** 57585 *(1 gray)*
F: **Bionicle Tooth with Axle Hole** 41669 *(6 red)*
G: **Bionicle Tooth with Axle Hole** 41669 *(4 white)*
H: **Bush** 3713 *(9 red)*
I: **Bush ½ smooth** 32123 *(11 yellow)*
J: **Liftarm 3x0.5** 6632 *(1 gray)*
K: **Axle Joiner Smooth** 59443 *(3 red)*

L: **Connector with Axle Hole** 32039 *(2 gray)*
M: **Axle Joiner Perpendicular** 6536 *(8 red)*
N: **Cross Block 1x3** 32068 *(4 gray)*
O: **Cross Block 3x2** 63869 *(2 gray)*
P: **Connector Toggle** 87408 *(1 black)*
Q: **Axle Joiner Perpendicular 3L** 32184 *(17 red)*
R: **Axle Joiner Perpendicular with 2 Holes** 42003 *(14 red)*
S: **Axle Joiner Perpendicular Double** 32291 *(2 red)*
T: **Axle Joiner Perpendicular Double Split** 41678 *(4 red)*
U: **Cross Block 2x2x2** 92907 *(2 gray)*

The LEGO EV3 kit has a variety of accessories that can couple with axles. Angle connectors are used to support axles or to alter the angle of an axle. Each connector has an identity number that is used to identify the angle.

Bushes are used to secure axles in place. Although the full bush has more surface area, the half-bush exerts about the same friction on an axle because the axle hole is smaller. If you need to make your robot extra secure in places, try using two half-bushes in place of a full bush. The red and white Bionicle teeth are mostly decorative, but they can act as bushes if you run out of standard bushes.

WARNING: *Bushes are some of the weakest parts in the EV3 kit and sometimes crack along the top or bottom rim. They still work as separators but they don't exert very much friction on the axle after they crack. If a bush is not holding firm, make a careful examination to see if it is cracked.*

Often you need a longer axle than provided in the EV3 kit. Axle joiners allow you to connect two axles together. It might appear that the kit only contains three axle joiners to extend axle lengths. However, the angle connector #2 works almost as well as an axle joiner. If you are really pressed, there are other ways to extend axles (see Figure 5-2).

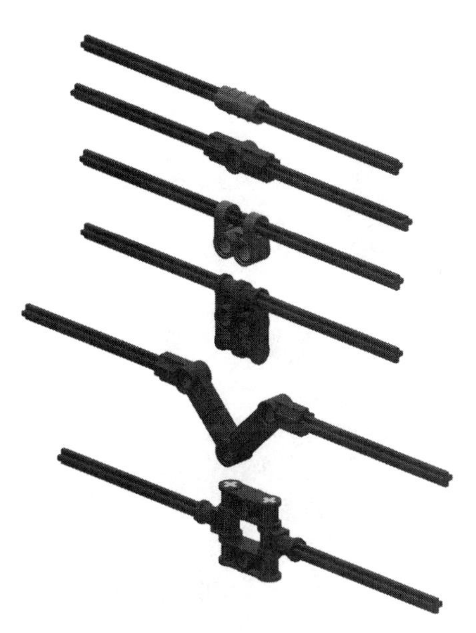

Figure 5-2: Extending axles.

My favorite parts are the double-perpendicular axle joiners. Together, these parts allow you to attach beams in a number of ways. There are two different double-perpendicular axle joiners that are complimentary to one another and fit together with an axle (see Figure 5-3: A and B). You can also fit a double-split perpendicular axle joiner with a dual-perpendicular pin joiner to create a door hinge. You can use a variety of other parts with the double-split perpendicular axle joiner to create other parts (see Figure 5-3: C through G).

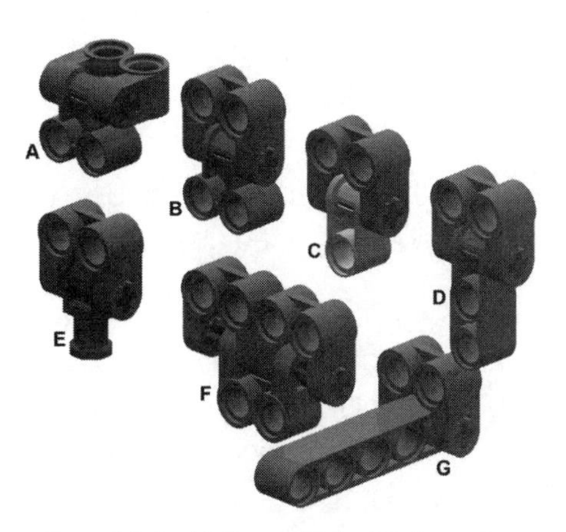

Figure 5-3: Double Perpendicular Axle Connectors.

Tires and Wheels

A: **Tire 43.2 x 22 ZR** 44309 *(4 black)*
B: **Wheel 30.4 x 20** 56145 *(4 black)*
C: **Wedge Belt Wheel Tire** 2815 *(3 black)*
D: **Wedge Belt Wheel** 4185 *(3 gray)*
E: **Tire 30x11** 50951 *(2 black)*
F: **Wheel Hub with Center Groove** 42610 *(4 gray)*
G: **Caterpillar Track with 36 Ridges** *(2 black)*

There is a moderate choice of tires in the EV3 kit, in fact better than the NXT kits. On top of tires, there are a pair of tank treads. The two treads are rubberized and contain 36 ridges. These ridges mesh perfectly with the black wheel hubs and allow your robot to crawl over varied obstacles.

It would have been nice to have a pair of larger diameter wheels, such as the motorcycle wheels that were once included in the RIS kit. These wheels are better suited for outdoor vehicles, allowing them to crawl over larger obstacles, and of course they are perfect for motorcycles.

> ***Website!***
>
> An Omni-wheel is a special wheel that allows a robot to move in all directions. If you want to try out these wheels, you can order them online (and view videos) from the following website:
>
> www.holonomicwheels.com

Gears

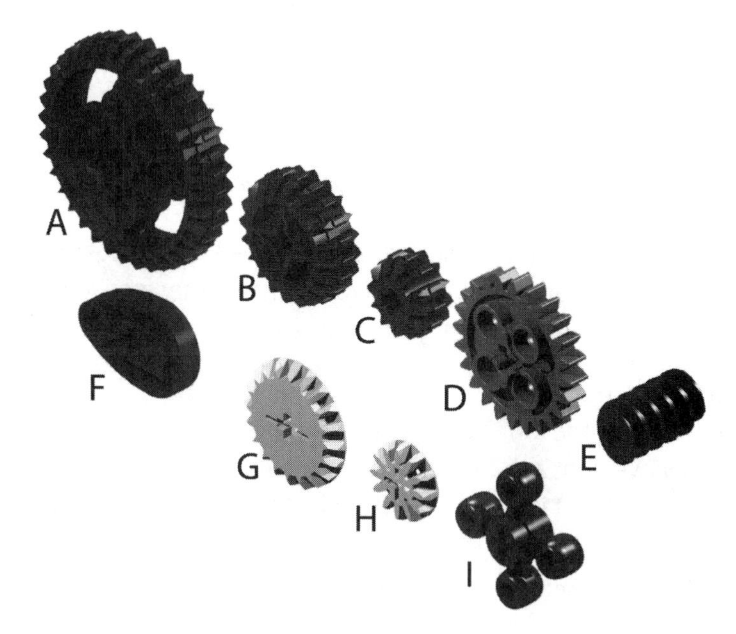

A: **Gear 36 Tooth Double Bevel** 32498 *(5 black)*
B: **Gear 20 Tooth Double Bevel** 32269 *(4 black)*
C: **Gear 12 Tooth Double Bevel** 32270 *(2 black)*
D: **Gear 24 Tooth** 3648 *(2 dark gray)*
E: **Worm Screw** 4716 *(2 gray)*
F: **Cam** 6575 *(2 black)*
G: **Gear 20 Tooth Bevel** 32198 *(1 tan)*
H: **Gear 12 Tooth Bevel** 6589 *(2 tan)*
I: **Knob Wheel** 32072 *(4 black)*

Torque is force in a rotational motion. A very powerful motor can produce a lot of torque, and likewise a weak motor produces low torque. Gears are used to transfer torque from one axle to another, often increasing or decreasing torque in the process. When two or more gears decrease speed, torque increases. When gears increase speed, torque decreases (torque and speed have an inverse relationship).

The smallest gears in the kit are the two 12-tooth bevel gears. Using these with the other compatible bevel gears, we can modify torque with the following ratios (see Table 5-1).

Gear	Ratio with 12-tooth gear
12-tooth	1
20-tooth	1.67
36-tooth	3

Table 5-1: Determining gear ratios.

Gear reduction occurs when a smaller gear connects to a larger gear. For example an 12-tooth gear on one axle will rotate three times for every one time a 36-tooth gear rotates. This is a 3:1 gear reduction.

Gear reduction can be used to make your tachometer more accurate. If it takes 60 counts to rotate the arm into a position, and your count is off by 2, this will reduce arm movement accuracy by 3% If, however, it takes 600 counts to move the arm, and the count is off by two, the loss in accuracy is only 0.3%.

NOTE: *In Chapter 3 we built a drag racer that used the opposite of gear reduction called a gear multiplier. This gear train used two sets of gears. The first set is 12 to 20, and the next set is 12 to 36. This speeds up the axle 5 times (36/12 * 20/12).*

If you make a very long gear train, you'll notice the *backlash* effect. Gears have a tiny amount of space between the teeth when they mesh. However, if you connect many gears together the spaces are cumulative, which can result in a lot of dead-space between teeth. This means that when the motor starts turning, the final gear won't turn until the backlash is consumed.

Most of the gears in the EV3 kit are bevel gears, which appear fatter and rounder than regular gears. Bevel gears can be used to change the axis of rotation by 90 degrees. You can also use knob wheels, which look like tire irons. The bevel gears mesh with one another, but they must be spaced properly, sometimes using half-bushes (see Figure 5-4).

Figure 5-4: Meshing Bevel Gears. Note the half-bushes.

Knob wheels perform the same function as bevel gears by changing the axis of rotation. Knob wheels are superior to the bevel gears, which tend to slip when under large forces. If you are creating a gear transfer that has substantial force on it, such as an arm or rough-terrain vehicle, use knob wheels.

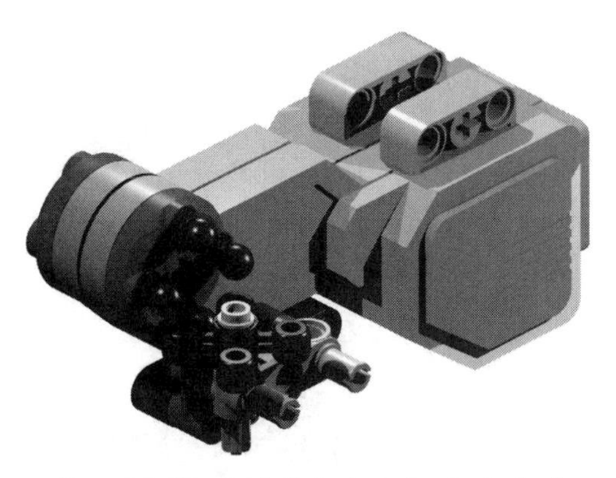

Figure 5-5: Using towball pins to replace knob wheels.

Worm screws are another part that can be used to change the axis of rotation. Worm screws can also be used for gear reduction. In fact, they are the most effective gear reducers; one complete rotation of a worm screw moves any gear by only one tooth. Rotation can only travel from the worm screw to regular gears (such as the dark gray 24-tooth gear), but regular gears cannot turn the worm screw (it locks if you try).

> ***Tip!***
>
> There are only 4 valuable knob-wheels in the kit, but you can stretch this number out by placing towball pins in the holes of the motor (see Figure 5-5).

In a combustion engine, cams are attached to an axle to drive the axle as the pistons move up and down. Conversely, in LEGO the cam is used to change a circular motion into a *reciprocal* motion (back and forth movement). In most robots in which I've seen the cam used it was to raise and lower a LEGO part, frequently an arm or a beam. The smooth egg shape of the cam allows it to slide easily along the underside of a hinged beam as it applies force to it.

Cables

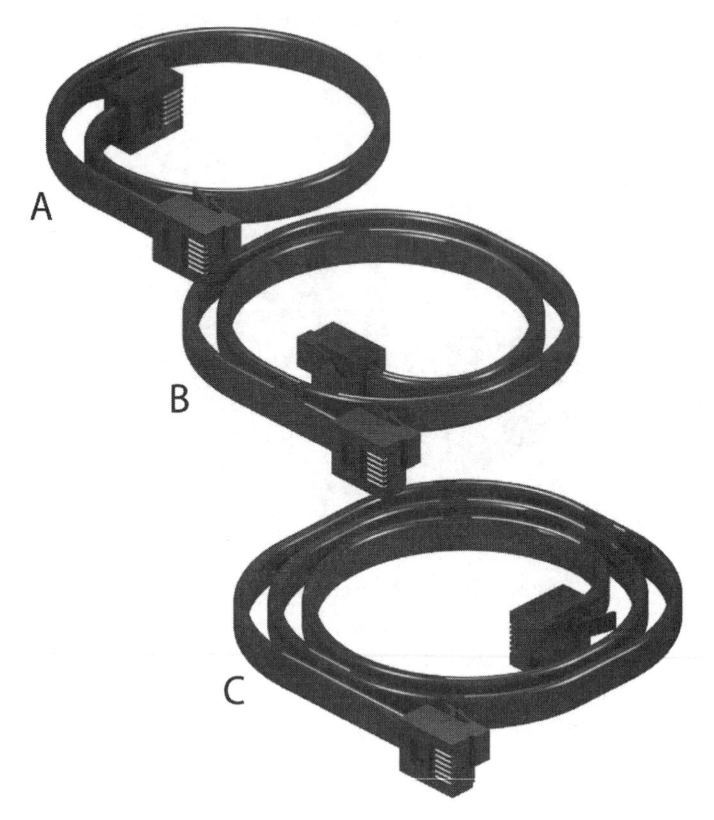

A: **Short Cable** 55804 *(4 black)*
B: **Medium Cable** 55805 *(2 black)*
C: **Long Cable** 55806 *(1 black)*

Standard MINDSTORMS cable lengths are 25 cm (10″), 35 cm (14.5″) and 50 cm (18.5″). One thing not immediately obvious is that if the cable is laid flat, the RJ12 connector faces up on one end and down on the other. This so the wires connect to the correct pins on the RJ12 connectors.

There is no part to connect two RJ12 connectors together into one longer-wire. However, different lengths can be purchased from other vendors for a modest cost. HiTechnic supplies a set of four wires: 12 cm (4.75″) - extra short, 16 cm (6.3″) – short, 70 cm (27.6″) – long, and 90 cm (35.4″) - extra long. Mindsensors. com sells wires that use ribbon cable. The advantage of ribbon cable is that they flex and twist easier than hard rubberized cables; the downside is they are not as rugged.

Other Parts

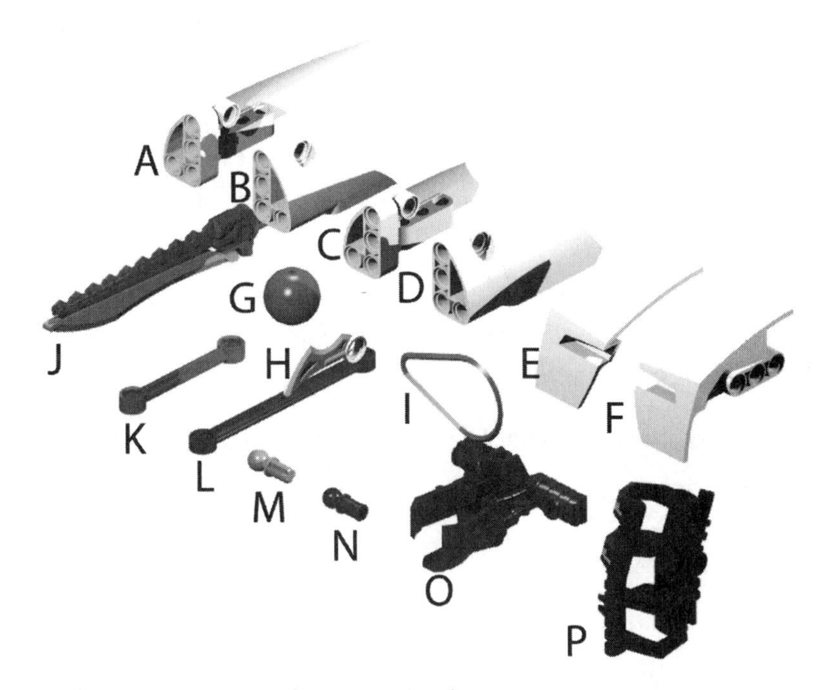

A: **Panel Fairing Long Right** 64393 *(3 white)*
B: **Panel Fairing Long Left** 64681 *(3 white)*
C: **Panel Fairing Medium Right** 64391 *(3 white)*
D: **Panel Fairing Medium Left** 64683 *(3 white)*
E: **Panel Fairing Left** 61071 *(1 white)*
F: **Panel Fairing Right** 61070 *(1 white)*
G: **Bionicle Zamor Sphere 16.5 mm** 54821 *(3 red)*
H: **Curved Blade** 98347 *(4 white)*
I: **Rubber Belt** *(1 red)*
J: **Sword with Silver Blade** 98568 *(6 red)*
K: **Steering Link Type 2** 2739B *(2 black)*
L: **Steering Link 9L** 32293 *(4 black)*
M: **Axle Towball** 2736 *(6 gray)*
N: **Friction Pin with Towball** 6628 *(6 black)*
O: **Bionicle Zamor Sphere Launcher** 54271 *(1 black)*
P: **Bionicle Zamor Sphere Magazine** 53550 *(1 black)*

The EV3 contains a large selection of white panels, which are generally used for decoration only. There are also several swords and blades, which again are generally used for decoration.

The Zamor Sphere canon is a fun addition to the EV3 kit and allows your robots to fire spheres with a decent velocity. To test out the cannon, attach the magazine to the launcher, load some balls into the magazine and stick an axle through the bottom hole.

Steering links and towballs are useful for steering mechanisms and tilting. The box cover robot uses these parts to shift the robot from side to side when it walks, and to move the arms and head.

The next chapter will examine different ways to use collections of parts from this chapter.

CHAPTER 6

Building 101

TOPICS IN THIS CHAPTER

- Design Patterns
- Laws of LEGO
- Engineering Goals

In the previous chapter we explored the individual parts contained in the EV3 kit and identified their function. We also described the versatility of the parts and how they can be used for more than one purpose. In this chapter we will look at collections of parts—mechanisms that consist of two or more parts but produce a single function. We will also look at some of the principles that govern the practical use of LEGO. The final part of this chapter will explore general goals that apply not only to LEGO, but to any design.

How to Invent

Before you start putting pieces together, it can help to have some idea of what you want to build. This inspiration can be an original idea, in which you are thinking about a problem and want to come up with a solution. This is invention in its purest form.

The second type of invention is when you are inspired by existing technology, and want to create your own version of the item. This is the Apple school of invention, where designers redesign existing products with their own spin.

Assuming you have an idea in mind, now you need to build it. Maybe these words will help you, maybe they won't, but the bottom line I want to get across to you is how design actually happens, as opposed to the clean, overly optimistic version you sometimes see in movies. Keep in mind, the real way design happens is a lot more uncertain than a 4-point checklist. With that said, here is a 4-point checklist.

Step 1: Slap something together that kind of works. In this step, you just want the basic parts in some sort of configuration that works a little bit. The robot is guaranteed to look ugly, but don't worry too much yet. Any combination of LEGO parts, no matter how convoluted or silly, will do in this step.

Step 2: Refine. By this time, you have some idea of what mechanisms your robot needs to work and how they fit together. In this step, you want to take all those convoluted kludges and turn them into something more presentable. Usually this involves replacing many parts with fewer parts.

Step 3: Identify problems. With pen and paper in hand, begin using the robot. Note down everything wrong, no matter how trivial. Many of the problems have to do with unintended flexing where it needs to be rigid, and insecure parts such as gears.

For example, with the SCARA robot in chapter 18, there were many small problems with the first version. The base was too unstable and bendy, the arm flexed too much, the claw joint was too weak, and the whole robot tipped over when the arm was in certain positions. Whew!

Step 4: Fix all the bugs from step 3. This one is pretty obvious. For the SCARA robot, I reinforced the base, added supports to the arm, redesigned the claw joint, and rebalanced the robot so it was more stable.

And then I repeated steps 2, 3, and 4 a few more times. With the obvious bugs out of the way, there were smaller bugs in the design I missed the first time. Also, each revision can introduce new bugs. As always, I made sure to step away from the robot for a day or two before looking at it again. It's surprising what a fresh pair of eyes and a little distance can do to make things more obvious.

Design Patterns

Design patterns are ways in which parts are put together to produce a specific function or mechanism. While much of the great challenge and fun of LEGO is discovering how to create mechanisms, there is no need to re-invent the wheel. This section will help you to familiarize yourself with common design patterns to give you a head start on your projects.

Caster Wheels

Two-wheeled robots are the most common platform for robotics because they are ideal for navigating in confined spaces. These types of robots usually require a smaller third wheel, called a *caster wheel*, to maintain balance. A caster wheel has the property of being able to rotate in all directions. You can often see these in the real world on desk chairs.

The LEGO EV3 kit includes both large and small wheels, but in general, you should use a smaller wheel since they swivel more easily than large, wide tires. In the EV3 kit, you can use the smallest wheel for a castor (see Figure 6-1).

Figure 6-1: A caster wheel.

The placement of the caster on your model is important. Casters form the third point of a triangle in your model, with the two drive-wheels forming the other two points. This gives your robot stability. Casters work best when they are in the same circle as the main wheels (see Figure 6-2 A). If you notice your vehicle jumps around while rotating, the caster is probably too far outside the circle (Figure 6-2 B). You should also position the mass of your robot over the center of the triangle for stability.

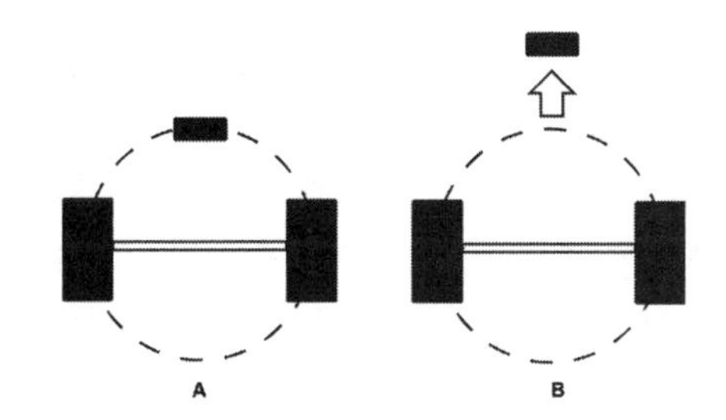

Figure 6-2: Proper and improper placement of caster wheel.

Lateral Motion

Lateral motion is movement in a straight line. Some devices that require lateral motion are scanners and photocopiers. Normally these machines use a gear rack, which is a beam with many gear teeth. EV3 kits do not contain gear racks, requiring that you find an alternate method for producing lateral motion.

Lateral motion can be difficult to achieve because a motor moves in a circle. Technically a vehicle achieves lateral motion by driving forward. However, there is a way to produce constrained, lateral motion without using bulky tires.

Figure 6-3 demonstrates a mechanism using EV3 parts to create lateral motion. It relies on a worm screw and a half-bush on an axle. The worm screw engages the half-bush and moves it a small distance. To achieve greater lateral movement, more half-bushes can be added (at spaced intervals) to a longer axle.

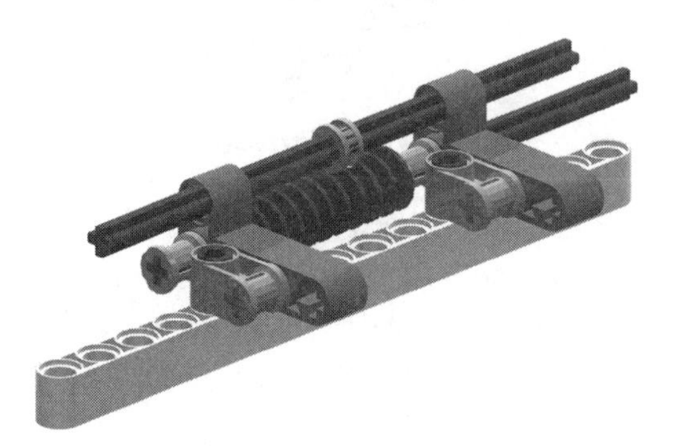

Figure 6-3: Lateral motion using a worm screw.

Periodic Movement

Sometimes you want a robot to perform novel periodic movement (such as turning its head or moving arms) without using an extra motor. You can create periodic movement by placing a gear on a main drive axle using one of the off-center holes on the gear (see Figure 6-4). Although regular gray gears are shown, the same principle applies to the black double bevel gears. The off-center gear borrows torque from the drive axle, periodically rotating another gear at intervals thereby generating periodic movement.

Figure 6-4: Movement at intervals with an off-center gear.

You can also use a variation of this mechanism to sequentially rotate four axles one after another using only one motor! This can be used to move four legs in sequence. To accomplish this, place up to four gears around the off-center gear.

Attaching Motors

Sensors are very easy to attach to a frame, whereas motors can pose a greater challenge. LEGO EV3 motors have four basic anchor points, plus the axle hole which can also serve as an anchor point (see Figure 6-5). Surprisingly, motors don't need an especially secure attachment—one pin at either of the connection points, plus the axle is good enough in most cases.

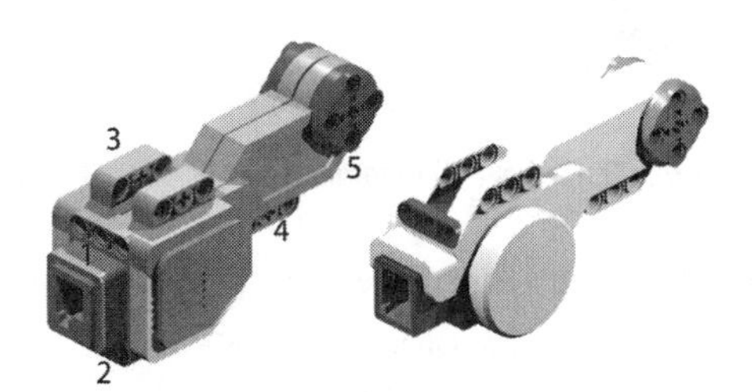

Figure 6-5: EV3 motor anchor points.

The unusual shape of the motors may also appear to be a hindrance, but the design actually works in your favor. For example, when building a robot arm, the motor extends the length of the arm. As a result, you can build robots using fewer parts.

Although the motor looks similar to the NXT motor at first glance, there are actually several important differences (see Figure 6-5 above). First, there are three more connector holes near the connector port. This allows stable connections of the motor to the robot chassis using one of the frames included in the EV3 set (see Figure 6-6).

Figure 6-6: Securing the motor to a frame.

Also, the motor axle hole is now in line with the most common/stable connection point (see Figure 6-7). This makes it easier to design some things where you want the joint to be in line with a beam, such as with an arm for example.

Figure 6-7: Alignment of the motor axle with rear connectors.

Another difference is the size. The part where the physical electrical motor resides is more bulky than the NXT motors. With one of my tank designs I wanted to put the wheel hub right next to the motor. However, the size of the bulky part of the motor interfered with the wheel hub and I had to abandon that design.

Tip!

When building a robot, you might periodically want to test movement without writing a small test program. A simple method to do this is to remove one gear of the gear-train so the motor is no longer resisting, and then spin the part with your fingers. Using this method, you can gauge how hard the motor must work, which can focus your mind on making low resistance gear ratios.

Sometimes when you attach your motor, the forward function makes the motor rotate backwards. So which way is forward? Figure 6-8 indicates default forward rotation of the motor. This is not necessarily a problem, since you can easily alter your code. different.

Figure 6-8: The default forward direction.

Cable Management

If you built LEGO's Ev3rstorm model, you understand the challenges involved with managing cables. Sometimes it can be heartbreaking to have to plug cables into a clean robot design. Suddenly your neat robot looks like a Medusas head of tangled wires! To add to your troubles, the cables are semi rigid, not pliable like a rubber band. Due to this rigidity, they can interfere with the movement of your robot. When this happens, you need to think seriously about cable management.

The easiest method of cable management is to allow the cable to spring out from the back of the robot in an arc, clear of any obstructions. This gives your robot a Japanese Manga look. It also protects the EV3 brick, since the wires can cushion the impact if your robot falls over.

If the cables continue to get in the way, the standard means to remedy this is to use two long pins with bushes (see Figure 6-9 A). To trap three cables, use long-pins spaced one unit apart with a 3-unit beam to cap it off (see Figure 6-9 B).

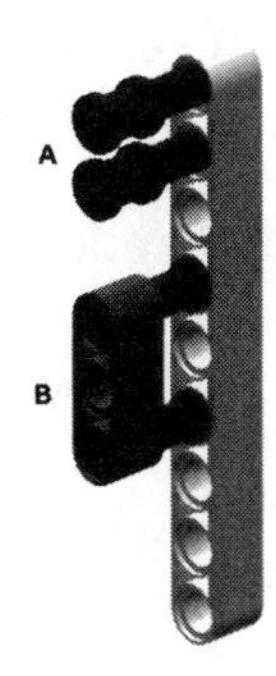

Figure 6-9: Cable management

Sometimes the selection of cables in the EV3 kit is not appropriate for your design. In this case, you can purchase different lengths from LEGO, Hi-TECHNIC or Minsensors.com.

Laws of LEGO

When most people think of LEGO, they think of little square bricks with studs that lock onto other bricks. Upon opening an EV3 kit, these same people may wonder what happened to the LEGO. They might be hard pressed to find any bricks with studs.

EV3 kit contains a menagerie of unusual parts that give LEGO a hi-tech look. For example, EV3 parts include TECHNIC beams with a length of holes. The EV3 brick and sensors only have holes for pins and axles. This is a unique paradigm for building, and this section will get you acquainted with this approach.

When you are faced with limitations, problems are solved by being creative. For example, creativity seems to flow in constructing a Haiku when one is limited to only a few syllables. Likewise, LEGO parts are simple structures with limits to their application. You are restricted by the number of parts, the type of materials, or how the parts can be attached to one another. Rather than impeding creativity, I think these limitations enhance creativity.

Bricks are good for building houses, but in the real world bricks are not used to build robots. You need rotating axles and structural beams for most robot projects. For this reason, the emphasis in the EV3 kit is on beams.

In the world of engineering, machines require screws, nuts, bolts, nails, glue, or welding to hold parts together. LEGO does not use these types of joints. Instead, everything is held together by friction. Friction, however, does not produce the most secure joint, especially when large forces are subjected to your models by powerful servo motors. Let's explore how to build functional models that won't fall apart.

Take a good look at the pins in the previous chapter. These are the main components for holding your robot together. The black friction pins are especially plentiful in the kit because these are the most versatile connectors.

The three-hole structure is everywhere on the main EV3 parts—on sensors, motors, and the EV3 brick. This is the key to understanding how to connect different parts.

At first, you may feel that you need precognitive abilities to predict whether or not the beam-holes are going to line up. Rest assured that LEGO knows that we don't all have extrasensory powers. They have designed these parts so that beams and holes will almost always line up at regular intervals.

Building something new is exhilarating. There are times when you will have your model working, yet you see a way to make it better. In order to make that change you need to rip apart what you already have and risk losing a good design in the process. But, you are driven to take the chance to turn a good design into a great design. It's a scary moment that can pay off with success or lead to frustration.

Stacking Beams

To stack studded bricks, you simply press the bottom of one brick into the top of another to gain elevation. With studless design the solution is not as obvious. So how do we do it?

You could try perpendicular connecting beams (Figure 6-10A), however, this not structurally stable and will wobble. To create a stable structure as depicted in Figure 6-10B, you can use L-beams, although this approach can be somewhat bulky. Another option is to attach axle joiners perpendicularly to the main beams (Figure 6-10C). I favor these parts when I need to stack one beam on top of another.

Tip!

A digital camera can be your best friend when you are building. Take a photograph of your robot if you are about to try out something new that requires disassembling a large part of your robot. If something goes wrong you can always use the photograph to help you rebuild your original design.

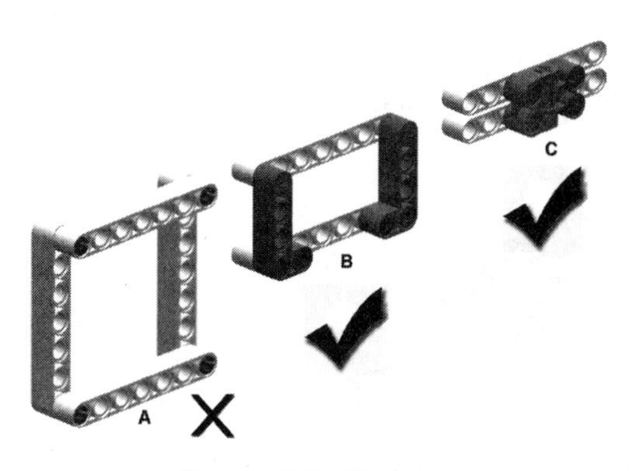

Figure 6-10: Stacking beams.

Tip!

To avoid *LEGO finger* (a harmless reddening of your fingertips) avoid pulling on small parts when trying to remove them from the structure. Axle-pins tend to be the worst offenders. Using an axle to push parts out from the other side will make it easier to pull small items out.

Changing the Axis of Beams

Often when building a robot, you need a row of holes to line up along a beam, but as luck has it, they are in the wrong orientation. This prevents you from attaching an axle or another beam along the proper line (one of the weaker aspects of building with LEGO). To overcome this problem you must transform the orientation from one axis to another—for example, from the x-axis to the y-axis.

There are several ways to do this. Figure 6-11 shows two beams with holes aligned along different axis. The red 2x2 cross blocks provide a sturdy connection at a 90-degree angle. Simply insert a 2-axle to connect the cross blocks then attach them to the beam using friction pins.

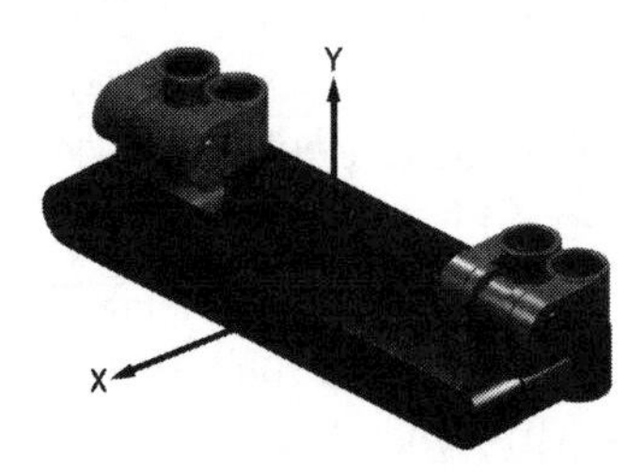

Figure 6-11: Changing the axis of beams.

There are other parts that can change the axis of beams, such as the dual perpendicular joiner (see Figure 6-12). However, because this part is symmetrical along the middle of the holes, it can change the spacing, and the holes along the new beam may not line up with the rest of the parts on your model.

Figure 6-12: Other methods for changing the axis of beams.

Securing Gears

Although it isn't good practice, it is possible to get away with poor design in some aspects of robot construction. Gears however, do not tolerate flimsy design. They need to be properly supported or they will slip while under torque.

Gears that are not properly supported make a telltale clicking sound. This occurs when two gears push away from each other, likely because they are on different beams. When gears are not on the same beam (see Figure 6-13A) the beams will flex, causing the gears to slip. By having two gears on the same beam, this does not occur and the gears function properly (see Figure 6-13B).

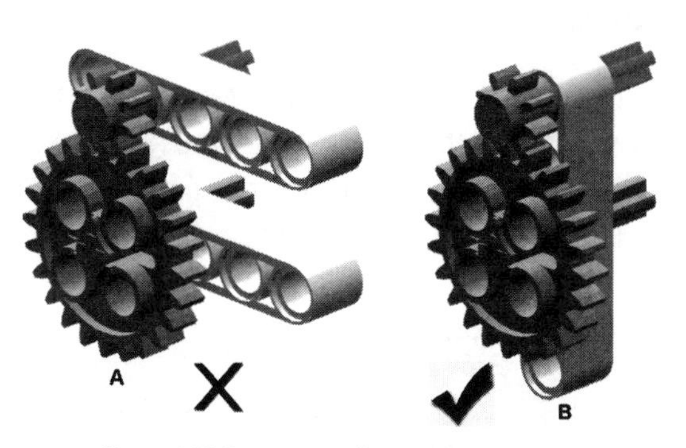

Figure 6-13: Improper and proper gear support.

It is also important to position the gear right next to the beam. If the gear hangs out from the beam, it can slip (see Figure 6-14A). You can force the gears to hug each other by adding a small beam to the ends of each axle (see Figure 6-14B). Even though the gears are hanging precariously in the open, they still work well.

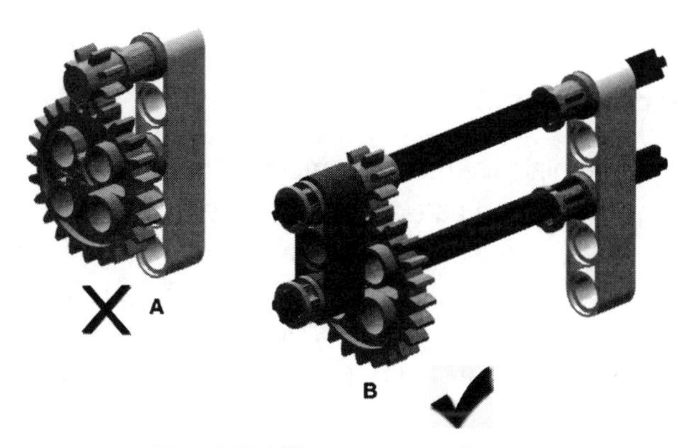

Figure 6-14: Adding extra support for gears.

Engineering Goals

Design goals that apply to the real world of engineering also apply to LEGO. LEGO, after all, also operates in the real world. When designing machines, there are steps that can help you refine your creation. In engineering, this process is called *requirements analysis*. With this method, you start with the requirements and design around them. This section will take a brief look at how requirements analysis can help you.

1. Define Functionality

It's important to know the primary function of your robot and focus on that. Ask yourself, what is the function of this robot? Sometimes it's merely to drive from one point to another. In this case, you need to determine the best steering method for your robot. Does it have to turn on the spot? Is analogue steering more useful?

2. Design around functionality

Once you've determined the functions of your robot, identify the most important part of your robot and build it first. For example, if you are making a robot arm, the claw is probably the most important feature, since every other part of the robot is in the service of moving the claw. Once the claw is done, you can add the supporting mechanisms, such as the arm that supports the claw. When the arm is done, build a base to support the arm.

As mentioned earlier, before trying to make it beautiful, just make it work. Once the core mechanism is working, if aesthetics is important to you, start refining the model by making it sleeker, lighter, and stronger.

3. Common design goals

Goals give you something to strive for. Sometimes the list of goals isn't apparent until you've played with the parts and constructed a prototype. For example, with arms, I didn't realize one of my goals was to make the arm move fast until I tried it out. Similarly, only after trying to have it lift an object did I realized one of my goals was to have it to lift objects of some mass. Ironically, sometimes you realize what all your goals were only when your project is done.

To help you pick your goals, think of the following properties most engineers want from their creations:

- Speed
- Strength
- Power
- Stability
- Accuracy
- Symmetry
- Compactness
- Minimalism
- Robustness
- Modularity

How fast do you want it to move? What sort of payload might your robot carry (arm or vehicle)? Is your robot going to race with other robots? Will it fight with other robots? Depending on the answers to these questions, you will have to provide your robot with the appropriate properties to attain your goals.

Strength

Strength is always valued, even if your robot won't experience rough use. Try dropping your robot from a height of 10 centimeters onto carpet and note what breaks off. Keep working on those joints until they don't break anymore. Some robots break easily when you try to pick them up, so try picking it up in different ways. Study the joint that failed and develop a means to strengthen that joint until it holds tight.

Speed

Speed is achieved in two ways: motor/gear/axle speeds and programming. The ARM9 processor inside the EV3 brick is fast. It processes faster than the motors are able to react. Take advantage of this. Don't have your robot pause using Thread.sleep() unless it is absolutely necessary. Try upping the motor speed as high as it can go. See if the LEGO motors can handle higher gearing, which can make your axle spin faster. You can also choose the medium EV3 motor over the larger one to attain higher speeds.

Power

The trade-off with speed is power. Since the specifications of LEGO EV3 motors are the same, you can have speed or power, but not necessarily both. A powerful robot with geared-down axles will be slower. You decide which is more important for your creation.

Stability

Stability is important to any design. Whether it is a stationary device or a moving robot, your creation should not be prone to tipping over. Chapter 27 on balance and walking explores this concept in more detail.

Accuracy

An arm that is only accurate to within five centimeters might not be very effective if it tries to pick up a marble. You can increase accuracy by gearing down motor movements or choosing the medium EV3 motor. Also, make sure your gears don't slip, otherwise your motor tachometers will no longer provide useful data.

Symmetry

Symmetry not only looks pleasing to the eye, but it can also aid other design goals, such as stability and compactness. There is a reason most animals in nature have symmetry.

Compactness

Making your robot mechanisms as compact as possible for the size of your robot. That is not to say you should make your robots small—it is perfectly valid to want or require a large robot. However, the mechanisms like gearing should be as compact as possible. Allow as little air as possible between parts.

Minimalism

Believe it or not, you want to use as few parts as possible. There's a saying in engineering, "You are done when you can no longer remove any parts". For starters, make sure there are no unattached pins sticking out. But, more importantly, optimize your design to use as few pieces as possible without compromising the essential features of your robot. For example, if your design uses four beams and you can come up with a design that does the same job with only three beams, the latter is superior.

Robustness

A robust robot is not only good on hard floors but also carpet. Your robot is robust when it deals well with diverse conditions. Making your robot easy to tune can also increase its robustness. For example, a robot that allows you to adjust the length of the legs can handle diverse situations.

Modularity

Modularity is important, especially with complex machines. The robot arm in this book is a prime example of modularity. It consists of a base unit that rotates like a crane, a lower arm, a forearm, and a claw. If you want to change any part of this arm, you can change one module without having to reconfigure the other modules.

To make modularity work, you need to know how the modules connect to each other. In practice, modularity is often inherent when you build something because modules tend to group together naturally.

CHAPTER 7

The World of leJOS

TOPICS IN THIS CHAPTER

- leJOS Overview
- The Oracle Java Virtual Machine
- Exploring the leJOS Platform

LEGO's programming language is great, especially for those just learning to program. However, if you are already a programmer and feel more comfortable with a professional language or if you think you have a future in programming, you may want to move to something more advanced.

This book uses leJOS, a software package that allows you to program the EV3 brick using Java. Why choose leJOS over the LEGO EV3 software? There are many answers but the most important is that leJOS allows you to do things you can't do with the standard LEGO software. For example, there are some very complex navigation classes in leJOS which would be difficult to program with the LEGO software. I'm not saying the LEGO software can't do this—it probably can. It just gets complicated.

If you are a Java programmer, you already know just how good the language is. This chapter will introduce what Java programmers have been craving—a Java environment complete with threads, arrays, floating point numbers, recursion, garbage collection, and total control of the EV3 brick.

Welcome to leJOS

The leJOS platform builds upon the previous version of leJOS, which ran on the LEGO NXT brick. The project developers took the leJOS code and ported it to the EV3 brick. Rather than keeping the code backward-compatible with the NXT, we decided to create a separate branch of code specifically for the EV3.

With previous versions of leJOS, we supplied our own JVM that was minimal enough to run in a small amount of memory of the RCX brick. This JVM was ported to the leJOS NXT brick and expanded. However, now that the EV3 has much more memory and runs Linux, we can use a fully featured JVM. As of this writing, we are using the Oracle JVM (more on this later).

NOTE: *The name lejos means 'far' in Spanish. In the name leJOS, the letters JOS are capitalized because those letters stand for Java Operating System. Since le means 'the' in several languages, this would mean* the Java Operating System. *leJOS now refers to the part of the package that is specifically for the EV3 brick. This includes Java packages such as lejos.hardware.motor (which will be covered later).*

A core part of leJOS, and perhaps the face of leJOS, is the menu system on the EV3 brick. It can almost be considered the GUI for the Linux operating system running on the brick. You can interact with your robot by running programs, running sample code, using tools for controlling motors and reading sensors, and inspect version information. You can also set up Bluetooth, WiFi, and sound options from the menu.

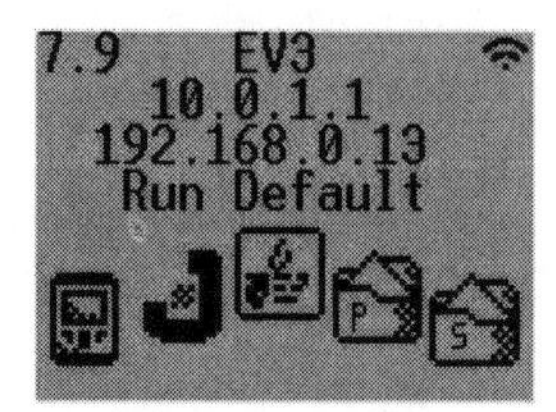

Figure 7-1: Viewing the leJOS menu

There are also tools on the PC side to compile and upload code to the EV3 brick. leJOS is multiplatform, and these days that means Windows, Linux, and Macintosh. leJOS is available for each of these platforms, allowing you to develop leJOS code under your favorite operating system. Among these tools is the EV3 Control Center, which allows you to communicate with the EV3 brick wirelessly or through USB (see Figure 7-2). More information on this tool can be found in the next chapter.

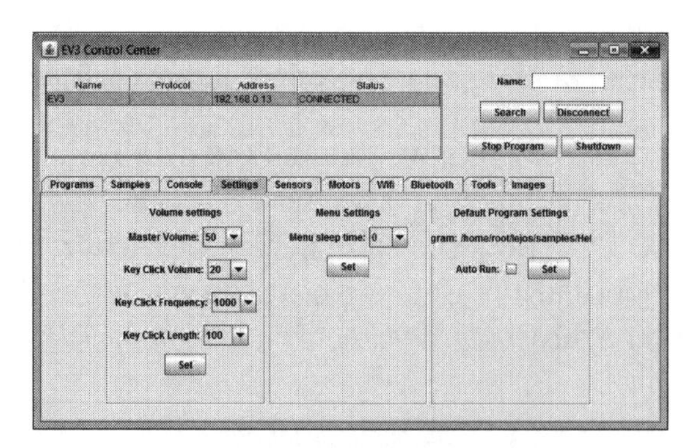

Figure 7-2: Using the EV3 Control Center.

We released the first alpha build of leJOS for the EV3 in January 2014, and since then it has been downloaded thousands of times. The Windows version is outpacing Linux and Macintosh combined by three to one. Germany is once again the top country for downloads. We also get an especially high number of downloads in September when students are going back to school.

Oracle's Java Virtual Machine

The Java Virtual Machine runs your Java code. A JVM is a virtual computer that can run the same code on multiple platforms, such as Windows, Linux, Macintosh, or of course the EV3 brick (because it runs Linux). For the first time ever, leJOS relies on a JVM supplied by an outside company, Oracle. Let's go over the JVM.

Speed is usually not a great issue for most robotics applications. Normally your robot can sit there for a few seconds while it thinks about its next move without anything terrible occurring. There are a few applications that require split-second reactions, such as robots that balance on wheels or on a ball (both of which are demonstrated in Intelligence Unleashed—ISBN 9780986832208).

As mentioned, the JVM runs much faster on the EV3. But how much is due to the EV3 hardware, and how much is due to the Oracle JVM? When we tried running our homemade JVM on the EV3's ARM processor, it ran about twice as fast as on the NXT. However, the benchmark speeds for the Oracle JVM indicate it runs about 25 times faster than our homemade JVM on the EV3! In total, your leJOS code will run about 50 times faster than it did on the NXT.

To put this in perspective, a Segway-like robot only needs to check the sensors and update the action every 8 ms, and the code to do this (each loop cycle) uses only 1/50 of a ms on the EV3. This means that technically the EV3 is capable of keeping up to 400 Segway robots balancing at the same time. In other words, you have plenty of computing horsepower for most of your needs.

Fun Fact!

According to ohloh.net, the leJOS project has taken 19 person years to develop and would cost over a million dollars to program from scratch (with theoretical salaries of $55,000 per year).

Tip!

There is something you can do to reduce delays. SD cards come in different speeds. Most SD cards are class 4, but there are also class 10 cards which provide faster memory transfer speeds. A higher class number will result in slightly faster boot times and faster executing time. However, it will only shave 5 or 10 seconds off the bootup time and a second or two off the time before a program begins executing. Whether or not this is worth it to you depends on how patient you are.

As mentioned earlier, the EV3 Linux is slow to boot up with leJOS, and slow to begin executing a program. In fact, because there is around a five second delay before launching a program, we created a wait screen with the official Java mascot (see Figure 7-3).

Figure 7-3: Meet Duke! You'll be spending some time with him.

JSE Classes

One other huge improvement that comes with using leJOS on the EV3 is that the standard Java classes (Java Standard Edition) are much more complete than they were previously. The standard Java classes include all the base Java packages, such as java.lang, java.math, java.net, java.util and so on.

With leJOS on the NXT, we created our own homemade Java classes. Most often, these classes were scaled down versions of the real thing in order to save memory. We only had about 200 KB or program memory on the NXT, but now there are gigabytes available. With memory no longer a limiting factor, you get all the classes with all the methods and all the functionality.

Exploring the leJOS Platform

The leJOS API includes the package lejos.hardware just for controlling the EV3. This section is a brief overview of what you can expect from the leJOS classes. There are basically three main packages in leJOS: lejos.hardware.*, lejos.remote.*, and lejos.robotics.* (plus a single package lejos.utility). Chapter 10 will explore how to use these packages, but for now, let's have an overview of what you can expect.

Motors are Everything

Every device has a primary function; the main reason you use that device. When you boil away everything else, the main reason for owning an EV3 is to control motors. Even the sensors are there to serve the motors. So leJOS lives or dies depending on how well it can control motors. This is where leJOS has an advantage over the LEGO firmware, in my opinion.

With the LEGO firmware, if you tell the motor to rotate forward 40 degrees, it rotates, slows down, overshoots the mark, then backs up and finally stops within five degrees. With many projects, this is not ideal. We wanted to improve the function so that it decisively moved to the destination, and when it got there, stopped without overshooting.

leJOS gives you a very refined algorithm that took a lot of time to perfect. The people behind these algorithms are Professor Roger Glassey of Berkely and Andy Shaw (a long-time leJOS developer formerly of Oracle). Their algorithm rotates the exact number of degrees, stopping exactly at the desired number. Whether you request a 45 degree rotation in leJOS directly on the brick or via Bluetooth, you will receive accuracy.

Furthermore, Andy thoroughly tested these algorithms under load by hanging a weight on the end of the motor to ensure accuracy under almost any condition. (The testing mechanism was based on a design by Philippe Hurbain, shown in Figure 7-4).

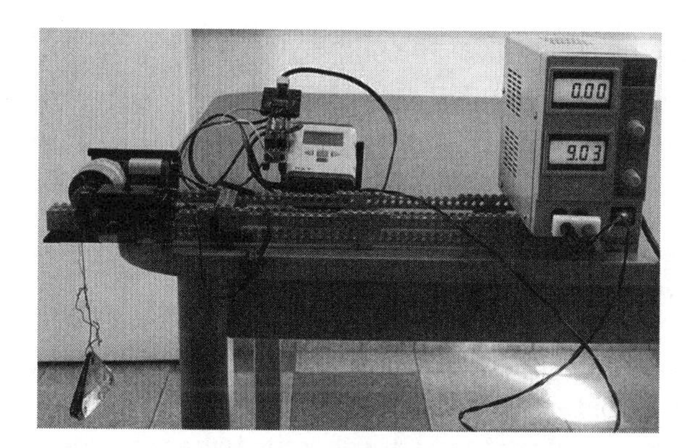

Figure 7-4: Testing motors under load

You can also control the motors by speed. You tell it the speed to maintain and the program will monitor the motors and change the power level to maintain a constant speed. If a robot is going up a hill or down a hill, it will maintain the same speed. You can even check to see how fast the motor is rotating, in real time.

Sensors

It was easy to develop sensor code for the RCX. They all used the same basic Sensor class because they expressed their measurements as analog values between 0 and 1023 (even the touch sensor). The EV3 sensors are far more complex and they provide much different output. We decided, therefore, to write unique classes for each EV3 sensor. For you, this makes using sensors in leJOS far simpler than before.

Buttons

All six buttons on the EV3 can be reprogrammed under leJOS. You can even use events to listen for button presses and react accordingly when one is activated. This makes it easy to separate the user interface portion of your code in an object-oriented style.

System Time

Time is kept on the EV3 brick as the number of milliseconds that have elapsed since it was turned on. Think of this as a time sensor for your robot. This can be useful for time-stamping when events have occurred. In fact, you can even accurately go down to the nanosecond level for even more accuracy.

Battery Power

leJOS is capable of checking battery charge. You might think this would give you a percentage of battery power remaining, but it actually returns a number representing the voltage of the batteries. If the battery charge starts to fall rapidly it usually means the EV3 is "running on empty".

LCD Display

leJOS allows you to take complete control of the LCD. You can draw lines, shapes and bitmap images to the display. leJOS even includes a complete character set so you can output text and numbers.

Sound

The leJOS API allows you to play simple sounds. In chapter 11, we will explore the sound functions. You can even upload and play prerecorded sound files. The sound quality has been greatly refined for the EV3, allowing larger wav files to play on the brick, resulting in much clearer sound.

Bluetooth Communications

Communications can take place between the EV3 brick and other Bluetooth devices. Not only can you communicate with your EV3 brick via your PC, but the EV3 brick can also communicate with mobile phones, GPS units, controllers, and other EV3 and NXT bricks.

Robotics Classes

The leJOS developers believe that if a class is useful for general robotics tasks, it should be part of leJOS. For example, navigation is a big part of robotics so the leJOS platform includes a variety of packages dealing with navigation. It also has a package for Rodney Brooks' subsumption architecture (Chapter 19). These classes allow you to jump over the low-level robotics programming tasks and go straight to programming more interesting behavior.

As of this writing, leJOS is the only Java API for the EV3 brick. If you like Java, this is the environment for you. The developers think it holds up well against other platforms for the EV3, such as C language or the standard LEGO software. However, don't hesitate to try other languages for the EV3 brick, such as Robot-C. C language is important in the programming world, and often goes hand in hand with Java.

CHAPTER 8

Control Center

TOPICS IN THIS CHAPTER

- leJOS Tools
- EV3 Control Center
- Building a Grabby Robot
- Remote Motor Control

The leJOS platform includes a number of useful PC utilities to aid robotic development. These utilities are located in the bin directory of your leJOS EV3 install. There are over half a dozen total, but we will focus mainly on the core utility, called EV3 Control Center. But first, let's briefly examine the PC utilities available with leJOS.

leJOS PC Tools

As mentioned in the intro, all the PC tools can be found in the bin directory of your leJOS install. In Windows, these files have a .bat extension. It's worth noting that many of these individual utilities are also incorporated in EV3 Control Center, which the majority of this chapter focuses on.

EV3Console

EV3Console connects to port 8001 in the EV3 menu and displays System.out and System.err. It needs a parameter -n in order to work with the network.

EV3Image

EV3Image converts images to the leJOS EV3 image format.

EV3MapCommand

EV3MapCommand lets you load a line map and drive a robot around a mapped area (e.g. a floor of your house), by setting waypoints. It can plot paths to a destination, avoiding obstacles, and then follow the path. Line maps are in svg format.

EV3MCLCommand

EV3MCLCommand is similar to EV3MapCommand but drives a robot with a distance sensor and does localization using the Monte Carlo Localization technique. It displays the particle set so you can see your robot homing in on its true position and maintaining it as it moves around.

EV3ScpUpload

EV3ScpUpload is used by the Eclipse plugin. It uploads a runnable jar file to the EV3 and optionally runs it.

EV3SDCard

This utility is used to burn a the leJOS image to your EV3. During the Windows installation this already popped up for you.

Now let's move onto the main attraction, the EV3 Control Center!

EV3 Control Center

EV3 Control Center is a comprehensive utility containing many of the other utilities in the bin directory. This mature user interface allows you to control motors, read sensors, change EV3 settings and more, all from a PC. Running the utility is simple, assuming you installed Eclipse.

Running EV3 Control Center

1. Click the left leJOS icon in the toolbar (see Figure 8-1), or you can double click ev3control.bat from the leJOS bin directory.

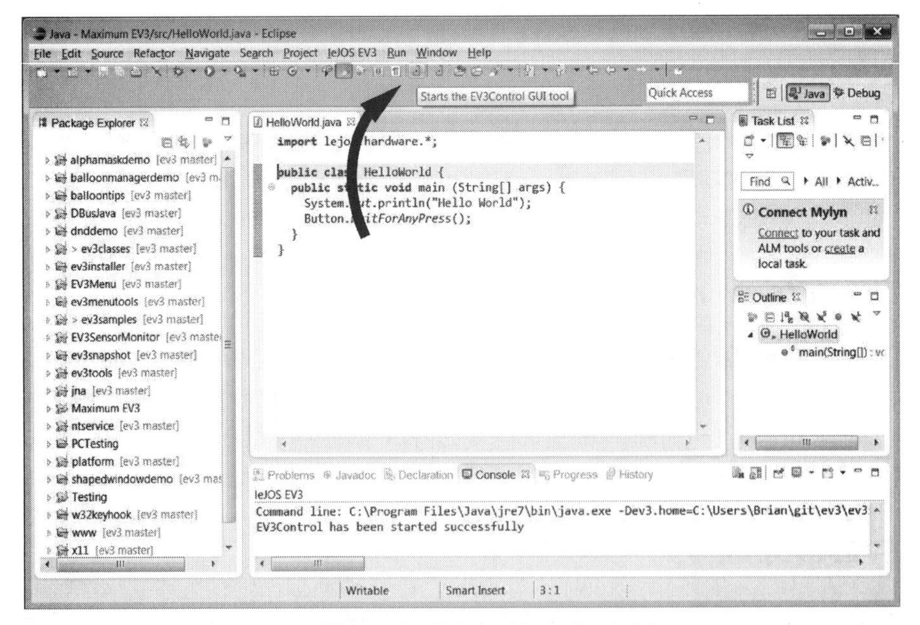

Figure 8-1: Clicking the EV3 Control Center icon in Eclipse

2. The main screen has options to connect to an EV3 brick (see Figure 8-2). Turn on your brick, connect the USB cable to the slave USB port on the EV3, and plug in the other end to your PC. Alternatively, you can set up a WiFi connection or enable Bluetooth (more on this in chapters 12 and 13).

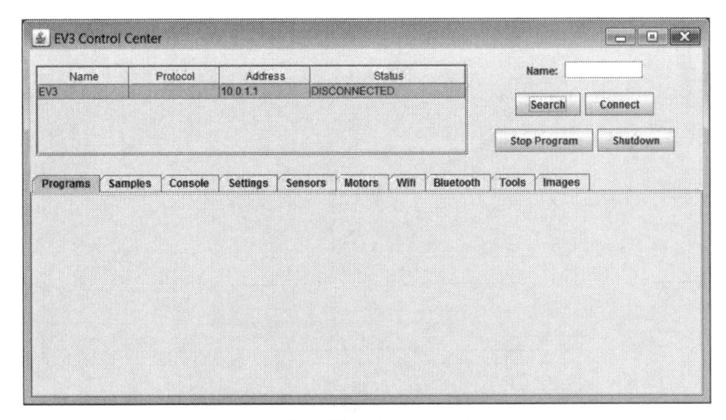

Figure 8-2: Displaying the connection screen

3. Click the search button and the program will search for all available EV3 bricks. Chances are only one brick showed up in the search. When you click connect, it will connect to the first available brick. If you want to connect to a specific brick, enter the name of the brick in the text field (the name is in the first column of the list of bricks it found).

Once your PC is connected to the EV3 brick, you can manually stop programs that are currently running with the Stop Program button. You can also shut down the EV3 brick by clicking the Shutdown button. Let's examine the different tabs available.

EV3 File Browser

The first tab contains a file browser of sorts for the programs directory. This is the main utility for file management. It's easy to compile and upload code, but often you want to upload data files to the EV3 brick, such as images, sounds, or map data. This tab allows you to do all that and more.

The file browser displays the contents of your EV3 Programs directory (see Figure 8-3). There are buttons at the bottom of the screen to delete files, upload files to the EV3 brick, and even download files to your PC in case your Java program was recording data for later analysis on a PC. To delete a file, first place a check-mark next to it before hitting the delete button.

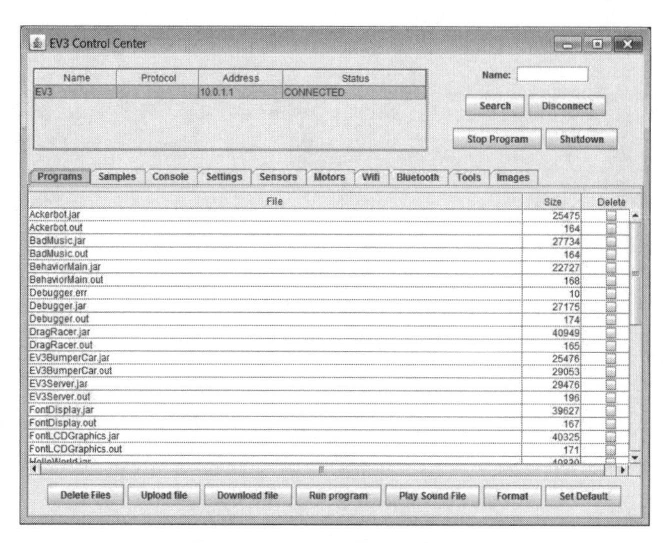

Figure 8-3 Browsing files on the EV3

There are also additional buttons. You can start a program on the EV3 brick. Select it from the list by highlighting it (you do not have to place a check-mark next to it), then click Run program. You can also play a sound file through the brick, such as a wav file. The format button erases the entire file system (this does not delete the Linux or leJOS).

Samples

Samples is similar to the Programs tab, only it allows you to manage which samples you want to keep on the EV3. leJOS comes with several samples, but it's possible you might want to erase some from here. You can also run samples from this section.

Console

The Console tab is perhaps the best part of EV3 Control Center. One handy way to view console output from System.out or System.err streams is to view the output on your PC. In order to do that, simply connect to EV3 Control Center (as described at the start of this section) and then any program you upload to the EV3 will echo its output to the Console window.

But that's not all. You can also use the GUI buttons on the Console window instead of the physical EV3 buttons (see Figure 8-4). You can also change the LED pattern of the EV3 buttons by changing the setting to one of 10 different values. Finally, the best part of the console window is the live screen capture of the EV3 brick. Together, this window allows you to physically interact with the buttons and LCD without having to chase your robot!

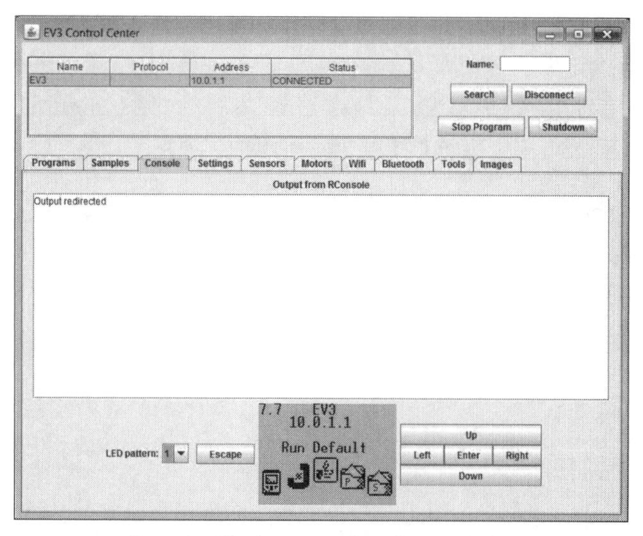

Figure 8-4: Viewing output from the console tab

Settings

The settings section allows you to change the EV3 volume and the Menu sleep time, which determines the interval of inactivity to automatically power off the EV3 brick (see Figure 8-5). If you want to play around with EV3 Control Center without the brick accidentally powering off, now would be a good time to raise this interval, to twenty minutes for example.

NOTE: *The EV3 brick will turn off after the default time if there is no activity. If you lose connection, your brick has probably turned off. You can also extend the automatic power-off time from the EV3 menu. Select the System menu, then select Sleep Time.*

Try it!

Try using the remote buttons by navigating through the EV3 menu. Simply press the GUI buttons for up, down, left and right and the actual menu on the EV3 will change. You can also see the screen capture of the EV3 echoed to your PC. Finally, try changing the LED pattern to see the light under the EV3 buttons flash different colors.

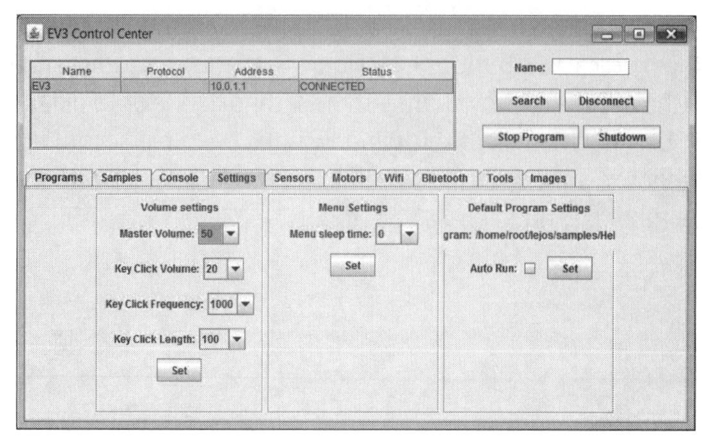

Figure 8-5 Changing the system settings

You can also check the default program that will run when you select "Run Default" from the EV3 menu. If you select Auto Run and then click the Set button, this program will run when you first turn on the EV3 brick.

Try it!

1. Connect an IR sensor to port 4.

2. In the lower right corner of the Sensors tab, select EV3 IR, and Distance (see Figure 8-6).

3. Place your hand in front of the the IR sensor and click update. You can see the raw value displayed near the bottom of the panel.

4. By clicking continuous you get real-time updates. Click Stop when done.

Sensors

The Sensors tab allows you to view real-time data from sensors connected to your EV3 brick, as well as the battery (see Figure 8-6). Instead of explaining it, let's give it a try! Try using the remote buttons by navigating through the EV3 menu. Simply press the GUI buttons for up, down, left and right and the actual menu on the EV3 will change. You can also see the screen capture of the EV3 echoed to your PC. Finally, try changing the LED pattern to see the light under the EV3 buttons flash different colors.

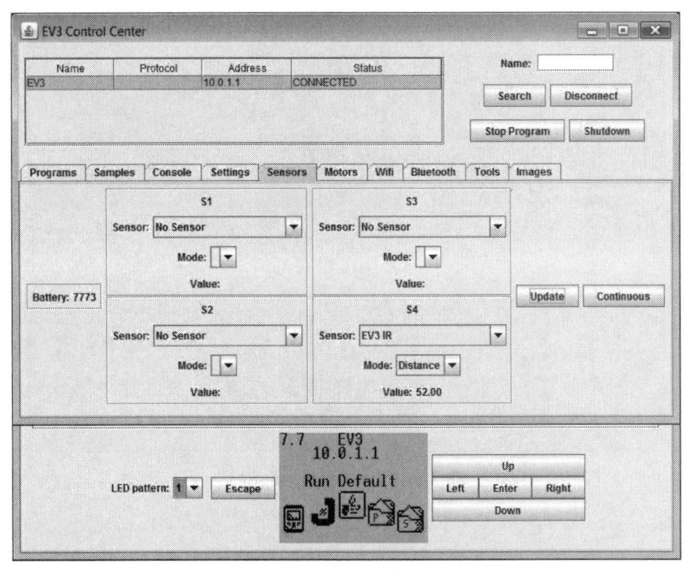

Figure 8-6 Monitoring sensors

Motor Control

A useful feature of EV3 Command Center is the ability to control motors without uploading Java code to the EV3 (see Figure 8-7). This can help you when constructing a new model, allowing you to test different motors without writing a single line of code.

The motors will rotate for as long as you hold down one of the direction buttons (Forward or Backward). The Reverse check-boxes allow you to reverse the direction of one or more motors. The Limit field allows you to select the amount to rotate every time you hold down the button. For example, if you type 90 into the field, every time you hold down forward it will rotate up to 90 degrees and then stop.

You can also control a differential robot by selecting two motors, and then pressing Forward, Backward, Turn Left or Turn Right. The latter two buttons will cause the two motors to rotate in opposite directions of one another.

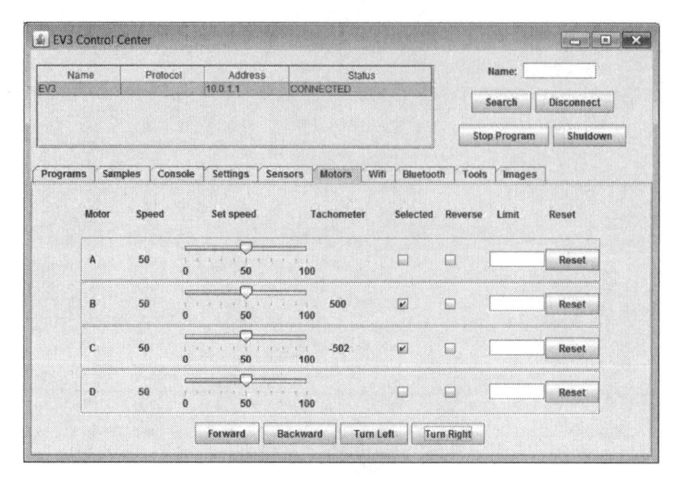

Figure 8-7 Controlling motors

Grabby Robot

In this section, we will use the EV3 Control Center to control a robot. The direction buttons will be used to control the steering of the robot, and we will employ the limit field to control a third motor to control the claws of the robot. But first, let's build the robot!

WARNING: *I recommend connecting a medium cable to the medium EV3 motor before beginning construction.*

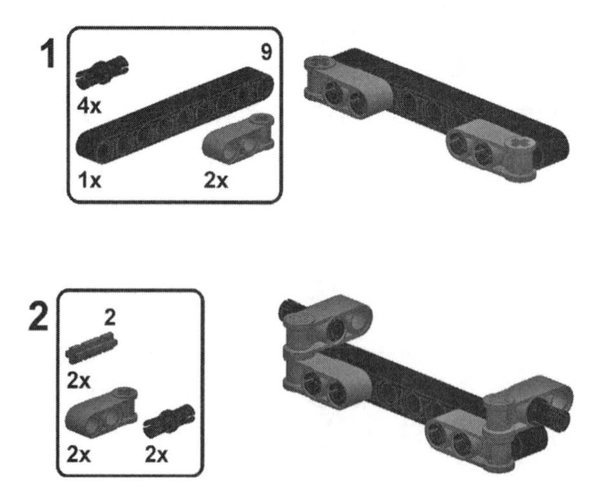

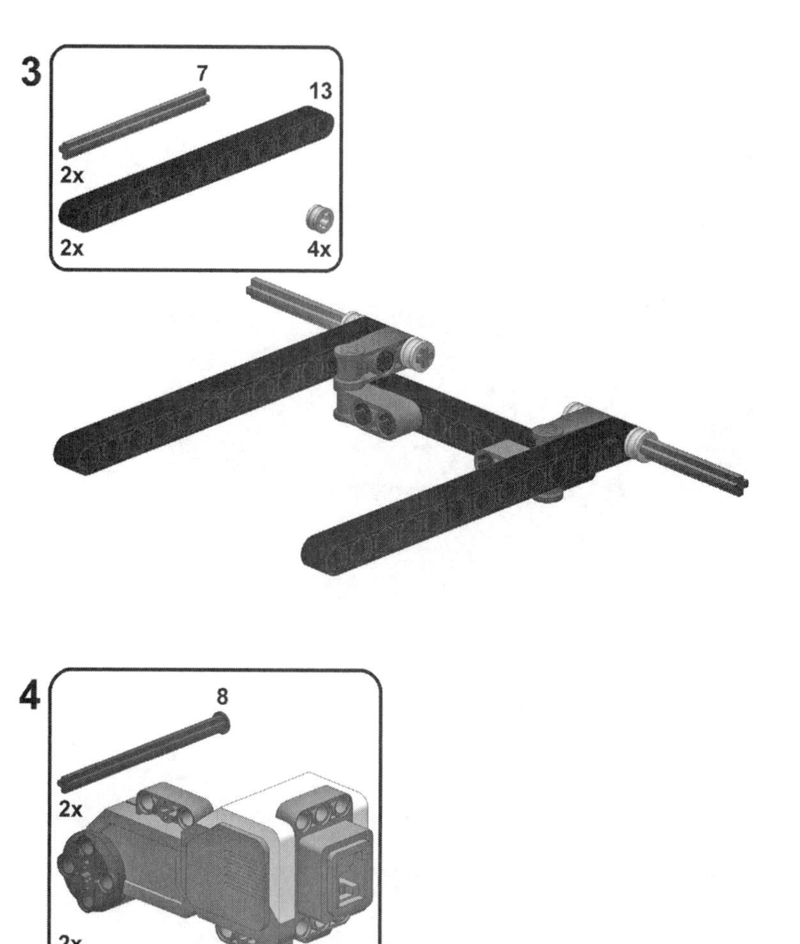
3
7
13
2x
2x
4x
4
8
2x
2x

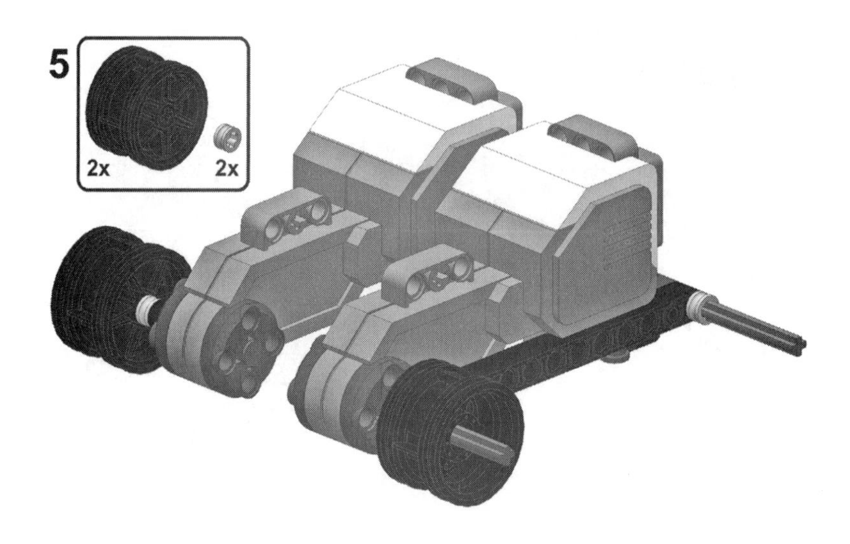
5
2x
2x

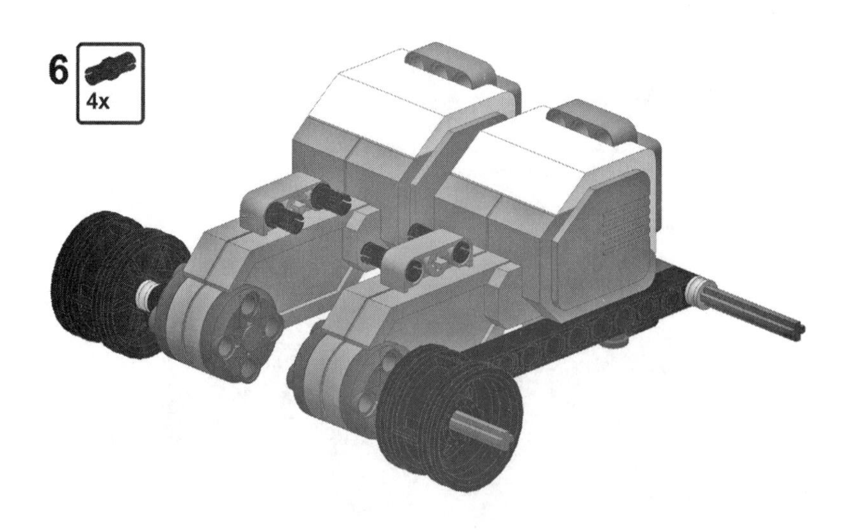
6
4x

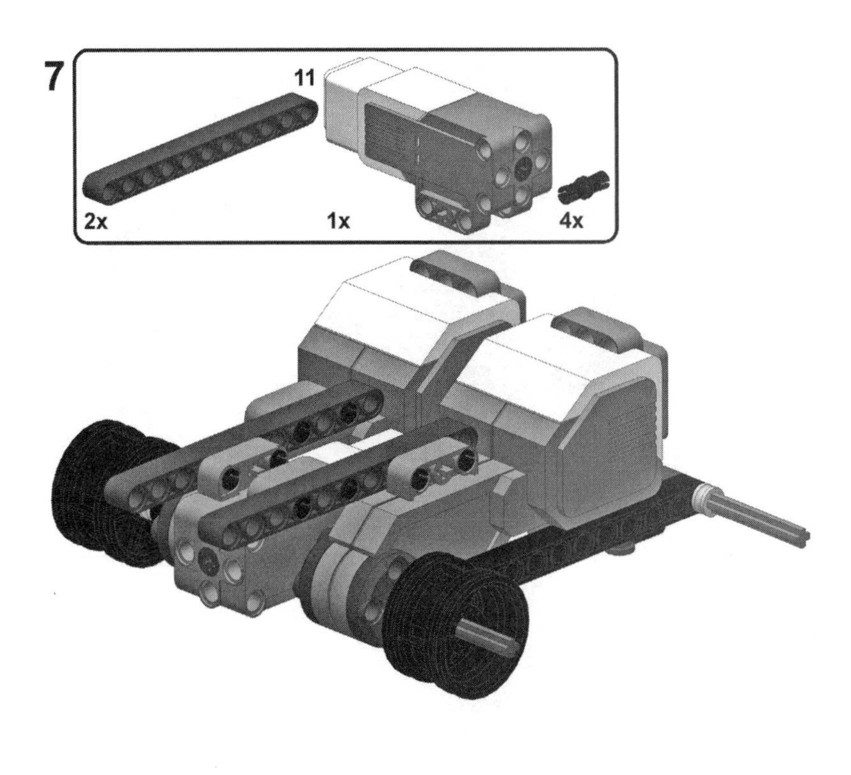
7
11
2x
1x
4x

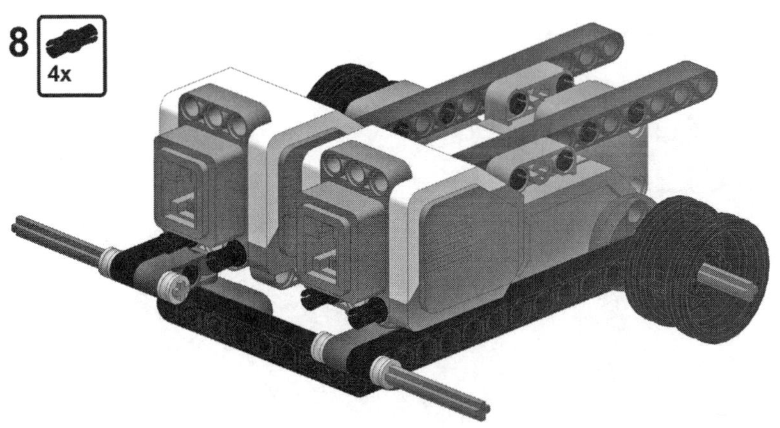
8
4x

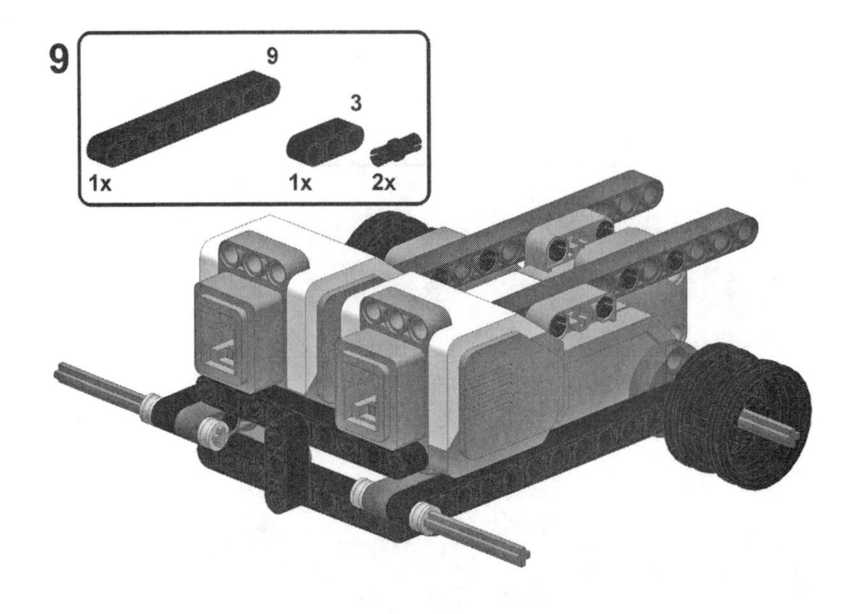
9
9
3
1x
1x
2x

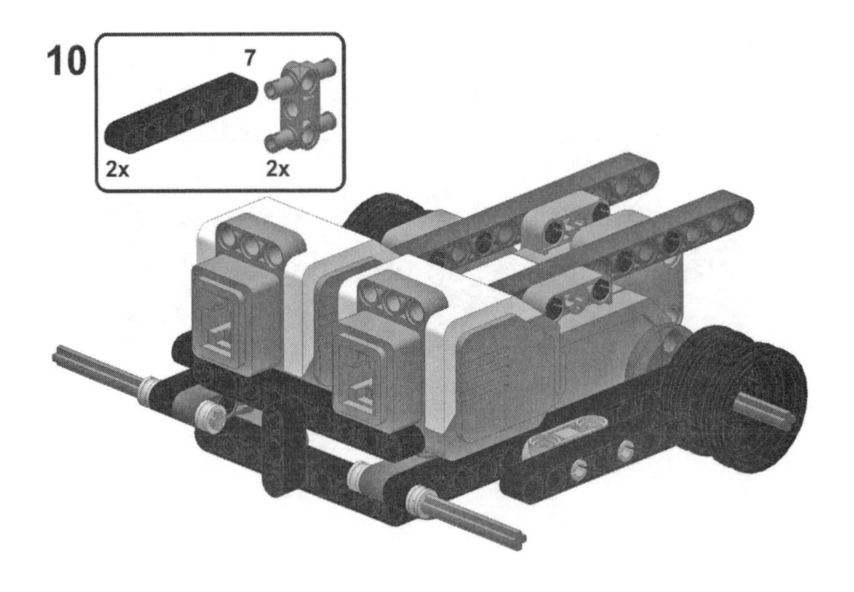
10
7
2x
2x

11

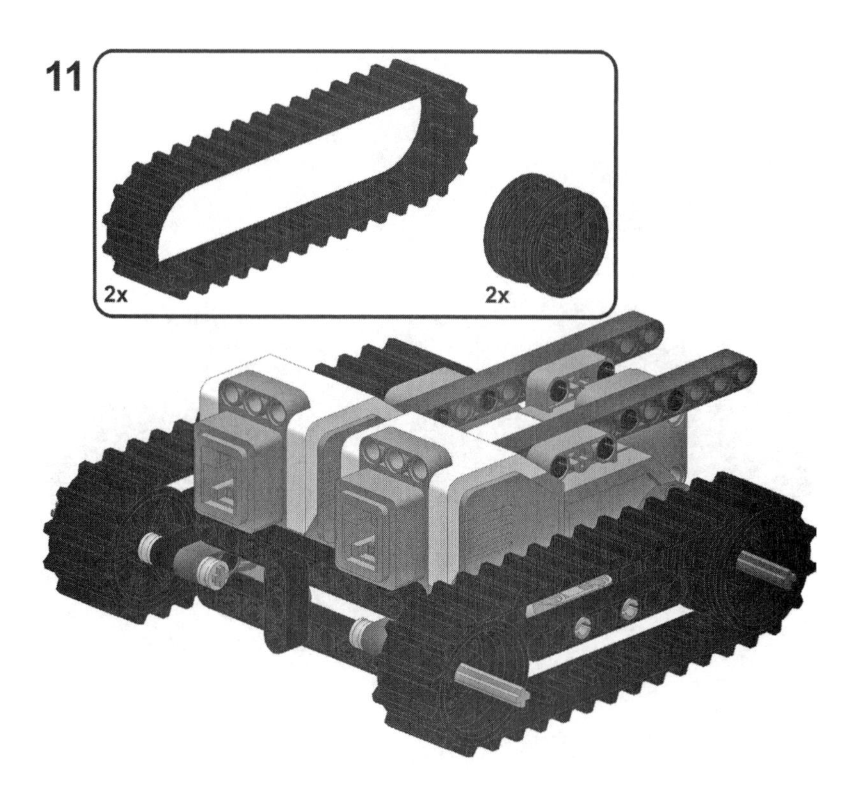

12

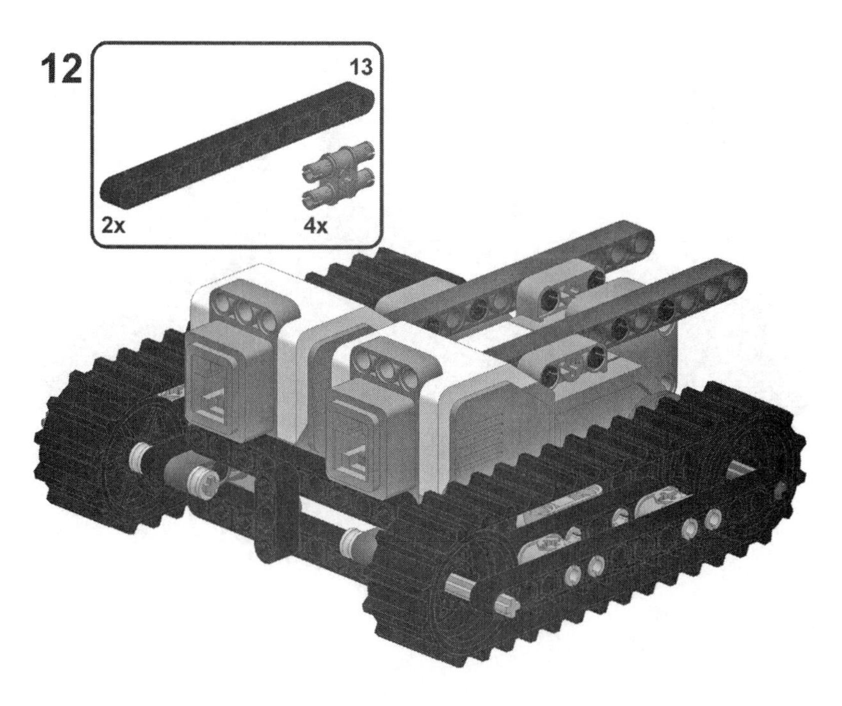

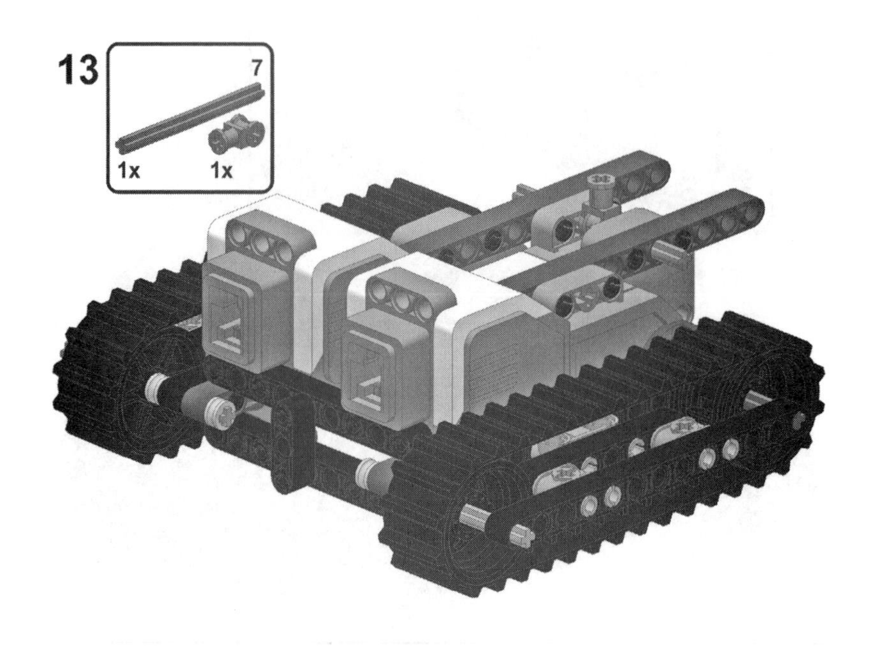
13
7
1x
1x

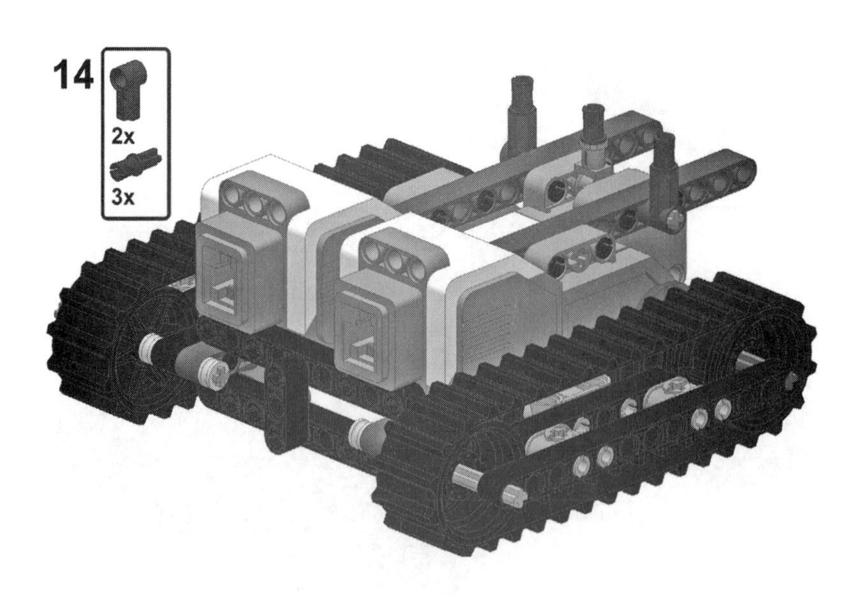
14
2x
3x

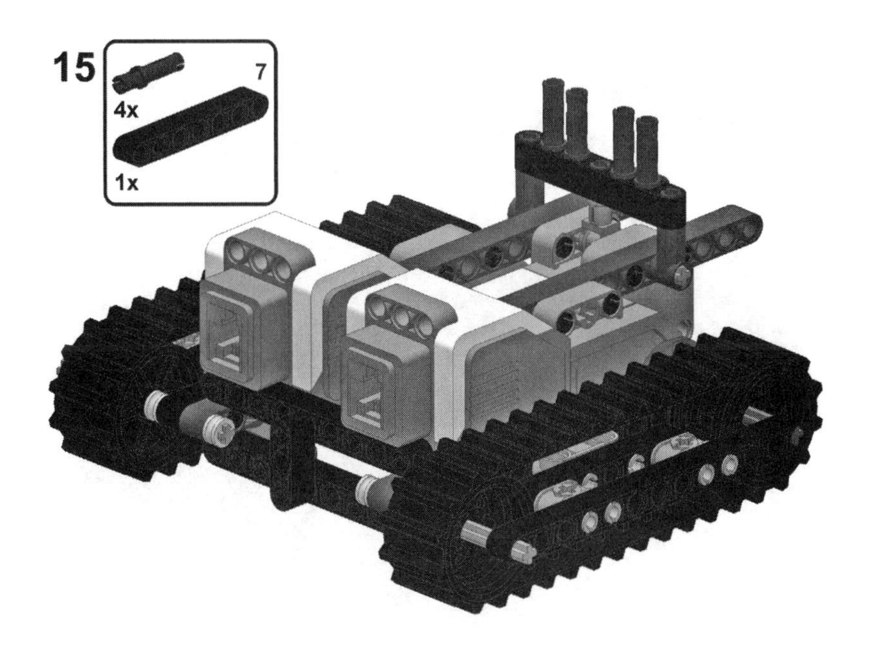
15
7
4x
1x

16
2x
2x

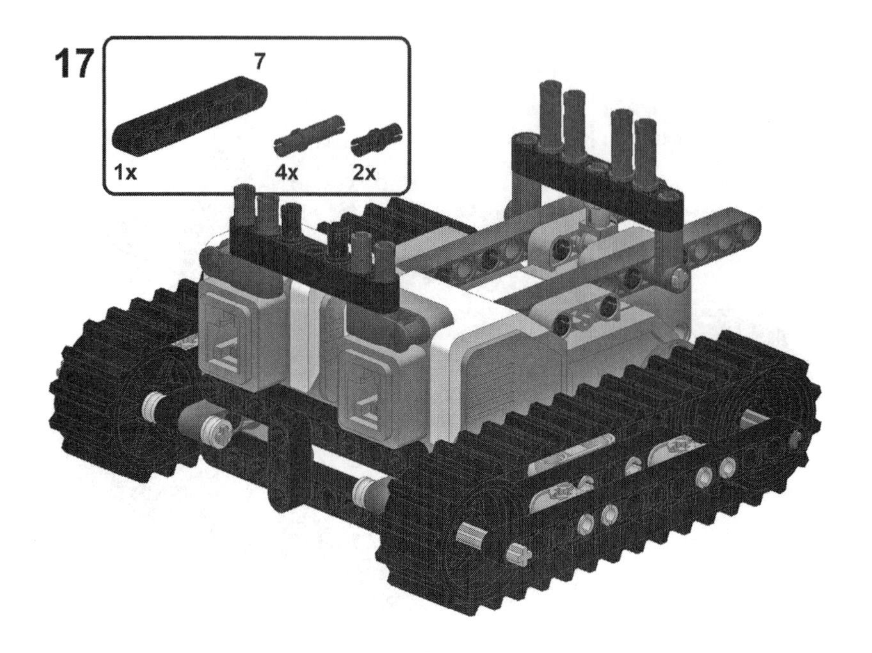
17
7
1x
4x
2x

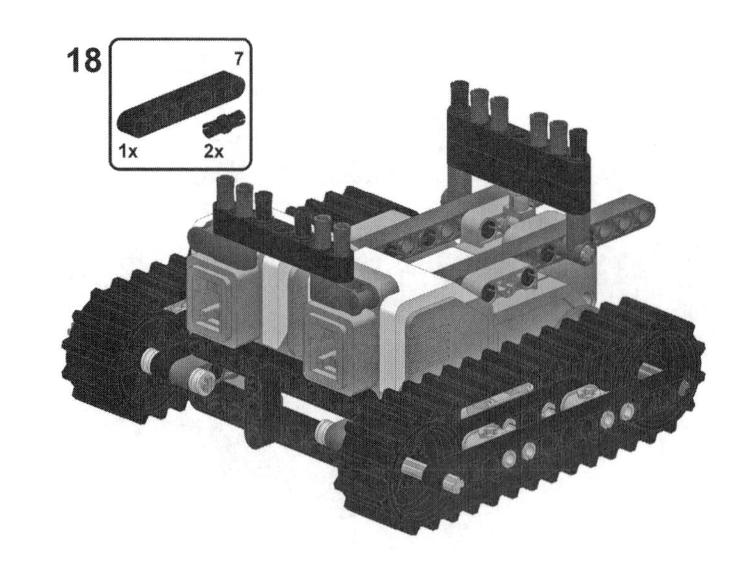
18
7
1x
2x

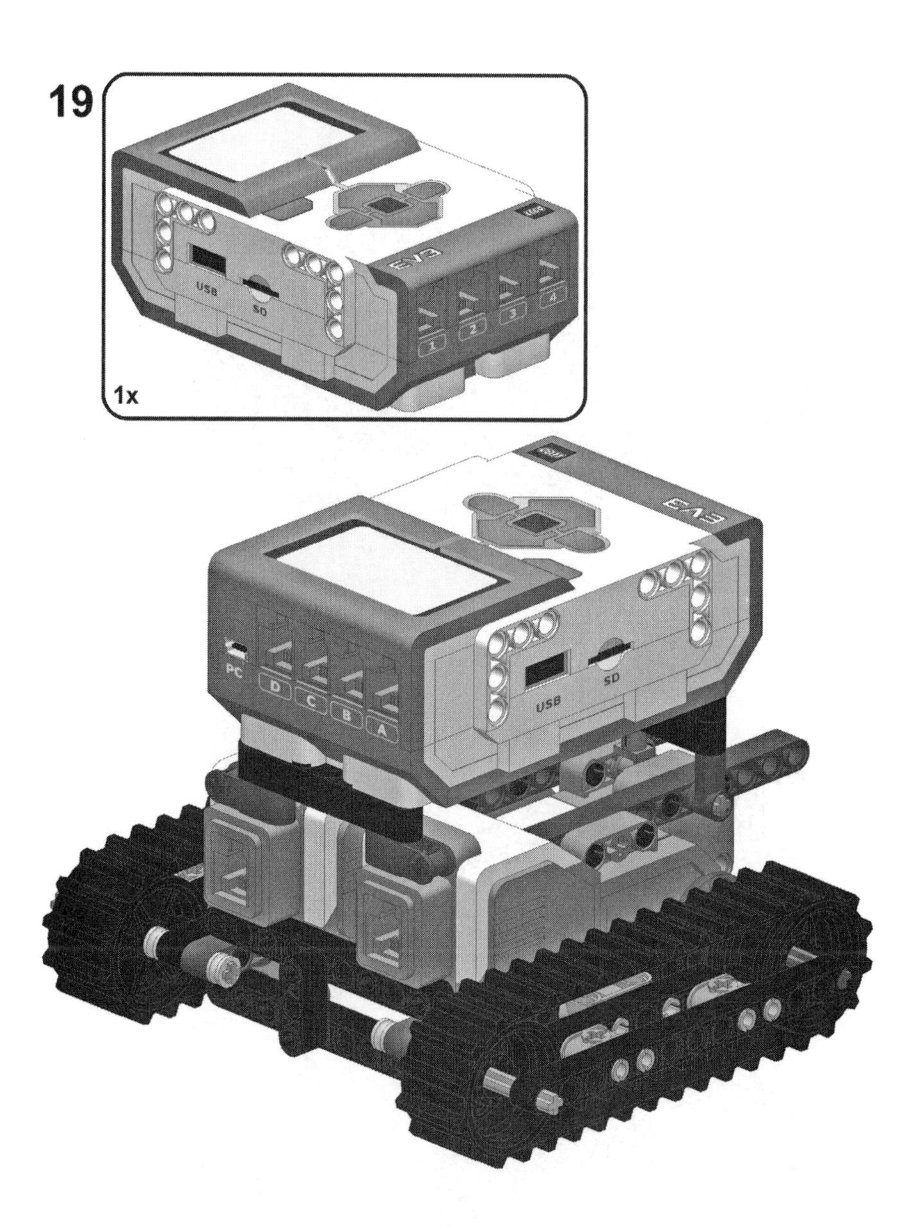
19
1x
USB
SD
EV3
1
2
3
4
PC
D
C
B
A

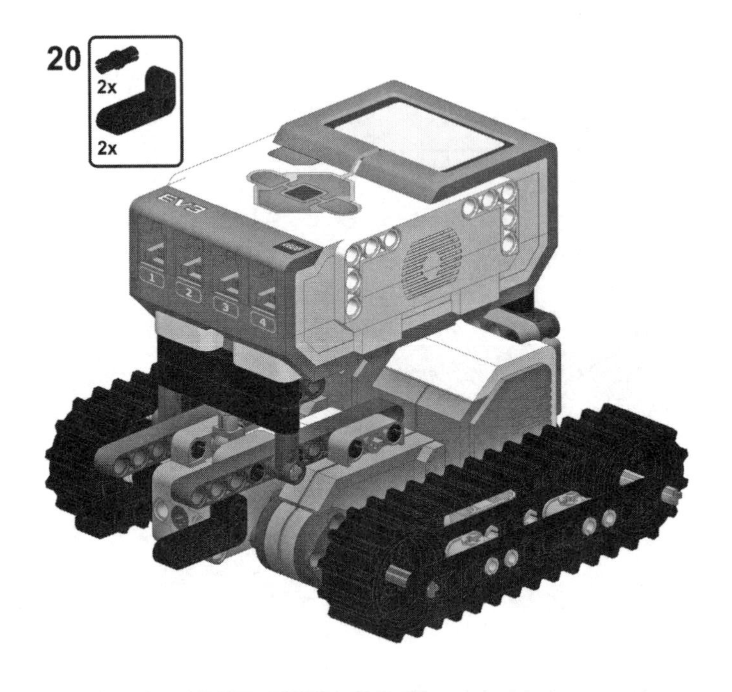
20
2x
2x
EV3
1
2
3
4

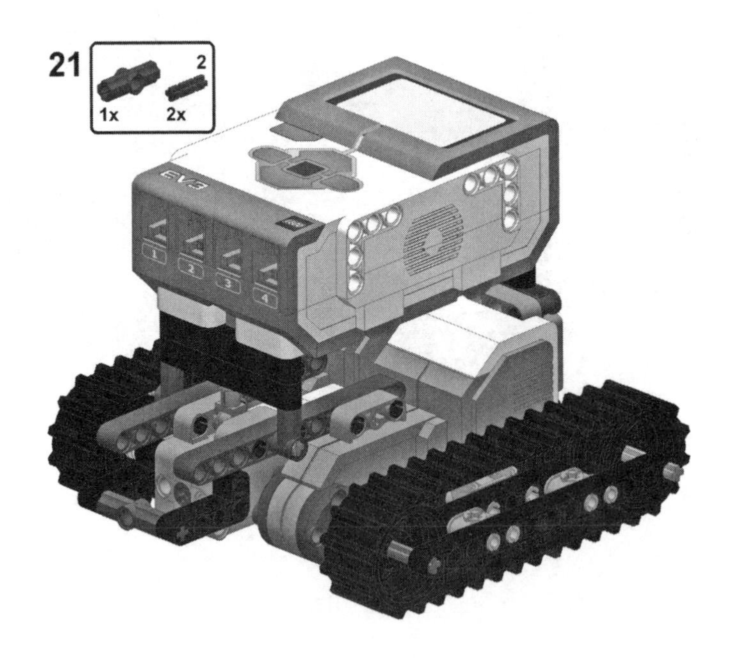
21
2
1x
2x
EV3
1
2
3
4

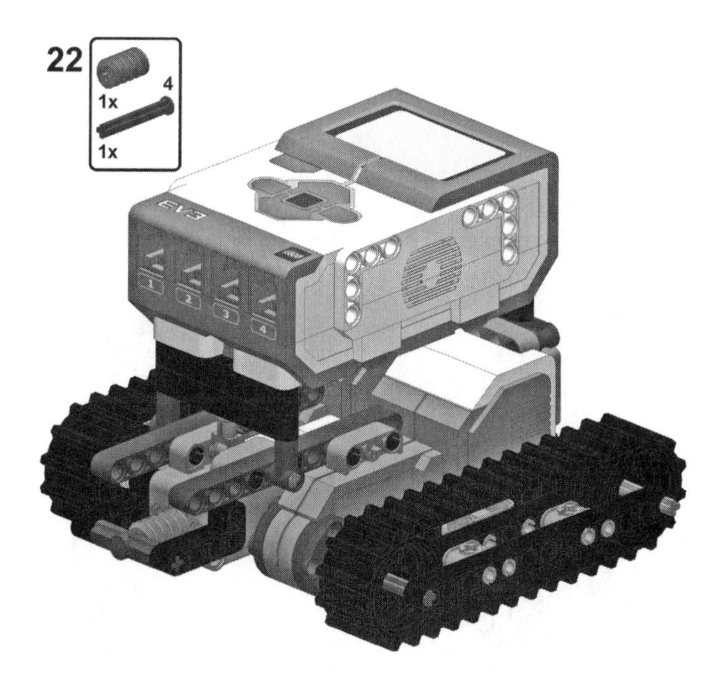
22
4
1x
1x
EV3
1
2
3
4

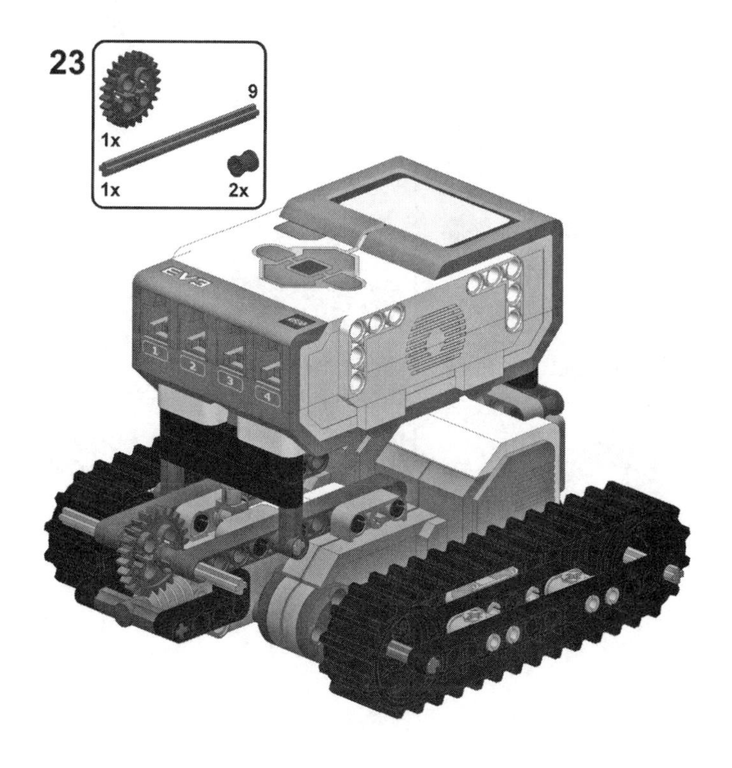
23
9
1x
1x
2x
EV3
1
2
3
4

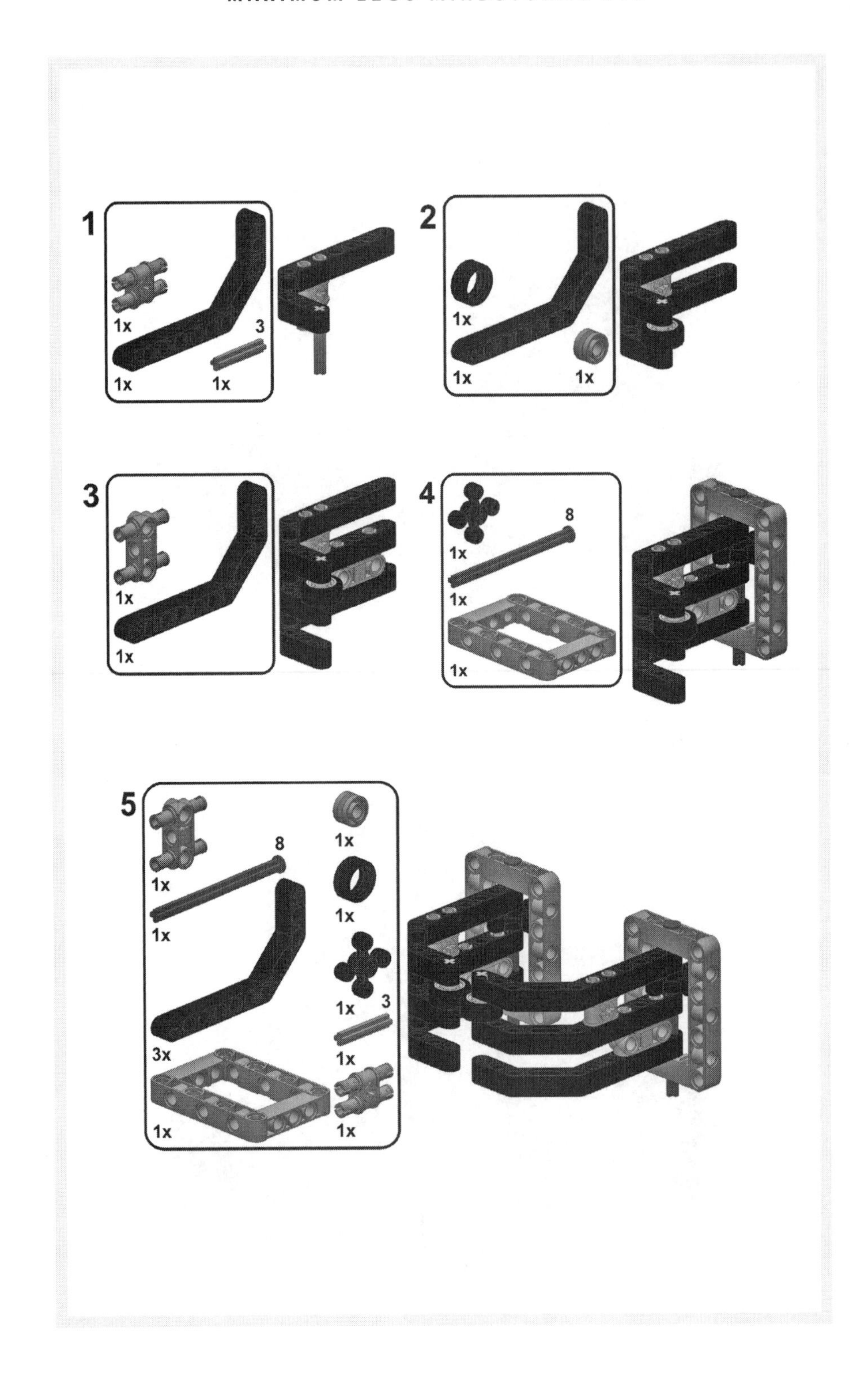
1
1x
3
1x
1x
2
1x
1x
1x
3
1x
1x
4
1x
8
1x
1x
5
1x
8
1x
1x
1x
3x
1x
1x
3
1x
1x
1x

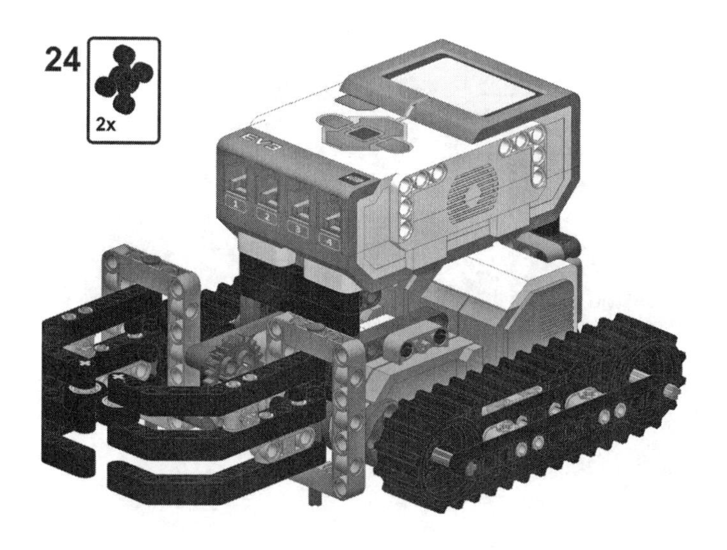
24
2x

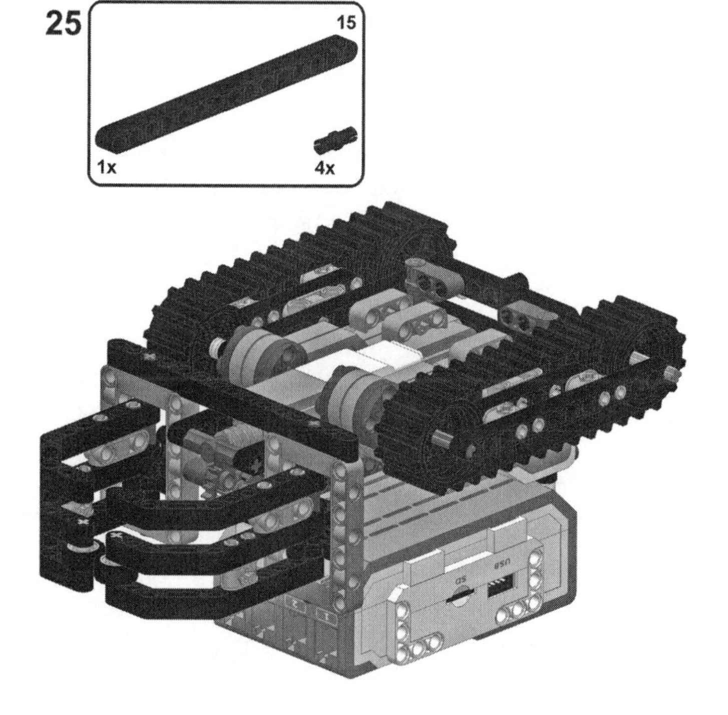
25
15
1x
4x

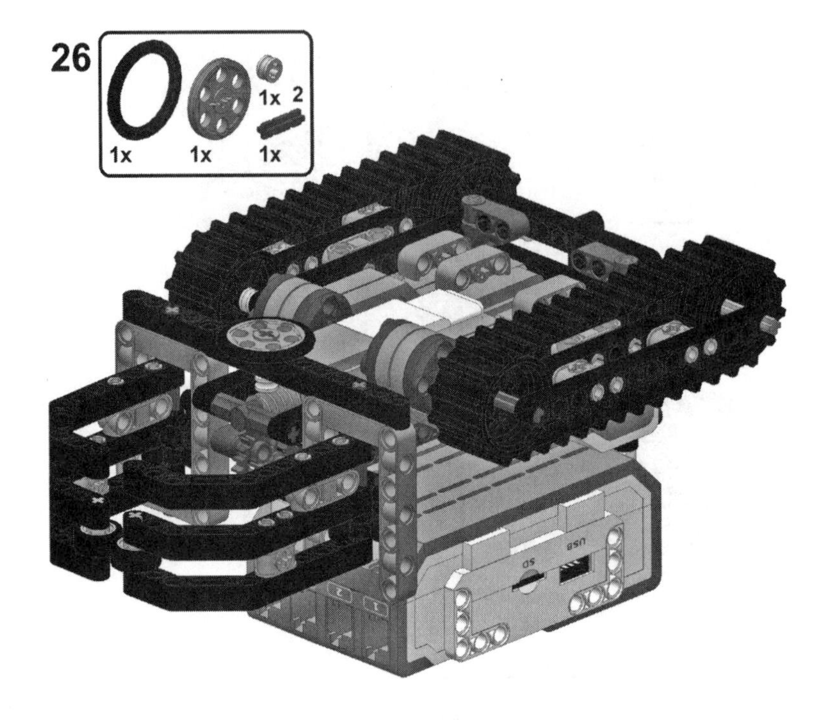

Cables

Insert a short cable from motor port B to the left motor. Use another short cable to connect port C to the right motor. (The cables will crisscross each other.) Insert a medium cable from the medium motor to port A.

Controlling Grabby

Now that we have a robot built, we can connect to it and then control it with the Motors tab. The robot is good at picking up empty pop cans or small light containers.

1. Follow the instructions above to connect to the EV3 Control Center. Select the Motors tab.
2. Select motors B and C by placing a check mark next to each of them. Now you can click the 4 direction buttons on the bottom of the screen to move Grabby. The motors will move for the duration that you hold down the button. You can also adjust the speed of the motors to make Grabby move faster or slower.

3. To open the claw, unselect B and C, then select A. Press backward until the claw is open.
4. Now drive towards an empty can of pop. To close the claw, click forward until the claw closes and lifts the can.
5. You can also experiment with limiting the rotation. However, keep in mind that the limit value has a positive or negative sign. No matter if you push forward or backward button, it will rotate according to the limit sign. Also, you must hold down the button until the limit is reached.

As you have probably just found out, EV3 Control Center is great at controlling robots! It's a breeze to adjust speed values on each motor and drive around. It's also easy to control secondary motors, such as the claw motor on Grabby.

> ***NOTE:*** *EV3 Control Center can be used for any robot. If your robot turns left instead of right, you can unplug the two motor cables and switch them with each other. Likewise if the robot goes backwards instead of forwards, check the Reverse option for those motors.*

WiFi

Normally you can configure WiFi to connect to a particular WiFi router with the EV3 menu, or you can even configure it manually by editing a config file. However, it is difficult to enter passkeys using the EV3 buttons. A keyboard makes it much more practical.

Therefore, a third way to configure WiFi is available from the WiFi tab (see Figure 8-8).

1. Make sure your WiFi adapter is plugged into the EV3 and enabled.
2. Select Scan to find a list of available routers.
3. Select your router in the list, then select Configure. A dialog box will pop up asking you to enter the password for your router. Enter it and click OK.

That's it! Your WiFi is now configured.

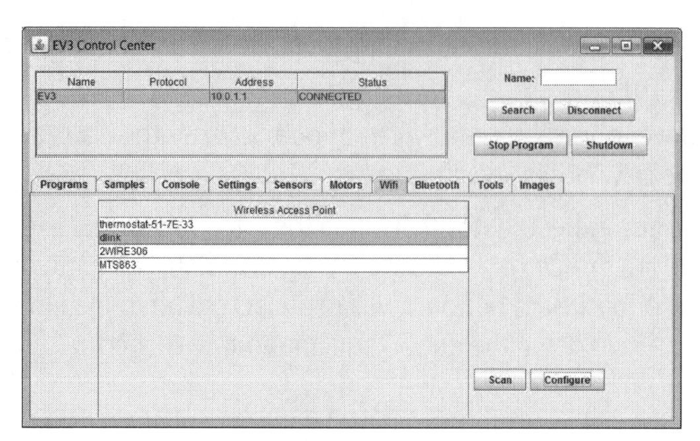

Figure 8-8: Configuring the WiFi from a PC

NOTE: *See chapter 12 for more information on WiFi.*

Bluetooth

It's also easier to configure Bluetooth from your PC, due to the presence of a PC keyboard.

1. Click the Search Button to find available Bluetooth devices (see Figure 8-9).
2. Select the device and hit Pair. This will allow you to enter the PIN code from your keyboard.

NOTE: *See chapter 13 for more information on pairing.*

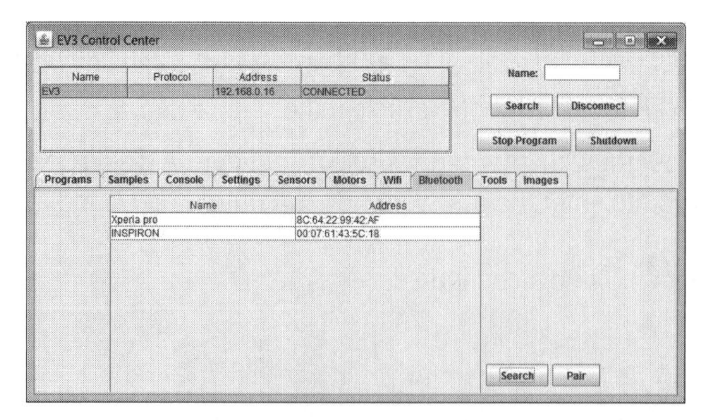

Figure 8-9: searching for Bluetooth devices

Tools

The Tools tab has functions for playing a tone, changing the EV3 brick name, and even sending test commands to I^2C sensors (see Figure 8-10). This is useful if you are programming a new sensor and would like to see how different commands behave (these commands are usually documented in the product literature).

NOTE: *Most NXT sensors are I^2C sensors. EV3 sensors do not use this protocol.*

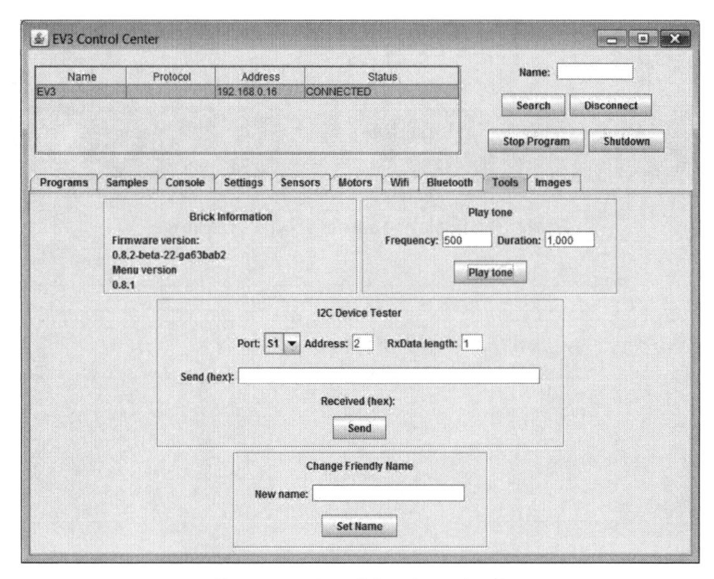

Figure 8-10: Examining the tools tab

Images

The leJOS API is capable of displaying images to the screen using the class lejos.hardware.Image. An image is a series of bytes that define whether pixels are on or off. In order to display images, the EV3 Control Center has the Image tab, which allows an array of bytes to display on the screen (see Figure 8-11).

For example, in order to display a solid 8x8 block, first click Edit. Then enter the following in the box:

```
(8,8) "\u00ff\u00ff\u00ff\u00ff\u00ff\u00ff\u00ff\u00ff"
```

Now when you click execute it will draw the image.

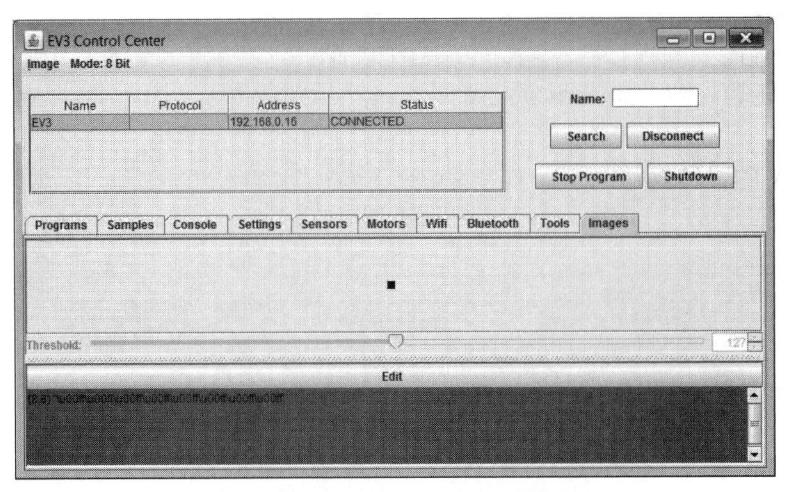

Figure 8-11: Displaying a series of bytes as an image

NOTE: *As of this writing, some of these tools are a work in progress. For example, the Images tab does not currently offer a lot of options.*

CHAPTER 9

Instant Java

TOPICS IN THIS CHAPTER

- Java Fundamentals
- OOP
- Core Java Language
- Classes
- Objects
- Interfaces
- Methods
- Primitive Data Types
- Operators
- Flow Control

There is a popular Java programming book called *Teach Yourself Java in 24 Hours*. This chapter, Instant Java, attempts to one-up that book by simplifying things even more. If this is your first time programming Java, you are in for a rewarding experience. Java has consistently been the most popular programming language out there, beating both C and C++ year after year. Java is a popular and well loved language for a reason.

Although this chapter will not teach you everything you need to know about Java, it will teach you enough to program your EV3 robots using leJOS. Everything you learn here also applies to programming Java applications on any platform, such as PC applications. After all, you are using the same Java Standard Edition as you use on a PC.

NOTE: *This chapter is an introduction to Java. If you intend to perfect your knowledge of Java, there are many good books on the market that teach the Java language in greater depth, such as* Head First Java *published by O'Reilly.*

Java Fundamentals

Every programming language uses keywords and symbols to provide core functionality. This functionality includes such things as declaring classes, methods, and variables. It also includes manipulation, such as adding two variables.

A programming language would be boring if it could only run commands one after another, so to give a program a branching behavior you need to be able to control program flow—in Java by using *if, for, while* and *do* statements.

In this section we will cover each of these fundamentals and show examples in actual Java code. Before getting into Java it would be a good idea to touch on a concept fundamental to Java: Object Oriented Programming.

OOP

Object oriented programming (OOP) is a simple but powerful concept. In order to appreciate OOP it should be compared to structured programming. Structured programming uses one body of code with multiple methods. This type of program is difficult to manage because methods and global variables are constantly added to the same program. As a program grows larger it becomes messy and error prone.

OOP has a significant difference that has powerful implications: code can be segregated into discrete units. These units, or objects, are usually defined by a theme. For example, imagine you are trying to program a robot to compete in a chess tournament. The robot needs to be able to move pieces, and also to play chess. One object would contain functionality centered on the theme of piece moving, and another object would contain functionality for thinking up chess moves (Figure 9-1). These units are defined by classes in Java. The ChessArm class would contain all the methods to move to chess pieces around the chess board, and the ChessPlayer class would contain all the methods for choosing which piece to move. Most likely a third class would call methods from the ChessArm and ChessPlayer classes in order to function.

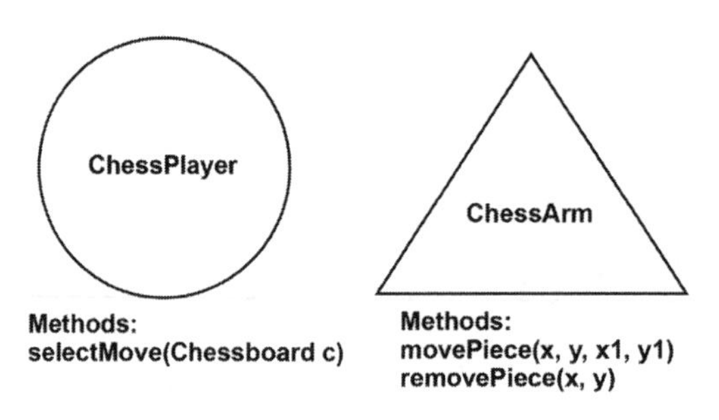

Figure 9-1 Objects in a robotics application

Classes are created using methods (functionality) and fields (data). Once a class is defined, it can be *instantiated*, which means an object is created which can be manipulated in code. One class can be used to create many objects. Another powerful feature of OOP is *type extensibility*. This is the ability of classes to inherit data and functionality from other classes. These topics will be fleshed out with actual examples in the following sections.

Source Files

In Chapter 2 you were shown how to set up your development environment and compile a simple program for the EV3. Let's now analyze the contents of a Java source file:

```
import lejos.hardware.LCD.*;
class Hello {
 public static void main(String [] args) {
```

```
  LCD.drawString("HELLO", 0, 2);
  Button.ENTER.waitForPressAndRelease();
    }
  }
```

Enter this code in a new class file named Hello.java, upload it to the EV3 and run it (refer to Chapter 3 if you are not familiar with how to do this on your system). You should see the word HELLO appear on the LCD display. In order to end the program, press the square enter button. The program is uninspiring, but it demonstrates some basic concepts.

This is about as simple as a Java source file can be. For now, we'll ignore the import statement in the first line. The second line contains a statement that defines the class name, in this example, Hello. Java is case-sensitive which means it notices if letters are capitalized. To a Java compiler, the word "Public" is different from "public". Note the curly brace located after the class declaration and the matching curly brace located at the end of the file. All of the methods and variables contained in the class must lie within these two curly braces, otherwise it will not compile.

Line three contains a method definition. As you can see, methods also use curly braces. This method is named main()—it is the main method that starts the ball rolling. All leJOS programs are started using a main() method. There are other keywords in the method definition: public, static, and void. For now, the only thing you should know is that these must be present for the main() method to function properly. There are also some words in parentheses after the method name, the significance of which will be explained further in the chapter.

The statements within the main() method are the heart of the program; they provide the functionality of the class. The first line outputs the word "HELLO" to the LCD display, and the next line waits for you, the user, to press the square enter button. Without this final line, the program will flash "HELLO" for a brief millisecond then go back to the regular menu system.

Classes

Classes are programming structures that define objects, sometimes called *meta-objects*. They are like templates that are used to create objects, typically by using the *new* keyword. One class can make a multitude of objects, much like a rubber stamp can create copies of the same design. For example, String is a class contained in the java.lang package. In order to create several String objects, we can use the following code:

```
String s1 = new String("String A");
String s2 = new String("String B");
String s3 = new String("String C");
```

Once an instance of an object is created it is possible to call methods on that object or access variables. The following line of code uses the toCharArray() method to retrieve a character array of the String object created above:

```
char [] name = s1.toCharArray();
```

This is the essence of object oriented programming. Objects contain all the data and methods they require in one place. This makes the code easier to understand, especially when compared to structured "spaghetti" code where methods and data are thrown together.

Until now we have seen only some very basic class definitions, but classes can also be modified by a number of keywords. For example, the following class contains a declaration using many keywords:

```
public abstract class Hermes extends Navigator
implements SensorListener {}
```

Let's examine each of the keywords available for class definitions.

Class Access

Classes can be either public or default (no keyword). A public class is visible to all other classes, meaning another class may interact with the class, doing such things as creating an instance of the class and accessing its methods. Default (sometimes referred to as package access) classes are only visible to other classes within the same package.

Extending Classes

One of the most useful concepts of object oriented programming is that a class can extend another class. Imagine that you program a class that controls a robot arm to move up and down. You decide to call this class Arm. Let's examine a very simplified version of what this class might look like:

```
import lejos.hardware.motor.*;
class Arm {
  public void armUp() {
    Motor.B.forward();
  }
public void armDown() {
    Motor.B.backward();
  }
}
```

Now imagine you create an enhanced robot arm that still moves up and down, but also has an attached claw that opens and closes. It would be preferable to reuse the code in the Arm class instead of starting from scratch. It is also cumbersome to copy and paste the Arm code because if you ever change or improve the Arm code you would have to recopy it.

Object oriented programming provides an elegant solution by extending the Arm class to inherit all the functionality of Arm, as well the ability to add new methods and data for the claw. In our example we will call this new class ClawArm. ClawArm extends Arm so Arm is the *superclass* and ClawArm is the *subclass* (Figure 9-2). All classes in Java automatically extend the Object class, as you can see in the diagram. If you follow the hierarchy of any Java class, the class at the top is always Object.

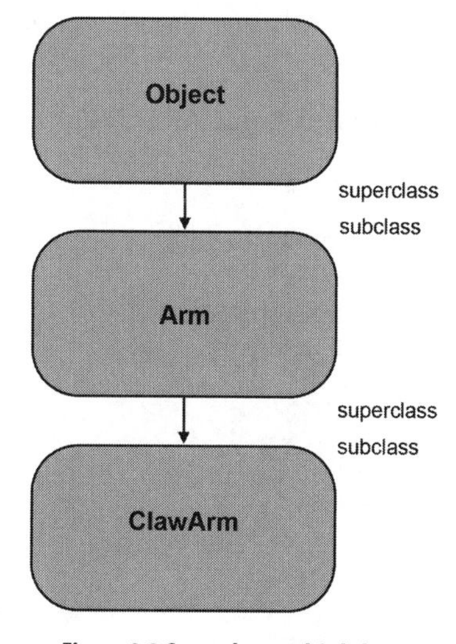

Figure 9-2 Superclass and Subclass

When a subclass extends another class, it inherits all of the functionality of the superclass. In Java code, the syntax for extending another class is as follows:

```
import lejos.hardware.motor.*;
class ClawArm extends Arm {
  public void openClaw() {
    Motor.C.forward();
  }

  public void closeClaw() {
    Motor.C.backward();
  }
}
```

The *extends* keyword is used above to indicate the superclass. Now anyone using the ClawArm object will be able to call the two methods of Arm, as well as the methods introduced in ClawArm:

```
ClawArm myRobot = new ClawArm();
myRobot.armDown();
myRobot.openClaw();
```

Abstract Classes

An abstract class (possibly) contains some functional methods, but also declares method names with no functional code. In a way, an abstract class is like a half-finished class; some methods are already provided, but others are just defined and must be filled in later. Since an abstract class is not complete it can not be instantiated. Let's examine an abstract class to see what they are about. We'll use the Arm class again:

```
import lejos.hardware.motor.*;
abstract class Arm {
  public abstract void spinArm();
  public void armUp() {
    Motor.B.forward();
  }
  public void armDown() {
    Motor.B.backward();
  }
}
```

In line two, the Arm class is now declared to be Abstract. It contains the same two functional armUp() and armDown() methods as before, but it also defines a third abstract method called spinArm(). Notice this method has semicolons at the end of the definition, but no curly braces? If we tried to instantiate this class in code, the Java compiler will respond with "Arm is abstract; cannot be instantiated".

The purpose of an abstract class is to be a higher level superclass that will be used by other subclasses. The subclasses all share the same code, making it more efficient and logical to program. So in this case, Arm is the general class, and the subclasses are more specific types of Arm (Figure 9-3).

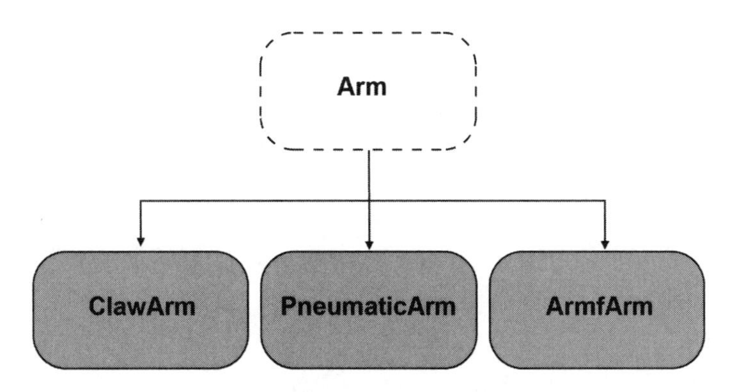

Figure 9-3 Hierarchy of Arm classes

Objects

Objects form the basis of object oriented programming. An object is an instance of a class that contains variables and methods. Put another way, they contain the data and functionality of a class.

A class is not a fully fledged object, but a programmer can use a class to create an object by using the *new* keyword to call a constructor on the class. The new object can also be assigned to a variable if the object needs to be referenced in later lines of code. The following lines of code show an object being initialized, and another assigning an object to a reference variable:

```
new ControlGUI(); // initialized but not assigned
String name = new String("Maximillion");
```

NOTE: *The text after // is called an inline comment. You can make a comment in the code (that does not affect the code) by typing two forward slashes before your comment.*

All Java classes are extended from the class Object. This class contains methods common to all Java objects, such as toString() and equals(). The toString() method provides a string representation of an object, and the equals() method compares two reference variables to see if they refer to the same object.

In Java, objects can be accessed by reference. This means two or more variables can refer to the same object, as the following pseudo-code demonstrates:

```
MyObject a = new MyObject();
MyObject b = a;
MyObject c = b;
```

As you can see above, all three reference variables (a, b, and c) are now referring to the same object. Imagine that a method is called using variable b, such as:

```
b.setValue(25);
```

This means the value will be changed for the object that a and c refer to as well, since they all refer to the same object. Now if you call getValue() on a, it will retrieve a value of 25:

```
int x = a.getValue(); // x = 25
```

Interfaces

Java does not allow a class to extend more than one class. Normally, with object oriented design, it is a good idea for a class to be responsible for only one type of behavior. If you try to make a single class perform all kinds of tasks it usually makes things unnecessarily complex.

But sometimes it is necessary to make a class with multiple behaviors. In these cases, interfaces are the solution. An interface can define methods and variables, but the methods do not contain any functional code. Since there is no functional code, an interface may not be instantiated (similar to an abstract class). The following is an example of an interface:

```
interface Steerable {
  public void turnLeft();
  public void turnRight();
  public void driveForward();
  public void applyBrakes();
}
```

As you can see there is no functional code here, only method definitions. So how does this help you as a programmer? Imagine a class that contains a method to steer any type of robot around a racetrack. We'll call this class RaceCar. Imagine that the RaceCar class has a method that steers a car around the racetrack by using the steering and driving methods contained by Steerable objects. We'll say the method definition looks like this:

```
public void drive(Steerable myCar)
```

By having your robot code implement Steerable, it can be used by the RaceCar class. It can also extend another class if need be (but this is not necessary). In the following example, we will also have the class extend the Thread class:

```
import lejos.hardware.motor.*;
class MyCar extends Thread implements Steerable {
  public void turnLeft() {
    Motor.B.backward();
    pause(200);
    Motor.B.stop();
  }

  public void turnRight() {
    Motor.B.forward();
    pause(200);
    Motor.B.stop();
  }

  public void driveForward() {
    Motor.A.forward();
    Motor.C.forward();
  }

  public void applyBrakes() {
    Motor.A.stop();
    Motor.C.stop();
  }
```

```
    public void run() {
      // Sensor watching code here
    }

    private void pause(int mSeconds) {
      try{
        Thread.sleep(mSeconds);
      }catch(InterruptedException e){}
    }
  }
```

As you can see, there are now complete method definitions for turnLeft(), turnRight(), driveForward() and applyBrakes(). Now the RaceCar class will be able to use the MyCar class and guide it around the track using a few lines of code to get things going:

```
MyCar speedy = new MyCar();
RaceCar.drive(speedy);
```

RaceCar now has an instance of MyCar, called speedy. RaceCar knows it has a Steerable, so it can now call the proper methods to drive the car around the track.

Import and Package Statements

You may have noticed the following statement at the top of the MyCar class:

```
import lejos.hardware.motor.*;
```

This is an import statement responsible for allowing the MyCar class access to all classes contained in the lejos.hardware.motor package. Keep in mind that this statement does not cause all these classes to load into memory. Only the classes used in your code are loaded into Java memory.

In the next chapter (Chapter 10) we will examine some of the packages in the leJOS API, such as lejos.robotics.navigation and lejos.hardware.ev3. These packages contain an assortment of classes that perform functions according to the theme of the package. The main theme of lejos.hardware.port is that all the classes in the package give you access to the input and output ports, such as MotorPort and SensorPort.

These classes can not be accessed by your code until the import statement is used, however. The advantage of the package system is that other classes can be hidden as well as sorted, but called out of hiding when you need to use them. You can also import just a single class from a package. For example, it we only want to use the Sound class we can use the following syntax:

```
import lejos.hardware.Sound;
```

Conversely, we could also access a class in a package without using an import statement by drilling down to it. The following line could be used to call a method directly from Sound:

```
lejos.hardware.Sound.beep();
```

NOTE: *The java.lang package is always imported automatically by Java. This is because the classes used in this package are central to the use of the Java language. It contains such important classes as Thread, System, String and Math which are used frequently.*

So how can you make your own packages? In Eclipse it is easy. Simply select File > New > Package and enter the name of the package. You can then create new classes and interfaces within this package.

Using the command line it is slightly more complicated. First, you must create a directory structure that matches the package name. This directory structure must start at a classpath directory. For example, you may want to make a package called robot.flying. If you have included c:\Java in your classpath, make the following directory:

```
c:\Java\robot\flying
```

Adding classes to your package is easy. Let's say you are making a class called AirShip and you want it to appear in the package robot.flying. Just include the package declaration at the top of your source file:

```
package robot.flying;
import lejos.hardware.motor.*;
class AirShip {
  /// Rest of code...
}
```

The package statement must appear before any other code otherwise the compiler will produce an error. Once this class is created, in order to import it, the class file must appear in the directory c:\Java\robot\flying. Usually I just store the source code file in the same directory as the package classes so that when it compiles it automatically dumps it in here, but you could move the class file here on your own.

NOTE: *The main purpose of packages is to organize code. On large projects with hundreds of classes, this can be very important. When creating your own leJOS programs you might not use very many classes. For this reason it might not be useful to create your own packages, but you will still need to know how to import classes from other packages in the leJOS API.*

Methods

A method gives a class its functionality. All methods must be contained within a class, and there is no such thing as a global method in Java. Methods have two important defining characteristics: return types and arguments. Let's examine a typical Java method definition:

```
public int readDistance() {
  return motorA.getTachoCount();
}
```

The first line of this method declares a return type of the primitive integer (int), so by definition the code within the curly braces must return an integer value. If a method declares it will return a value but the code does not return anything, the compiler will produce an error. Methods can just as easily return objects too. As you can see above, the parentheses are empty and contain no arguments. Let's examine a method that uses arguments:

```
public void setMotors(BasicMotor left,
BasicMotor right) {
  leftDrive = left;
  rightDrive = right;
}
```

The above method has no return type, as indicated by using the keyword void. This method uses two arguments, left and right. These arguments can be objects (in this case, two BasicMotor objects) or primitives.

NOTE: *Recursion is allowed in leJOS. This is an advanced programming technique of having a method call itself (often several hundred times) before a criterion is satisfied and the methods all return. Keep in mind, recursion can be very memory intensive depending on how many levels deep it goes.*

Constructor Methods

A constructor method is a special method used to initialize an object when it is created. The code within a constructor can contain absolutely anything, but generally it is limited to setting variables and preparing the object so it will work when a programmer starts calling methods. The following class shows a properly defined constructor method:

```
class Kangabot {
  int jumpCentimeters;
  public Kangabot(int jumpDistance) {
    jumpCentimeters = jumpDistance;
  }
  public static void main(String [] args) {
    Kangabot roo = new Kangabot(5);
  }
}
```

The constructor method starts at line 3. As you can see, it has the same name as the class name, and it has no return type (not even void). In this case the constructor is simply used to initialize a programmer-defined variable jumpCentimeters. All objects use a constructor method, but if you don't specifically create one then the object will have a default no-arguments constructor implicitly defined by Java.

NOTE: *When you call the constructor method from another body of code, that body of code will stop its execution until the constructor returns. This means that if your constructor starts running methods that loop forever, your program will effectively freeze. This is one of the classic errors new programmers make.*

Static Methods

A static method is a method that can be called without creating an instance of a class. To contrast static methods with member methods, lets examine the following code.

```
MyRobot merl = new MyRobot();
merl.attack();
```

The code above demonstrates how a method is normally called from an object. With static methods, however, there is no need to create an instance first. If we changed the attack() method above to a static method, we could call it straight from the class as follows:

```
MyRobot.attack();
```

There are many Java classes, such as the java.lang.Math, that contain exclusively static methods. Since no recurring data needs to be kept in an object for Math methods to work, there is no reason to go to the trouble of creating an instance of Math. So methods are called directly from the class instead, as follows:

```
int result = Math.sin(0.5);
```

It is easy to write a static method, but there are two rules that must be obeyed in order for it to work properly:

- A static method may not directly call an instance (non-static) method
- A static method may not directly use an instance (non-static) variable

---

NOTE: *If you're worried you won't remember all this, don't worry. When you make a mistake, Eclipse does a good job of pointing out what you did wrong.*

Overriding Methods

As we saw, when one class extends another class the subclass inherits all the methods from the superclass. But sometimes it is beneficial to change one of the existing superclass methods in order to provide the subclass with altered or enhanced functionality. For example, when you create a two jointed robotic arm, you also need to program a class with various methods to control the arm. Imagine that the simplified API for the arm looks like this:

```
class RoboArm {
  public void goToPoint(int x, int y, int z) {
    // Code to move hand to 3-D coordinate…
  }
  // More methods...
}
```

Assume that your code is able to move the hand to any coordinate in 3-dimen-dional space. Now imagine that you add an extra joint around the wrist area of the robot arm. This means you need to alter the program code in order to accommodate the physical change.

The object-oriented way to do this is to extend the RoboArm class and then replace the goToPoint() method with new code. Replacing an existing method defined in a superclass is called *overriding* a method. There's not much to it, as the following example demonstrates:

```
class SuperArm extends RoboArm {
  public void goToPoint(int x, int y, int z) {
    // New code to move 3-jointed arm
  }
}
```

Primitive Data Types

The leJOS project fully supports all eight primitive types. There are 4 integral numbers, 2 floating-point numbers, a character type and a boolean type. Table 6-1 shows these types and their sizes.

Keyword	Minimum	Maximum	Default	Bits
byte	-28	28 - 1	0	8
short	-216	216 - 1	0	16
int	-232	232 - 1	0	32
long	-232	232 - 1	0	64
float	varies	varies	0.0	32
double	varies	varies	0.0	64
boolean	none	none	false	2
char	'\U00000'	'\U65535'	'\U00000'	16

Table 6-1 Java Primitives

In some languages, such as Visual Basic, it is possible to declare a variable without specifying its type. Java, on the other hand, is a *strongly typed language*, which means all variables must be declared as belonging to a specific data type. The following class shows each of the primitive variables declared and initialized with a value:

```
class Primitives {

  byte b = 127;
  short s = 32767;
```

```
    int i = 2147483647;
    long l = 2147483647;

    float f = 100.123f;
    double d = 100.123;

    char c = 'M';

    boolean boo = true;
  }
```

All of these variables are declared at the class level. Note in line 8 it is necessary to place an 'f' after the literal number. This is because literals are treated as double values by default, and it is illegal to assign a double value to a float variable.

NOTE: *Primitives are passed into methods (as arguments) by value. When a variable is passed in this manner it is just as if a copy of the variable was passed into the method. If the value is changed within the method it will not be reflected outside of the scope of the method.*

```
import lejos.hardware.lcd.*;

class ByValue {
  public static void main(String [] args) {
    float height = 72f;
    float metricHeight = getCentimeters(height);
    LCD.drawInt((int)height, 0, 0);
  }

  public static float getCentimeters(float inches) {
    inches = inches * 2.6f;
    return inches;
  }
}
```

The code above passes the value of variable height to the method getCentimeters. Inside the method, the value of the variable is multiplied by 2.6, but this does not mean the actual variable height is also multiplied. Since a copy of the variable is passed to the method, height is not affected.

Sometimes it is necessary to convert from one type of primitive to another. This can be done by casting, as when a ceramics artist casts plaster into a shape by using a mold. In Java, casting can convert a smaller primitive to a larger number type, or conversely, a larger value to a smaller primitive type. Let's examine how both of these look in code:

```
short x = 500;
int y = x; // implicit cast
int a = 500;
short b = (short)a; // explicit cast
```

Notice above that to convert a short number to an int requires no special syntax. Why? Because a smaller 16 bit primitive (short) will always fit within a larger 32 bit primitive (int). (If you have a 16 gallon jug and a 32 gallon jug, the contents of the 16 gallon jug will always fit in the 32 gallon jug no matter how full the 16 gallon jug is.) However, in lines 3 and 4, in order to convert the int to a short it must be explicitly stated that the conversion is occurring. If the number is too large to store in the smaller primitive then the number will be truncated.

It is also possible to cast floating-point numbers into integrals and vice versa. When converting a float to an int, the decimal places will be chopped off out of necessity. The following code demonstrates this:

```
float a = 555.555f;
int b = (int)a; // explicit cast
int x = 25;
float y = x; // implicit cast
```

Arrays

An array is a collection of objects or primitives that can be accessed using an index number. This index number starts with zero for the first element in the array. An array is an object in every sense of the word. It contains all the methods of the Object class such as equals() and toString(). It also has a variable called length that indicates how many elements it contains. The only difference is that it is initialized in a different manner from a regular object or primitive as the following examples demonstrate:

```
int [] x = new int[20];
x[5] = 2000; // changes the 6th value in the array
boolean [] boo = {true, false, false, false,
true, false};
```

If you pass an array into a method as an argument and change one of the variables in the array, this will also be reflected outside of the method. So, like any object, an array is accessed by reference.

Naming Rules

As a programmer it is up to you to give your classes, methods, and variables unique names. There are several rules about naming these elements in code:

- The name must begin with a letter (upper or lower case), an underscore '_', or a dollar sign '$'
- The name may contain only characters, numbers, underscore or dollar sign.
- The name may not be a keyword (such as true, false, or null). Table 6-2 shows a complete list of keywords.

abstract	default	if	private	this
boolean	do	implements	protected	throw
break	double	import	public	throws
byte	else	instanceof	return	transient
case	extends	int	short	try
catch	final	interface	static	void
char	finally	long	strictfp	volatile
class	float	native	super	while
const	for	new	switch	
continue	goto	package	synchronized	

Table 6-2 Java Keywords

Operators

Operators lie at the very core of programming because they perform the actual calculations, which is what a computer is designed to do. Operators essentially perform low level math calculations involving bits.

Mathematical Operators

The basic mathematical operators usually function the same across all programming languages, so most people are quite familiar with them. They are:

- + addition
- - subtraction
- * multiplication
- / division
- % remainder (modulo division)

> **Tip!**
> If you have a hard time remembering whether to use the backslash \ or the forward slash / for division, remember the division symbol tilts in the same direction as the % symbol on the keyboard.

The most unfamiliar operation here for most new programmers is the modulo operator. It produces the remainder of integral number division; that is, the amount that was not able to divide evenly into a number and was thus left over. Modulo can be used for both integer numbers and floating point numbers. Examine the following code:

```
class MathTest {
  public static void main(String [] args) {
    int result1 = 15 % 2; // yields 1
    float result2 = 15 % 2.3f; // yields 1.2
  }
}
```

NOTE: *There is no power operator in Java, such as the calculation* 3^9*. In order to perform a power calculation you must use the Math.pow() method, as follows:*

```
Math.pow(3, 9)
```

Comparison Operators

Comparison operators are used to compare two variables. All comparison operators return a boolean value to indicate if the comparison is true or false. Comparison operators are most often used in looping constructs, such as if-then and while (see below). The following comparison operators are used in Java:

```
==   // equals
>=   // greater than or equal to
<=   // less than or equal to
>    // greater than
<    // less than
!=   // not equal
```

Comparisons can be made as follows:

```
boolean a = (25 > 24); // a = true
boolean b = (25 == 24); // b = false
```

Boolean Operators

Some operators are specifically for numbers (greater than, less than), some are for boolean comparisons only (&& and ||), and some can be used for both numbers and boolean (!= and ==). The following can be used with boolean operands:

```
&&   AND
||   OR
^    Exclusive OR (XOR)
!=   Not equal
==   Equal
```

Let's examine these in some code:

```
boolean a = (25>24)||(12==13); // true
boolean b = (25>24)&&(12==13); // false
boolean c = a == b; // false
boolean d = a != b; // true
boolean e = a ^ b; // true
boolean f = b ^ c; // false
```

Most programmers are familiar with AND, OR, EQUAL and NOT EQUAL. Exclusive OR (XOR) is different from OR, however. Let's first compare this with OR. OR is true if one or the other, or both operands are true. XOR is only true if one or the other are true. If both are true, then XOR produces false.

Program Flow Control

A program that executed statements one after another in the same order each time it was run would be rather boring and predictable. It is the branching quality of a program that gives it power and flexibility. This quality is known as program flow control.

If Statements

If statements are very easy to use and very powerful. A typical if statement examines a boolean value and executes a block of code if the boolean value is true. The following example is typical example of an if-statement in leJOS code:

```
if(x == 10) {
  Sound.beep();
}
```

Notice that no then keyword is used. All conditional code must appear in the curly braces. Alternatively, if there is only one statement to execute, the braces are not required:

```
if(x == 10) Sound.beep();
```

You can also use an else statement to execute a block of code should the boolean value equal false:

```
if(x==10) {
  x = 0;
  Sound.beep();
} else {
  Sound.buzz();
}
```

Any code can be placed within the conditional code, including other if statements. This type of code construct is called a nested if-statement.

```
if(x==10) {
  if(y == 5);
    Sound.beep();
}
```

Switch Statements

Switch statements allow the code to check for a large number of comparisons. The code consists of a switch statement to indicate the variable being checked, and then a series of case statements to decide what to do if the variable matches the case constant. A typical switch-case statement looks like this:

```
int command = 1;
String output;
switch (command) {
  case 1: output = "Forward";
    break;
  case 2: output = "Backward";
    break;
  case 3: output = "Forward";
    break;
```

```
    case 4: output = "Backward";
      break;
    default: output = "Stop";
  }
  System.out.println(output);
```

There are a few things to notice in the above code. First, there is a default choice at the end in case none of the cases match the switch variable. Second, after each case statement is the keyword break. This is optional, and is used to prevent the code from checking through the rest of the case statements, and especially the default case statement which is executed by default unless a break appears.

For Loops

Conditional loops are used to repeat a code block a number of times until a condition is met. One of the most popular loop constructs is the for-loop. This loop repeats a block of code a predetermined number of times until a condition is satisfied. A for-loop consists of three main parts.

- Counter initialization
- Boolean condition
- Counter increment

These parts are stated in the following order:

```
for(counter initialization; boolean condition;
counter increment)
```

Let's examine a for-loop in some actual code:

```
import lejos.hardware.*;
class SoundLoop {

  public static void main(String [] args) {
    for(int freq=500;freq<1000;freq += 50) {
      Sound.playTone(freq, 30);
    }
    try{
Button.RUN.waitForPressAndRelease();
    } catch(InterruptedException e) {}
  }
}
```

The preceding example declares and initializes a variable called freq, checks if the variable satisfies the condition, executes the block of code, then increments the freq integer by 50 and rechecks the condition until the condition evaluates to false.

NOTE: *If the boolean condition is empty, it's assumed to be always true, so it will repeat in an endless loop as the following code demonstrates:*

```
for(;;) {}
```

While and do-while Loops

While loops are actually very similar to for loops, only they are not constructed specifically for incrementing a variable a set number of times. The while loop evaluates one boolean value, as follows:

```
while(boolean condition)
```

The following code uses a while loop to keep a robot moving forward until it gets to a dark area:

```
import lejos.hardware.port.*;
import lejos.hardware.sensor.*;
import lejos.hardware.motor.*;

class CockroachBot implements SensorConstants{

  EV3ColorSensor cs;
  SensorMode ambient;

  public CockroachBot() {
    cs = new EV3ColorSensor(SensorPort.S2);
    ambient = cs.getMode("Ambient");
  }

  public static void main(String [] args) {
    CockroachBot bot = new CockroachBot();
    Motor.B.forward();
    Motor.C.forward();
    while(bot.isBright()) {
      // Keep moving forward
    }
    Motor.B.stop();
```

```
    Motor.C.stop();
  }

  public boolean isBright() {
    float[] sample = new float[ambient.
sampleSize()];
    ambient.fetchSample(sample, 0);
    return (sample[0] > 55);
  }
}
```

It is also possible to make the while loop execute the block of code at least once, then evaluate the condition. This is done using the do keyword, as follows:

```
do {
  // code body
} while(boolean condition);
```

Exception Handling

Exception handling is one of the unique features of Java that makes it popular with programmers. With exception handling, the error checking part of your code can be segregated from the rest of the functional code. This makes code neater and easier to understand. Java accomplishes this by enclosing method calls that may throw exceptions within a try-catch block. The following pseudo-code shows how this works:

```
try {
  // method that may throw exception
}
catch(Exception type) {
  // code to deal with exception
}
```

When a method is prone to throwing an exception the try block must be called. The error is dealt with in the catch block. There is a third, optional part called a finally block. The finally block is executed once either the try or catch block has finished executing:

```
try {
  // method that may throw exception
}
catch(Exception type) {
```

```
    // code to deal with exception
  }
  finally{ // Optional
    // code that is always executed no matter what
  }
```

For the most part, under leJOS, the only time you'll really deal with exceptions is when using the Thread.sleep() method, as follows:

```
  try {
    Thread.sleep(100);
  } catch(InterruptedException e) {
    interrupted = true;
  }
```

The Java fundamentals in this chapter should be enough to get you programming your own code. There's still more to learn though! Chapter 20 will continue your Java education by exploring more standard Java features.

CHAPTER 10

The leJOS API

TOPICS IN THIS CHAPTER

- lejos.hardware
- lejos.hardware.sensor
- lejos.hardware.lcd
- lejos.hardware.motor
- lejos.remote packages
- lejos.robotics.navigation
- lejos.robotics.subsumption

Imagine a tiny robot that wanders around your house. In the past, such an activity was short lived because the robot invariably became stuck. It would tip over, run into a wall without the bumper activating, or the wheels would become stuck on some low-lying object.

Now imagine the following robot: it wanders around your house avoiding objects with the distance sensor. If the sensor misses an object, the robot can still tell if the wheels are stuck by monitoring decreases in rotation speed. If the robot tips over it uses a tilt sensor to identify the problem. It can even use speakers to emit a tiny voice that says, "I fell over on my side." It can then try to right itself. Such a robot could be left alone for hours, and when you return it would still be exploring your house. This is possible with leJOS. You just need to know where to find the classes and methods, which together form the Application Programming Interface (API).

The leJOS API has hundreds of classes, each with a multitude of methods. This does not include the classes that make up Java Standard Edition, just the original classes designed by the leJOS developer team. This chapter will focus on the main methods and classes, however it would be impossible to cover everything here. If you want to examine all the classes, check out the API documentation.

NOTE: *The leJOS developers periodically modify the API. Please refer to your install directory under lejos/docs/ev3/index.html for the latest API documentation (see Figure 10-1). You can also visit www.lejos.org.*

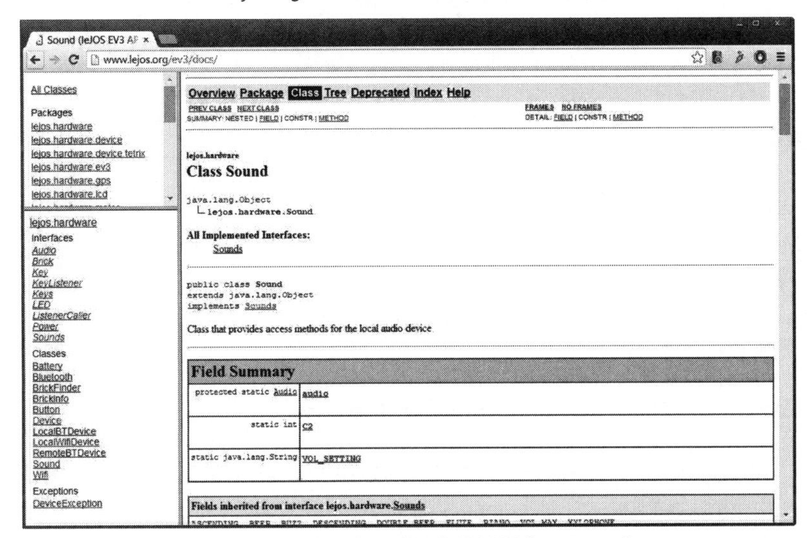

Figure 10-1 Viewing the online leJOS API documentation.

lejos.hardware

The classes and methods in the lejos.hardware package directly access most of the lesser functions of the EV3 brick.

Battery

The Battery class allows you to determine the voltage produced by the EV3 batteries. Rechargeable batteries provide approximately 7.4 volts, while alkaline batteries produce 9 volts. When the voltage level begins to fall below 6.5 it means they are almost expended.

lejos.hardware.Battery

```
static float getBatteryCurrent()
```

Returns the current being drawn by the EV3 brick.

```
static float getMotorCurrent()
```

Returns the current being drawn by the motors.

```
static float getVoltage()
```

Returns the battery voltage in volts.

```
static int getVoltageMilliVolt()
```

Returns the battery voltage in millivolts.

Button

The Button class contains static instances of the six buttons (much like the Motor class contains static instances of the motors). These four instances are ENTER, ESCAPE, LEFT, RIGHT, UP, DOWN. Many times we just want the code to stop until a button is pressed, so we can use the waitForPressAndRelease() method:

```
// Stops code until ENTER pressed
Button.ENTER.waitForPressAndRelease();
```

You can also use a simple while-loop to stop your code while it waits for the user to press a button:

```
while(!Button.ENTER.isDown()) {}
```

Java also offers event listeners. Java can initiate an action, or several actions, that are dependent on an event occurring (when a user presses a button, for example). There can be more than one listener waiting for an event to happen. When an event occurs, all the classes that are listening will be notified. The following example shows how to program an event listener:

```
import lejos.hardware.*;
class PlaySound implements KeyListener {
  public void keyPressed(Key k) {
    Sound.beepSequence();
  }
  public void keyReleased(Key k) {}
}
```

This class implements the KeyListener interface, which contains two method definitions: keyPressed() and keyReleased(). All interface methods must be defined in the class implementing the interface, even if the method is not used. When the button that is registered with this listener is pressed, the EV3 will play a series of beeps. Now let's examine a class that registers this listener:

```
  import lejos.hardware.*;
  class ButtonTest {
    public static void main(String [] args) {

Button.ENTER.addKeyListener(new PlaySound());

Button.ESCAPE.waitForPressAndRelease();
    }
  }
```

As you can see in line four, the ENTER button has an instance of the PlaySound listener registered with it (up to four button listeners can also be registered for each button). The next line halts the program until the escape button is pressed. We could also have opted to do this in one class by having ButtonTest implement the ButtonListener interface, and then add itself to the ENTER button.

lejos.hardware.Key

```
void addKeyListener(KeyListener aListener)
```

Adds a listener of button events.

```
boolean isDown()
```

Check if the button is pressed.

```
void waitForPressAndRelease()
```

Wait until the button is released.

***NOTE:** The buttons in leJOS all produce default tones automatically when pressed. If you want to disable the tones so you can use your own sounds in your code, use Button. setKeyClickTone(). By setting the frequency of a button to zero, the tone is disabled. For example:*

```
Button.setKeyClickTone(Button.ENTER.getId(), 0);
```

Sound

The Sound class is responsible for playing sounds through the EV3 speaker. The playTone() method is the most versatile. There are also convenience methods for playing basic sounds.

lejos.hardware.Sound

```
public static void playTone(int aFrequency, int
aDuration)
```

Plays a tone, given its frequency and duration. Frequency is audible from about 31 to 2100 Hertz. The duration argument is in hundredths of a second (centiseconds, not milliseconds) and is truncated at 256, so the maximum duration of a tone is 2.56 seconds.

Parameters:*aFrequency - The frequency of the tone in Hertz (Hz).*

aDuration - The duration of the tone, in centiseconds. Value is truncated at 256 centiseconds.

```
public static void beep()
```

Beeps once

```
public static void twoBeeps()
```

Beeps twice

```
public static void beepSequence()
```

Downward tones

```
public static void buzz()
```

Low buzz

```
public static void playSample(File file)
```

This method plays a wav file.

Sound will be covered more in Chapter 10.

lejos.hardware.sensor

There are more sensors supported for the EV3 than there are any other device, including motors. These sensors are mainly NXT sensors, which are backward compatible with the EV3. They come from companies like Mindsensors, HiTechnic, Dexter Industries and of course LEGO. All sensors are accessed in the same way using a sensor framework, which was introduced for the EV3. Chapter 16 describes how to access these sensors, so we will move on for now.

lejos.hardware.lcd

The lcd package contains classes and interfaces for manipulating graphics in the LCD. For th most basic output you can use LCD (or even System.out) but for more advanced displays you can turn to GraphicsLCD.

LCD

The LCD class offers some basic methods for drawing strings and numbers to the screen. In leJOS, the font size allows for eight rows of text, with each row containing as many as 18 characters (see Figure 10-2).

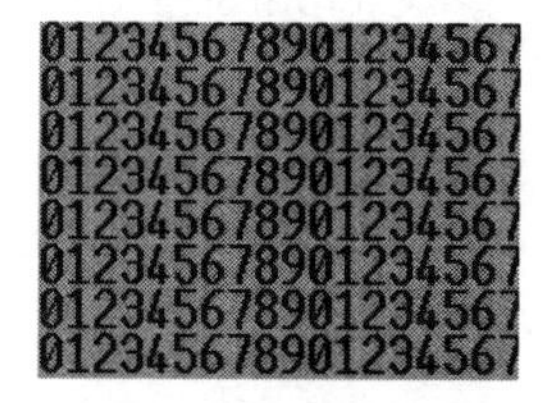

Figure 10-2: LCD displaying almost 18 characters across

```
public static void drawString(String str,
int x, int y)
```

Displays a string on the LCD at a specified x, y coordinate. The coordinates start in the upper left corner. Each coordinate is one character in size.

```
public static void drawInt(int i, int x, int y)
```

Displays an integer on the LCD at specified x, y coordinate.

```
public static void drawInt(int i, int places,
int x, int y)
```

This is a more advanced method for displaying integers. Characters from a previous call remain on the LCD. This method adds spaces to the front of the interger value, which will erase old characters. Displays an integer on the LCD at x, y with leading spaces to occupy at least the number of characters specified by the places parameter.

```
public static void clear()
```

Clears the display.

GraphicsLCD

GraphicsLCD is used to draw text and a variety of primitive shapes to the LCD screen, such as circles, ovals, arcs, lines, and rectangles. Unlike the official Java Graphics class, this one draws directly to the LCD screen, rather than to a Panel or other component. You must first create an instance of Graphics before using the class:

```
GraphicsLCD g = LocalEV3.get().getGraphicsLCD();
g.drawLine(5,5,60,60);
g.drawRect(62, 10, 25, 35);
g.refresh();
```

If you want to display text of different sizes, you must use the GraphicsLCD class. The following code uses the smallest font size to fit more text within the limited space of the LCD.

```
import lejos.hardware.*;
import lejos.hardware.ev3.*;
import lejos.hardware.lcd.*;

public class FontLCDGraphics {
```

```
  public static void main(String[] args) {
    GraphicsLCD g = LocalEV3.get().getGraphicsLCD();
    g.setFont(Font.getSmallFont());
    for(int y=0;y<128;y=y+8)
      g.drawString("012345678901234567890123456789", 0, y, 0);

    Button.ESCAPE.waitForPressAndRelease();
  }
}
```

Using the smallest font, this code produces 16 lines of text with 30 characters each (see Figure 10-3). You can also use the largest font, which produces much less text.

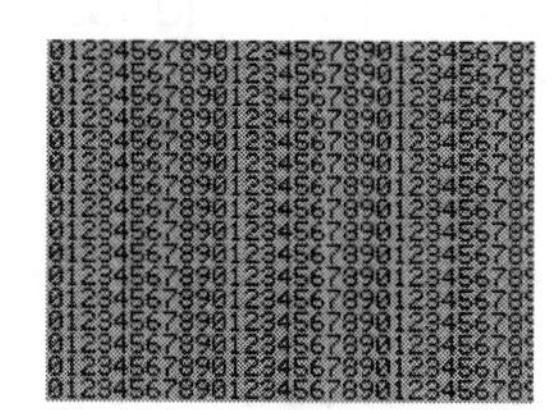

Figure 10-3: Viewing the smallest font size.

Let's examine the methods available to GraphicsLCD .

```
public void setPixel(int color, int x, int y)
```

The color parameter is obtained from Graphics, and represents either black or white. Use Graphics.BLACK or Graphics.WHITE.

```
public void drawLine(int x0, int y0, int x1, int y1)
```

Draws a line between the points (x1, y1) and (x2, y2).

```
public void drawRect(int x, int y, int width, int height)
```

Draws a rectangle. The left and right edges of the rectangle are at x and x + width. The top and bottom edges are at y and y + height (see Figure 10-4).

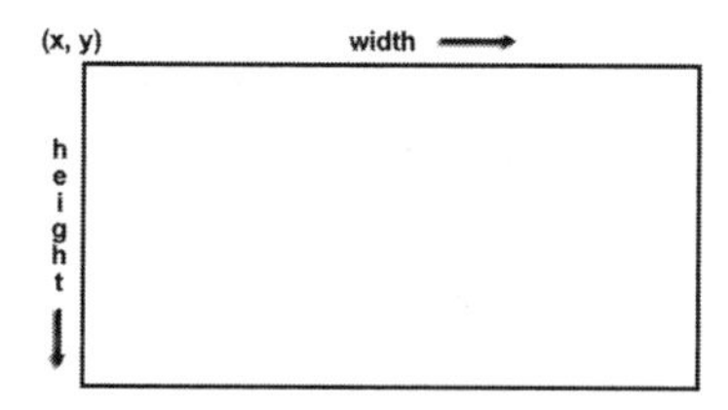

Figure 10-4 Drawing a rectangle.

```
public void fillRect(int x, int y, int width,
int height)
```

Draws a rectangle, as above, but fills it solid.

```
public void drawArc(int x,int y,int width,int
height,int startAngle,int arcAngle)
```

The drawArc() method is used for drawing circles, ovals, and of course arcs. It draws the outline of a circular or elliptical arc covering the specified rectangle (see Figure 10-5). The resulting arc begins at startAngle and extends for arcAngle degrees. Angles are interpreted such that 0 degrees is at the 3 o'clock position. A positive value indicates a counter-clockwise rotation while a negative value indicates a clockwise rotation.

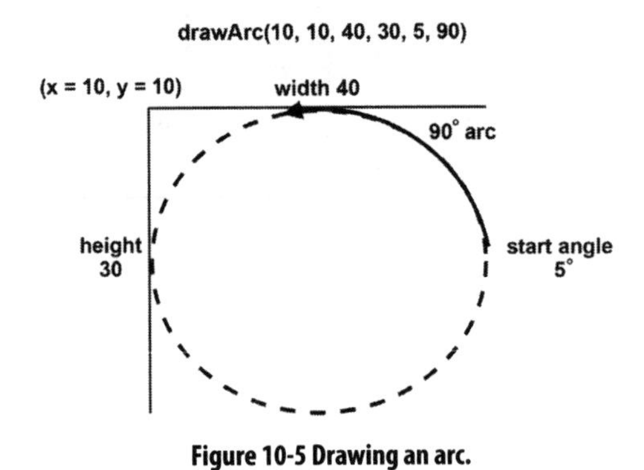

Figure 10-5 Drawing an arc.

```
public void fillArc(int x,int y,int width,int
height,int startAngle,int arcAngle)
```

The fillArc() method is used for drawing solid circles, ovals, and arcs. It fills in the area from the center of the arc to the outer line. It's useful for drawing Pac-Man.

```
public void drawRoundRect(int x,int y,int width,
int height, int arcWidth, int arcHeight)
```

Draws a round-cornered rectangle (see Figure 10-6). The left and right edges of the rectangle are at x and x + width, respectively. The top and bottom edges of the rectangle are at y and y + height. The arcWidth and arcHeight parameters determine the size of the rounded corners.

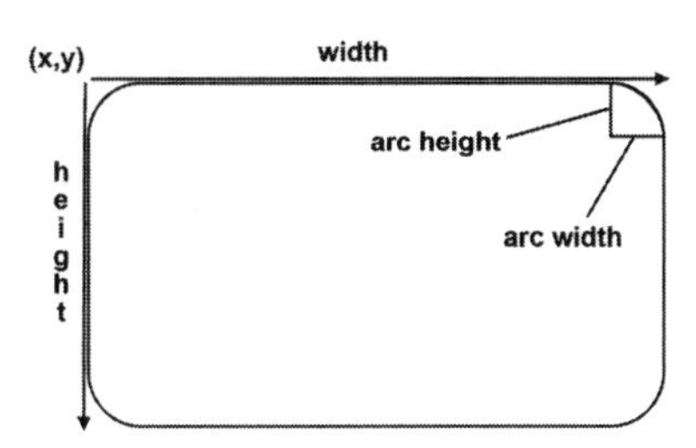

Figure 10-6 Drawing a round rectangle.

```
public void refresh()
```

Updates the LCD display. Nothing will appear on the LCD screen until refresh() is called.

```
public void clear()
```

Clears the LCD display buffer. Must call refresh() before this method shows up.

lejos.hardware.motor

The LEGO EV3 is capable of using a variety of different motors. There are two different motors available in the EV3 kit (large and medium) but outside of this there are legacy motors supported. The leJOS API contains classes for all these motors, which are found in the lejos.hardware.motor package.

There are two main types of motor classes. Unregulated classes allow you to control the motor via power only. Regulated motors, on the other hand, constantly monitor the tachometers to make sure speeds and rotations are accurate.

EV3LargeRegulatedMotor

The tachometer allows for robust motor movement of the EV3 motors. Speed is in degrees per second. The actual maximum speed of the motor depends on battery voltage and load. Let's examine the major methods of this class.

WARNING: *There is a big change in how to access motors compared to how it was done with the NXT. Avoid using the static instances of motors via Motor.A, Motor.B, etc. If you use this, all four motor ports are assumed to contain large EV3 motors, and are initialized as such. This is a problem if you use the static reference, then try to initialize an EV3MediumRegulatedMotor.*

```
Motor.A.forward();

// The next line throws an exception

EV3MediumRegulatedMotor b = new
EV3MediumRegulatedMotor(port);
```

The result is an exception because the motor port is already open. Make new motor objects instead.

---

lejos.hardware.motor.EV3LargeRegulatedMotor

```
public final boolean isMoving()
```

Returns true if the motor is in motion.

```
public final void stop()
```

Causes the motor to stop instantaneously. Once it is stopped, it resists any further motion.

```
public void flt()
```

Causes the motor to lose power and glide to a stop.

```
public void rotate(int angle)
```

Causes the motor to rotate the desired angle (in degrees).

```
public void rotate(int angle, boolean
immediateReturn)
```

Causes the motor to rotate a desired angle. The method returns immediately and the motor will stop by itself when the angle is reached.

Parameters:

immediateReturn - if true, method returns immediately.

```
public void rotateTo(int limitAngle)
```

Causes the motor to rotate to limitAngle. The tachometer should be within 2 degrees of the limit angle when the method returns.

```
public void rotateTo(int limitAngle, boolean
immediateReturn)
```

This is the same as the method above, except with the option of returning immediately after the method call.

`public void shutdown()`

Disables the speed monitoring functions of this motor.

`public void suspendRegulation()`

Removes this motor from the motor regulation system. After this call the motor will be in float mode and will have stopped. Note calling any of the high level move operations (forward, rotate etc.), will automatically enable regulation.

`public void setAcceleration(int acceleration)`

Adjusts smoothness of acceleration. Motor speed increases gently, and does not overshoot when smaller values used.

`public final void setSpeed(int speed)`

Sets motor speed, in degrees per second. Up to 900 is possible with fully charged batteries.

`public final int getSpeed()`

Returns the speed this motor is set to, in degrees per second. Does not return the actual measured speed - see getActualSpeed().

`public int getLimitAngle()`

Returns the angle (in degrees) that a Motor is rotating to.

`public final boolean isRotating()`

Returns true when motor is rotating toward a specified angle.

`public int getRotateSpeed()`

Returns actual speed, in degrees per second, calculated every 100 ms. A negative value means the motor is rotating backward.

`public int getTachoCount()`

Returns the tachometer count in degrees.

`public void resetTachoCount()`

Resets the tachometer count to zero.

```
public int getSpeed()
```

Returns the current target speed.

```
public void forward()
```

Causes the motor to rotate forward.

```
public void backward()
```

Causes the motor to rotate backwards.

```
public boolean isMoving()
```

Returns true if the motor is in motion.

Tip!

EV3 robots have a hidden sensor to detect when they've hit an obstacle: the motors. That's because the EV3LargeRegulatedMotor class has a getRotation-Speed() method. If the motor is turning slower than it's supposed to, there's probably an obstacle hindering the robot and slowing down the motors. You can even call isStalled() to see if the motor has stalled. The method setStallThreshold() is used in conjunction with isStalled().

EV3MediumRegulatedMotor

There is a class called EV3MediumRegulated motor which has basically the same methods as EV3Large-RegulatedMotor. The main difference is that the algorithms are tuned slightly to give better performance for the medium motor.

UnregulatedMotor

For some applications, such as Segway-like balancing robots, an unregulated motor is required in order to control the motor more precisely. The UnregulatedMotor class is used for these cases. It contains only the most basic commands, such as forward(), backward() and setPower(). It also has access to tachometer readings with getTachoCount(). This class works with all EV3 motors and the NXT motor.

lejos.remote.*

The leJOS communications API allows communications through a number of methods, including WiFi, Bluetooth and USB. Through these classes, your EV3 can connect to other LEGO Mindstorms devices and control them. You can read more about Bluetooth and WiFi in later chapters.

lejos.robotics.navigation

The navigation API controls steering functions and navigation for robots. Chapters 21, 22 and 23 explore the navigation API in detail.

lejos.robotics.subsumption

The subsumption package contains classes for implementing Rodney Brook's behavior control. It uses a Behavior interface and an Arbitrator class. Chapter 19 explores the basic subsumption architecture.

This takes care of the boring stuff! Now you can move on to more exciting things, like programming robots. Hopefully this overview will allow you to create the robot you have been dreaming of.

CHAPTER 11

Sound

TOPICS IN THIS CHAPTER

- Playing a tone
- Synthesizing notes
- Recording sound
- Playing prerecorded sound

Sound is not as vital to robotics programming as other topics, but it can aid with robot to human communication. For example, when the robot detects movement, it can say a meaningful phrase such as "I see you". However, you probably won't be having deep conversations with a robot by the end of this chapter.

This chapter will study the different ways you can use sound with leJOS. Some are useful while some are just for fun.

Playing a Tone

The EV3 can generate a simple tonal sound. In leJOS there are several methods in the lejos.hardware.Sound class to play different sounds. Sound.playTone() will play a tone by accepting an argument for frequency and duration. The following code plays an ascending series of tones.

```
import lejos.hardware.*;
import lejos.hardware.lcd.LCD;

public class ToneScales {
  public static void main(String [] args) {
    for(int freq=100;freq<11000;freq+=100) {
      LCD.drawString("FREQ: " + freq, 2, 3);
      Sound.playTone(freq, 500);
    }
  }
}
```

The sound fidelity of the EV3 speaker and synthesis is much better than the NXT. You will notice the frequency goes to levels only a dog could hear. I stopped at 11kHz, but you could raise the number to see how high you can detect the tone.

Synthesizing Notes

The tone sound is rather bland and robotic.We can also synthesize tones that are more sophisticated by creating an ADSR envelope. This stands for Attack Decay Sustain Release. Synthesizers use ADSR envelopes to create unique instrument sounds. The envelope is the shape of the sound being created over time (see Figure 11-1). Each of these letters stands for the time it takes to complete its portion of the envelope.

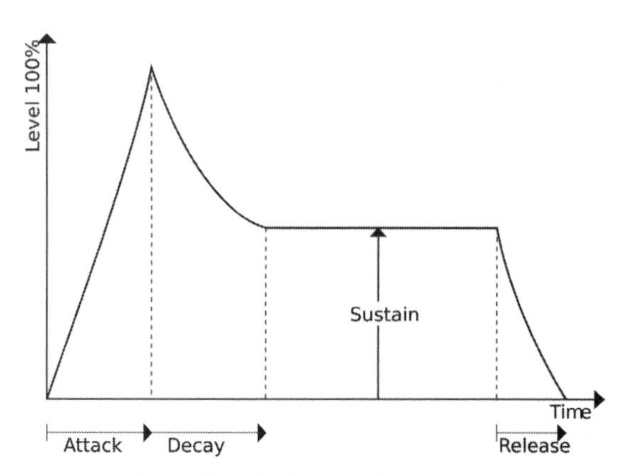

Figure 11-1: Viewing an ADSR envelope.

- Attack time is the time for the initial run-up from nil to peak.
- Decay time is the time for the subsequent run down from the attack level to the designated sustain level.
- Sustain level is the time during the main sequence of the sound's duration, until the note ends.
- Release time is the time for the level to fade from the sustain level to zero.

The Sound class has three predefined ADSR envelopes, which are called instruments. Each instrument is merely an array of five values: flute, xylophone, or piano. So how do we play these instruments? By using the method playNote().

```
import lejos.hardware.*;
import lejos.hardware.lcd.LCD;

public class NoteScales {
  public static void main(String [] args) {
      for(int freq=2500;freq>1000;freq-=100) {
    LCD.drawString("FREQ: " + freq, 2, 3);
    Sound.playNote(Sound.FLUTE, freq, 100);
    }
  }
}
```

Recording and Playing

EV3 can store and play sound files, with either the standard LEGO firmware or with leJOS. The LEGO software uses a proprietary sound format which uses files with the *RSF* extension. These sound files use less memory than MP3 and WAV file formats but they also contain less sound information. That is why they don't sound as clear.

leJOS does not use the RSF sound file format. Instead, it uses the more standard WAV file format. The sound file can be an 8-bit or 16-bit PWM (Pulse Wave Modulation) file, otherwise known as a WAV file. You can also play sample rates from 8K to 48K. However, it is not capable of playing stereo sound because there is only one speaker.

There are three steps to playing your own files:

- Download or record a WAV sound file
- Upload the file to the EV3
- Play it from the menu or with Java code

Just for Fun!

We're aware that the instruments don't really sound like the instrument constant name. Don't worry though, you can make your own instruments by creating an array of five integers. Here is the instrument definition for piano:

```
public final
static int[] PIANO
= new int[]{4, 25,
500, 7000, 5};
```

Try altering these values. Then, in the code above, substitute FLUTE with this array to see how it affects the resulting sound.

Downloading a Sound File

The easiest way to get a WAV file is to download one from the Internet. Luckily there are a large number of these available for free. You can easily find sites with a simple search, but here is a site with a good selection of working WAV files:

www.wav-sounds.com

Once you have located a file you like, right click it and save it to a directory. We will upload it to the EV3 later in this chapter. If you have no intention of recording a sound file, feel free to skip ahead.

Tip!

You can find a collection of RSF sound files in the directory where you installed the LEGO software:

```
C:\Program Files\
LEGO Software\LEGO
MINDSTORMS EV3
Home Edition\
Resources\Brick
Resources\Retail\
Sounds\files\
```

Recording a Sound File

To record a sound file, you need to connect a microphone or headset to your computer. Unfortunately, Windows Vista and higher versions no longer record files in the WAV file format. Instead, we will have to use another sound recording tool, such as All2WAV.

As previously mentioned, the sound file can be an 8-bit or 16-bit PWM (Pulse Wave Modulation) file, otherwise known as a WAV file. You can also play sample rates from 8K to 48K.

Windows

> ***Tip!***
>
> To make sure the microphone level is recording properly, go to Start > Control Panel > Sounds and Audio Devices. Make sure Sound is not muted, then click the Audio tab. Under Sound recording, click Volume… and make sure the volume is 100% and not muted. Sometimes, when a Bluetooth headset is added to Windows, it mutes the device by default, so make sure to check this setting.

1. Download All2WAV. It is available from a number of sources, but we will use CNet. You do not have to download the CNet Installer if you select the direct download link:

   http://download.cnet.com/All2WAV-Recorder/3000-2168_4-10267686.html

2. Install All2WAV by double clicking the downloaded file. Follow the instructions.
3. Run All2WAV. Under Quality, make sure to not exceed WAV 48,000 Hz; Mono; 16-bit (Figure 11-2). It is important that you select mono and not stereo.
4. Click the red button and speak into the microphone.
5. Click File > Save and save the WAV file to a project directory. Save it as hello.wav.

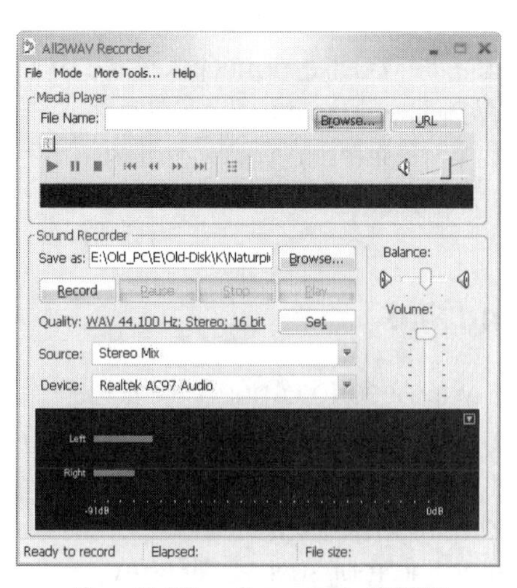

Figure 11-2 Recording sound in All2WAV

Linux

1. A popular sound recording program for Linux is KRec, but it is often not installed by default. Install this on your system and run it from the Multimedia program group.
2. Select File > New and use the default settings for this file, but choose Mono instead of Stereo.
3. When you are ready to record, click the red Record button and speak into the microphone (see Figure 11-3). Hit Stop when done.
4. Select File > Export... and save the file with a .wav extension. Save it anywhere for the time being.

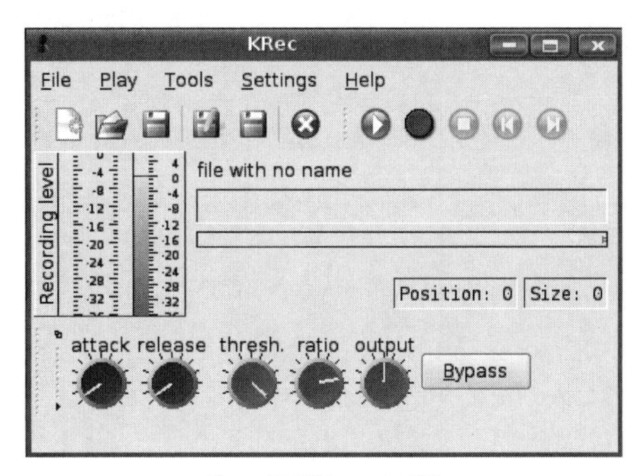

Figure 11-3 KRec under KDE.

Macintosh

1. OS X users can use GarageBand if it is installed on the system, or any version of Quicktime Player after 2006.
2. In QuickTime Player, select File, Record Audio...
3. When you are ready to record, click the Record button and speak into the microphone. Hit Stop when done.
4. Save the audio file in .wav format.

Uploading the File

At this point we assume you have either downloaded or created a WAV file. Now it is time to upload the file to your EV3 brick. We will use EV3 Control Center (see Chapter 8), although you could also use a SCP client to connect to the brick and transfer files, such as WinSCP (more on that in Chapter 28).

1. As mentioned earlier, the leJOS Eclipse plugin has an icon in the toolbar for EV3 Control Center. Alternatively, in the directory you installed leJOS is a subdirectory called bin. Navigate to that directory, and run a file called ev3control.bat.

2. If your EV3 is plugged in and turned on, press Connect (see Figure 11-4).

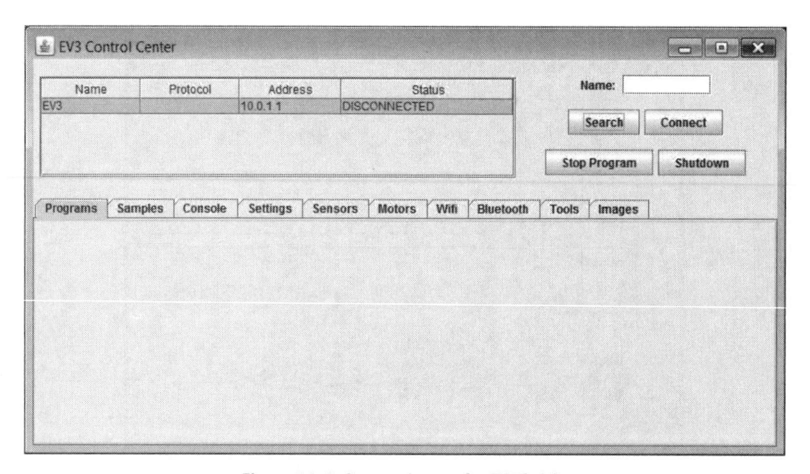

Figure 11-4: Connecting to the EV3 brick.

3. Shortly you will see a file listing of your EV3 brick (see Figure 11-5).

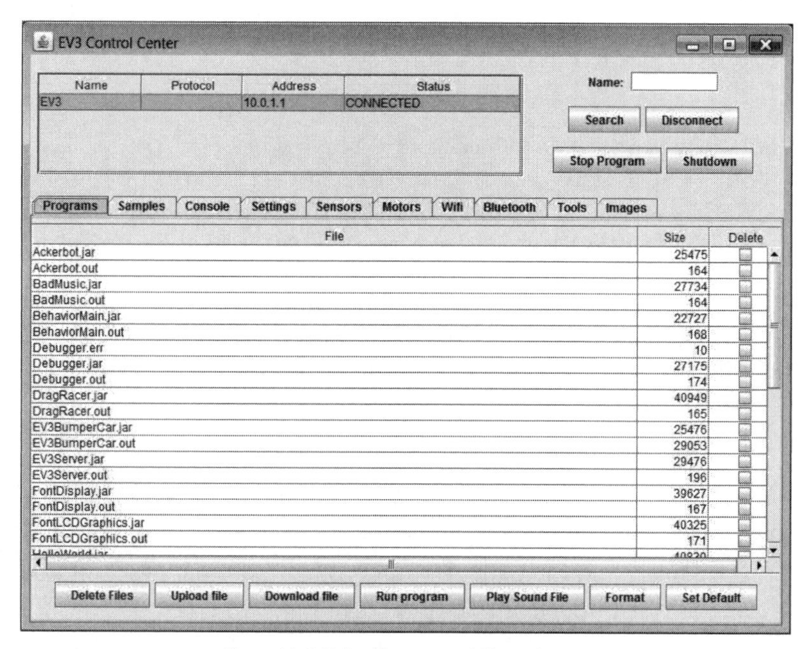

Figure 11-5: Uploading a sound file to the EV3.

4. Click on Upload file, and browse to the WAV file you saved earlier. Select the file and click Okay.

5. The EV3 will beep when the file is uploaded.

We are now ready to play the file!

Playing a Sound File

To play the file, we can browse to it using the leJOS EV3 menu. Simply browse to Programs, then you should see the WAV file along with your program files. You can select the file and then click play. If you obtained a proper WAV file, you should hear it from the EV3 speakers.

This is nice, but we really want our code to play the file while running a program. It will take a few lines of Java code to play a sound file. The code is self explanatory so I will skip an explanation:

```
import java.io.File;
import lejos.hardware.*;
import lejos.hardware.lcd.*;

public class PlaySample {
  public static void main(String [] args) {
    File myFile = new File("austinpowers.wav");
    LCD.drawString("Playing " + myFile
getName(), 0, 2);
    Sound.playSample(myFile, 100);
    LCD.drawString("Done hit ESC", 0, 4);
Button.ESCAPE.waitForPressAndRelease();
  }
}
```

That's it. Keep in mind that you will need to change the filename in the above example to the filename you downloaded. Make sure you put the sound file in the same directory as the code (if you used EV3 Command Center it will be there). If all goes well, your EV3 brick will play the file you recorded.

NOTE: *If the file does not play, chances are that it is a stereo file and not a mono file. You can use one of the many free audio recorders on the market to convert a stereo file to a mono file, such as Audacity.*

CHAPTER 12

WiFi Basics

TOPICS IN THIS CHAPTER

- WiFi
- Setting up WiFi
- Advanced Setup
- Using WiFi

Earlier in the book we looked at uploading programs to the EV3 brick using the USB slave port. However, this can quickly become tiring because you have to plug in a USB cable and unplug it every time you use a mobile robot. We need a better solution.

There are two wireless solutions available to the EV3 brick. The first is WiFi, which we will examine in depth in this chapter. The second is Bluetooth, which will be examined in the next chapter. Both have strengths and weaknesses, and both can do amazing things that the other is not capable of. For this reason, I recommend getting both a WiFi adapter for your EV3 brick, as well as a Bluetooth dongle for your PC.

WiFi

Most people are familiar with WiFi, so I will not go into great detail about it here, other than how it applies to the EV3 brick. To use WiFi, you will need an 802.11n WiFi adapter (see Chapter 1 for more details of ordering one). Once you have the adapter connected to the USB port on your EV3, it becomes part of your home network, and can then access the Internet.

If I had only one choice between WiFi and Bluetooth, I would take WiFi without hesitation. The speed of uploading programs to the EV3 brick is fantastic—it happens in the blink of an eye. The range of wireless communications is superior to Bluetooth. And once set up, the EV3 brick automatically connects to the network when it is booted up. (With Bluetooth, you have to manually connect each time through your operating system.)

And finally, WiFi allows the EV3 brick to connect to the Internet right away. With Bluetooth, it can't connect to the Internet (at least without setting up tethering). One of the first things leJOS does when the EV3 menu runs is to check a time server on the Internet and update the system time. This is only possible if you have a WiFi adapter.

WARNING: *When building a LEGO robot, it is prudent to remove the WiFi adapter from the brick. This is because the WiFi adapter is more breakable than plastic LEGO parts and can become damaged while building. I actually cracked the shell on my Edimax adapter and had to glue it back together. However, keep in mind you must keep the USB port area clear of LEGO parts if you intend to use WiFi after you are finished building the robot.*

Setting up WiFi

Setting up WiFi from the leJOS EV3 menu is easy. Simply search for your router's SSID (Service Set Identifier) and then connect to it by entering a passphrase.

1. Insert the SD card, insert your Wifi dongle (if you have one) in the USB host socket, and turn on the EV3.

2. You will need to connect to your Wifi access point. Scroll right in the EV3 menu until you select WiFi (see Figure 12-1), then hit the enter button.

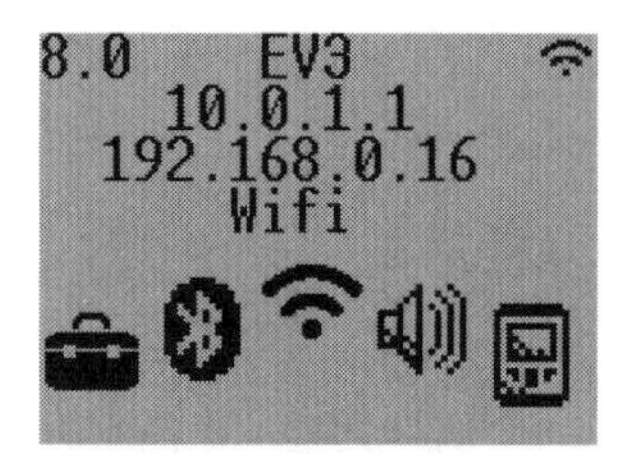

Figure 12-1: Selecting WiFi

3. You should see a list of WiFi access points, including your own home router (see Figure 12-2). Press down until the SSID for your Wifi access point is selected and then press Enter.

Figure 12-2: Selecting the SSID of your router

4. The next screen allows you to enter the passphrase (Figure 12-3). Use the left, right, up and down keys to select letters, numbers and punctuation and press Enter to add the character to the line of text that is shown on the bottom line of the screen.

NOTE: *There are special character selections near the bottom. "U" selects upper case letters, and "l" lower case. "x" deletes the last character you typed, and "D" stands for Done.*

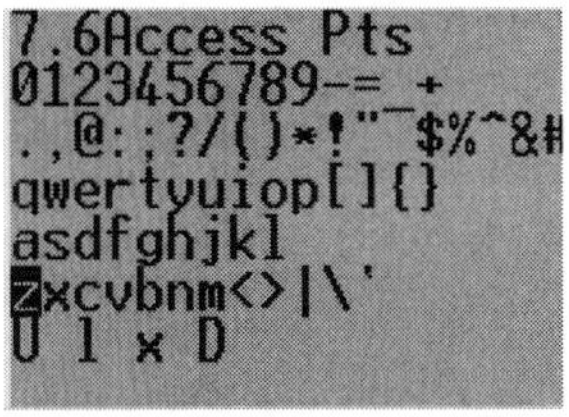

Figure 12-3: The EV3 simulated keyboard

5. Enter your router passphrase and then select D for Done.
6. You should then see some messages on the screen as the network connection is restarted. If all goes well, you will return to the menu with two IP addresses shown. The second is your WiFi IP address.

NOTE: *Next time you reboot your EV3, the WiFi will connect automatically as long as the EV3 is in range of the router.*

Advanced Setup

You can also manually configure the WiFi settings via a configuration file. Your reasons for doing it this way might be because the simple setup from the menu did not work and you need to tailor your WiFi setup according to how your router is configured. To do this, you will need to use either SSH or SCP to connect to the EV3 via the USB cable (see Chapter 28 for more information on this).

The leJOS menu creates a file called *wpa_supplicant.conf* in /home/root/lejos/bin/utils when it connects to a router. If a file does not exist there, it looks to the directory /etc/wpa_supplicant. Using the directions for editing a file in Chapter 28, create a file in /etc/wpa_supplicant called wpa_supplicant.conf. A basic configuration file has the following settings (quotes are needed below):

```
ctrl_interface=/var/run/wpa_supplicant

network={
  ssid="myssid"
```

```
    key_mgmt=WPA-PSK?
    psk=12345
}
```

The ssid is the name of your router, key_mgmt is the type of wireless security used by your router, and psk is the passkey. You can read more about WPA Supplicant online if there are more settings you need to change to connect to your WiFi network.

To do a quick test of your WiFi, try pinging it from your PC. Drop to a command line and type:

```
ping 192.168.0.xx
```

Replace the IP address above with the IP address displayed on the front of your EV3 brick. Hit ctrl-C to stop the pings.

Using WiFi

Now that you are connected to your WiFi network, you can upload program code via the Eclipse plugin. It works just like it did with USB—it will autodetect the EV3 on the WiFi network and begin uploading simply by pressing the green run button. Likewise you can connect to the EV3 via SCP or SSH (as mentioned, Chapter 28 has more information).

The next chapter shows how to connect to a Bluetooth PAN. The chapters after that applies to both WiFi and Bluetooth, and allows you to control an EV3 robot with code over a WiFi or Bluetooth PAN network.

CHAPTER 13

Bluetooth

TOPICS IN THIS CHAPTER

- Introduction to Bluetooth
- Pairing the EV3 with your PC

Bluetooth is a major feature of EV3. Not only can you upload code to the robot wirelessly, eliminating the need to plug and unplug the USB cable each time you upload a revision, but you can do so much more.

In this chapter, we will examine Bluetooth and what it really is. Then we will pair your EV3 brick with your Bluetooth adapter, which will allow you to upload code to your EV3 brick wirelessly. We will also connect your PC to the EV3's Personal Area Network (PAN). In subsequent chapters we will try other wireless projects.

NOTE: *Refer to Chapter 1 for instructions on purchasing a Bluetooth adapter.*

Meet Bluetooth

Bluetooth is a wireless protocol that allows devices to communicate with each other. The most popular Bluetooth devices include mice, keyboards, wireless headsets, and mobile phones. In fact, a Bluetooth equipped mobile phone can even control your EV3 brick. Likewise, your EV3 can control Bluetooth devices, including other NXT and EV3 bricks.

Bluetooth was introduced in 1998 and was initially slow to catch on, but soon vendors began rapidly adopting it. Many companies, including Logitech and Microsoft, have manufactured Bluetooth Wireless Keyboard and Mouse combos. Wireless headsets for mobile phones and even cars are becoming ubiquitous. Both Sony and Nintendo use Bluetooth for their console controllers. This mass acceptance led LEGO to adopt the standard for the NXT brick and continue its use for the EV3 brick.

But what is Bluetooth? The technology can be compared to USB in many ways. A single Bluetooth adapter effectively acts like many wireless USB ports. Bluetooth eliminates cables from keyboards, mice and game controllers.

As mentioned, Bluetooth is used by a lot of other devices. But what can you use Bluetooth for with a LEGO EV3 brick? The following lists some of the major uses for Bluetooth and the EV3.

- PC uploads program code and data to EV3
- PC controls EV3 robot
- EV3 sends data to PC
- EV3 sends data to another EV3
- Mobile phone controls EV3 robot
- EV3 sends data to mobile phone
- PC connects to EV3's PAN
- Serial Bluetooth keyboard paired with EV3 brick
- GPS paired with EV3 brick
- EV3 connects to Bluetooth game controller

That's an overwhelming number of uses, and there are probably even more! But don't worry, we'll cover the most important topics one at a time.

Many people wonder how Bluetooth is different from Wi-Fi (the 802.11 standard). For starters, Wi-Fi is designed for networking, usually with one or more wireless routers and many computers. You wouldn't imagine using an 802.11g network to control a mouse wirelessly—it would be unresponsive and costly. Bluetooth is for data communications between two devices only—usually a small device and a computer.

NOTE: *Bluetooth operates on 2.45 GHz, while WAP 802.11g is on 2.4 GHz (see Table 13-1). Although these frequency bands overlap, you probably won't experience conflicts—Bluetooth and Wi-Fi use frequency hopping, jumping from one band to another up to 1600 times per second.*

The only time my Bluetooth connection appeared unstable was in an area with five wireless networks (including the one connected) plus a cordless phone system. While I was streaming music through the network the keyboard became slow and unresponsive.

Device	Frequency
Bluetooth	2.45 GHz
Wi-Fi (802.11g)	2.4 GHz
Cordless Phone	900 MHz (older), 2.4 GHz, or 5.8 GHz
X-10 Camera	2.4 GHz
Mobile Phones	800 MHz
Headphones	924 - 928 MHz

Table 13-1 Wireless frequencies of common devices.

Bluetooth is capable of transmitting data at 24 Mbit/s, which is over 50 times the speed that the NXT's Bluetooth 2.0 transmitted data (see Table 13-2). These speeds are more than adequate for transmitting programs and files to the EV3 brick.

Wireless Standard	Typical Data Rate
IR	2.4 Kbit/s
Bluetooth 2.0	460.8 Kbit/s
Bluetooth 4.0	24 Mbit/s
Wi-Fi 802.11b	6.5 Mbit/s
Wi-Fi 802.11g	24 Mbit/s
Wi-Fi 802.11n	600 Mbit/s

Table 13-2 Comparing wireless speeds.

Multiple Bluetooth adapters can operate in the same area since Bluetooth uses frequency hopping to avoid conflicts. There shouldn't be much of a problem in a classroom setting, even with dozens of Bluetooth adapters. If interference occurs, USB cables can be used to transmit code to robots.

Pairing too many devices to the same Bluetooth adapter is a larger problem. I've found that other Bluetooth devices are affected when there is a lot of traffic over a single Bluetooth connection, especially with voice communications. My Bluetooth mouse behaved like it was drunk when I was using a Bluetooth headset.

Not all Bluetooth devices are compatible with the EV3. First, it must be Bluetooth 2.0 or higher, although 4.0 is definitely optimal. Second, some Bluetooth solutions, such as those built into Dell notebook computers, are not compatible. If you aren't sure and don't want to go through the trouble of researching this, you can order a Bluetooth adapter from LEGO. As long as Linux users can use their adapter with the BlueZ Bluetooth stack, there should be no compatibility problems.

Try it!

If you want to confirm that your Bluetooth stack is working, do the following:

1. Select Start > Control Panel > System and Maintenance > System.
2. Click the Device Manager button.
3. You will see a list of devices, including Bluetooth. Expand the Bluetooth selection and highlight the Bluetooth wireless hub (see Figure 13-1).
4. Click the Properties icon and you will see some general information. Click the driver tab to view the device driver (see Figure 13-2).
5. As long as you see "This device is working properly" In the device status, you should have no compatibility problems.

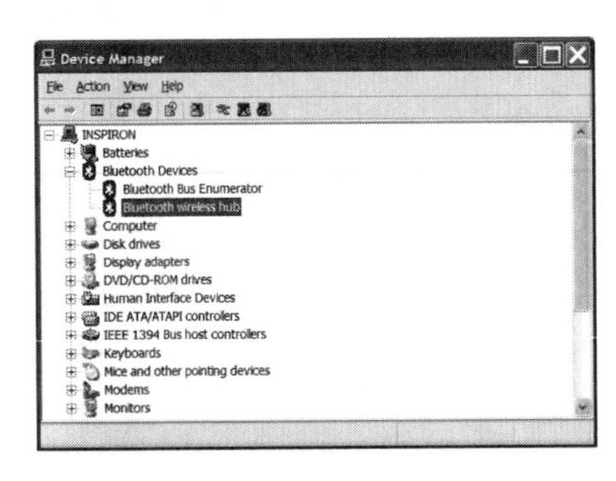

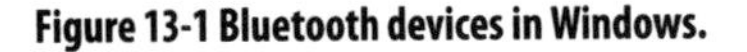

Figure 13-1 Bluetooth devices in Windows.

Figure 13-2 Device driver information.

Pairing the EV3 with a PC

Before you can use Bluetooth you must pair the EV3 brick with your PC. Pairing occurs when one device tells another device that they can be friends. The reason for pairing is to prevent strangers from accessing your Bluetooth devices without your permission. By pairing, you give the devices permission to interact (like plugging a USB cable into a computer).

Pairing is done by locating the Bluetooth device, then entering a four digit code on both devices. I recommend using the LEGO software to pair the devices. If there are any problems, the software will attempt to help you.

If you don't want to use the LEGO software, you can pair the devices using your operating system.

1. Turn on your EV3 brick. From the main menu, select Bluetooth (see Figure 13-3).

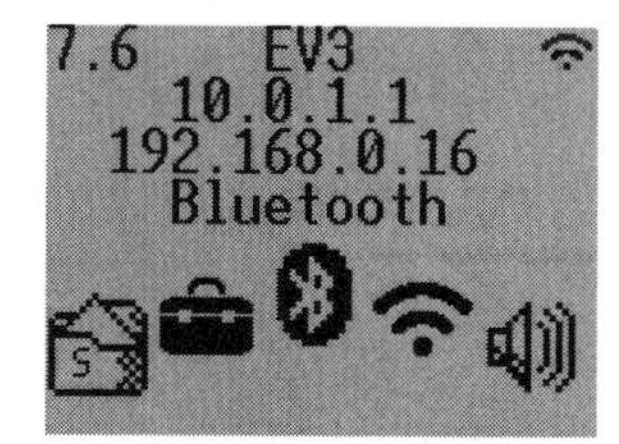

Figure 13-3 Selecting the Bluetooth menu

2. There are four options: Search/Pair, Devices, Visibility, and Change PIN. If visibility is off, select the visibility icon (an eye) and press the enter button (see Figure 13-4).

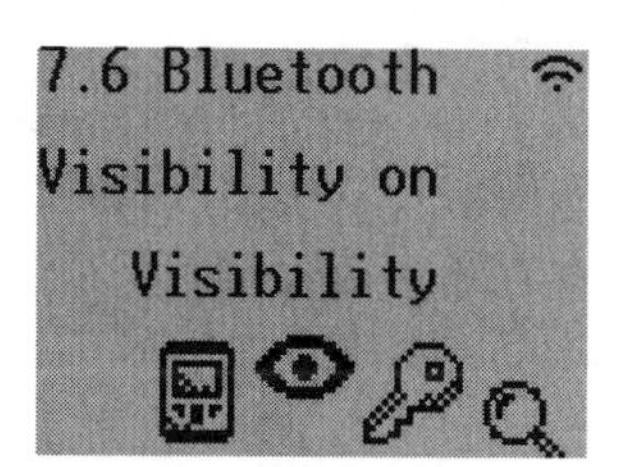

Figure 13-4 Turning on visibility

3. Now that the EV3 brick is visible, you can begin the pairing process. You can initiate the pair from either the EV3 or a PC. We'll use the PC here. In Windows, go to Control Panel > Hardware and Sound > Bluetooth Devices you should see a list of devices with some options at the top (see Figure 13-5).

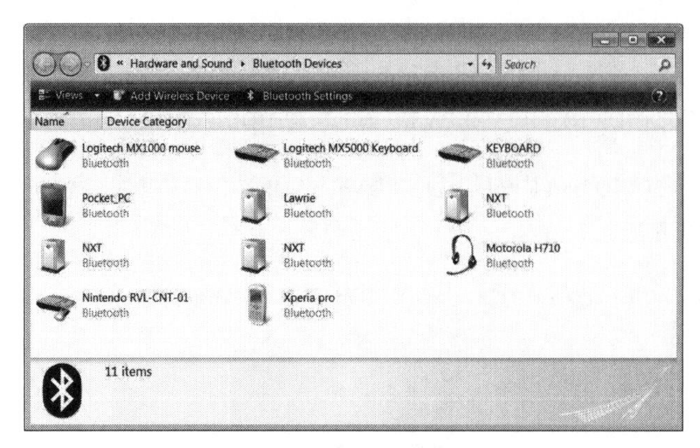

Figure 13-5: Initiating the pair sequence

4. Click the Add Wireless Device button. It will search and eventually the EV3 should come up. Click that and then click Next.

5. The next screen displays three options (see Figure 13-6). Click on the middle option, and then enter 1234 as the PIN (unless you changed the default PIN in the EV3 menu).

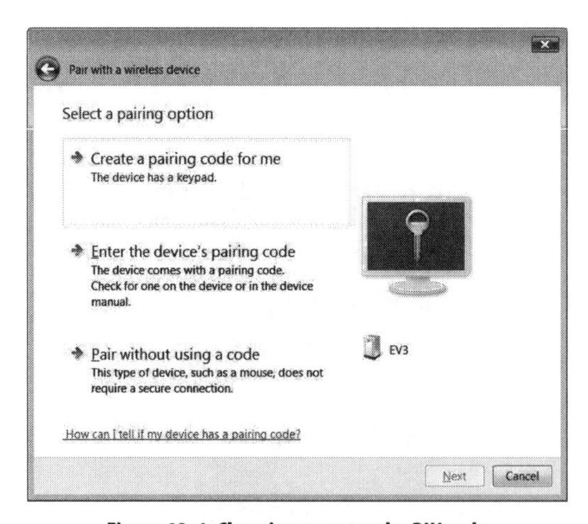

Figure 13-6: Choosing to enter the PIN code

6. Once you enter the correct PIN code on the PC it will complete the pairing process. Click close and you are ready to communicate.

You only need to do this procedure once. The two devices are now paired for all time, unless you remove the device from either Windows or the EV3 menu.

Connecting to Bluetooth PAN

A Bluetooth Personal Area Network (PAN) is similar to a Wide Area Network (WAN), which is what your computer devices connect to in your home using WiFi. Both networks use of an IP address to identify a device on the network. Connecting to a PAN automatically creates a TCP/IP connection between your computer and the EV3 brick. When connected to the PAN, you can upload files and Java programs to the brick, as well as control the brick using java.net classes.

Once you have your Bluetooth adapter installed, your PC can connect to the EV3's Bluetooth PAN. This is accomplished not through the Bluetooth settings, but rather in the networking settings. To do this, we will open Network Connections.

Windows

1. Click the Start button > Control Panel > Network and Internet > Network and Sharing Center. Then click Manage network connections on the left side of the window (see Figure 13-7).

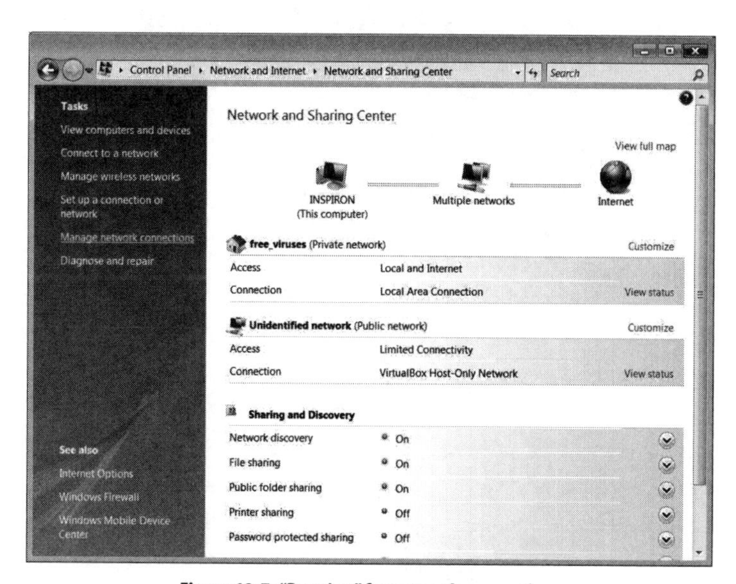

Figure 13-7: "Panning" for network connections

2. You should see "Bluetooth Network Connection" with a red X. Click this icon once to highlight it (see Figure 13-8).

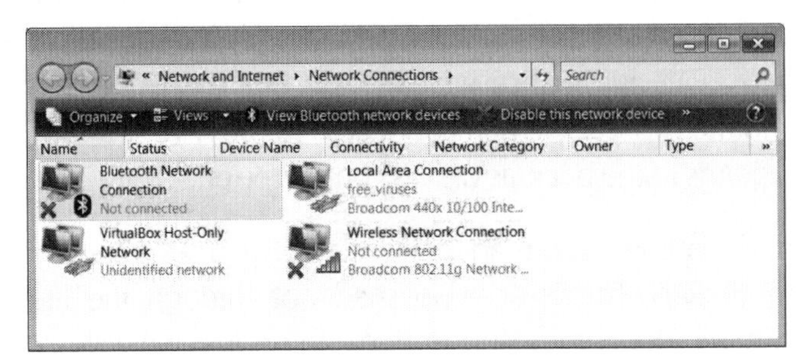

Figure 13-8: Selecting a Bluetooth connection

3. On the toolbar, click View Bluetooth network devices.

4. After a moment, one or more access points should show up (see Figure 13-9). One of them should say "EV3 Network Access Point BlueZ PAN Service". Click this once to highlight it and then click Connect.

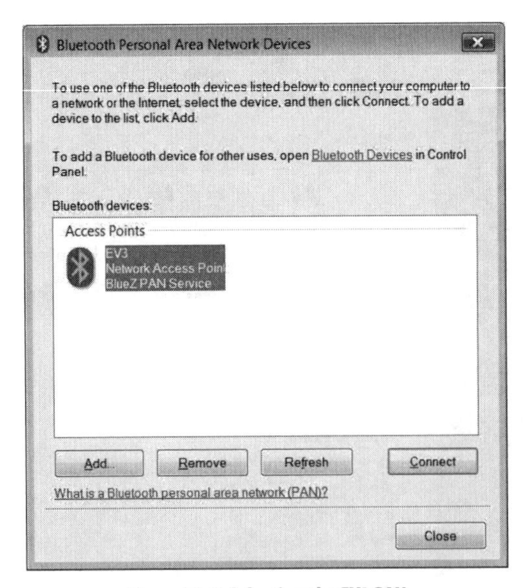

Figure 13-9: Selecting the EV3 PAN

5. After a moment it should say Connection Successful. The Network Connections window should now say "Connected to EV3" under the Bluetooth Network Connection.

NOTE: *Every time you turn on your EV3 brick, you will need to reconnect to the Bluetooth PAN using the above instructions.*

Now What Can I Do?

Now that you are connected, you can access the EV3 using the IP address 10.0.1.1. For example, if you drop to a command line and ping the EV3, you will get a reply.

```
ping 10.0.1.1
```

You can also use network tools to access the EV3 brick. This includes using SSH to connect using a terminal program (such as PuTTY) or using SCP to transfer files (see Chapter 28 for more on SSH and SCP). You can also upload files through the Eclipse plugin using Bluetooth (the uploader will detect Bluetooth automatically). Simply hit the green run button in Eclipse and it will upload the code to your brick via Bluetooth.

In subsequent chapters we will do some network programming using Bluetooth (alternately, you can use WiFi).

CHAPTER 14

Client-Server Robotics

TOPICS IN THIS CHAPTER

- Communications API
- Client-Server Coding
- Remote Control Car
- Telerobotics

Have you ever wanted to do more than exchange information over the Internet? How about something that allows you to reach right into your computer and physically manipulate something on the other side of the world? This chapter explores ways to perform physical movement around the globe, right from your computer.

Now that Mindstorms has networking with the EV3, we can easily program a robot that can be controlled through a network. This is the moment we've been waiting for! And not just over a home network. Any network, including the Internet, can be used to transmit commands.

Telerobotics with the Internet

The philosopher Marshall McLuhan described electronic media as an extension of our senses. For example, radio allows our ears to travel to a conversation at the top of Mount Everest. Television allows our eyes to see the bottom of the ocean. Electronic media extends the range of our senses.

McLuhan anticipated the Internet even before ARPANET. He saw it as a way for individuals to extend communications using their senses. McLuhan was very poorly understood in the fifties and sixties because no one understood what he was talking about. Today, with the Internet realized, McLuhan seems very straightforward.

Using the LEGO EV3, we can take McLuhan's concept of extending our senses even further. In this chapter, we will use the Internet (or any network) to explore telerobotics. If McLuhan were alive today, he would probably describe telerobotics as an extension of both our senses and our limbs. We will use your computer and network to control robots over any distance–in another room or another city. Let's make that happen.

Communications API

Since leJOS now uses the full Java Standard Edition classes, we can use complete versions of all the classes in java.net and java.io. These classes allow data exchange from the EV3 brick. The communications classes use streams, located in the java.io package, so anyone familiar with streams will find this chapter easy to understand. Java uses sockets to obtain input and output streams from a remote computer.

In this chapter, we'll use only the most basic streams relevant to sending and receiving data: InputStream, OutputStream, DataInputStream and DataOutputStream. Input/Output Streams are the foundation of Streams, and they are useful only for sending bytes. If you want to send other data types such as characters, Strings, integers, and floating point numbers you will need to use data streams (see DataInputStream and DataOutputStream below).

java.io.InputStream

InputStream is the superclass representing input streams. Input streams transfer bytes. It is an abstract class so it can not be instantiated on its own. An instance of InputStream can be obtained from a Socket class.

```
public int read() throws IOException
```

Reads the next byte of data from the input stream. The value byte is returned as an int in the range 0 to 255. This method blocks (waits) until input data is available, the end of the stream is detected, or an exception is thrown.

```
public int read(byte[] b) throws IOException
```

Reads a number of bytes from the input stream and stores them into the buffer array b. The number of bytes actually read is returned as an integer. This method blocks until input data is available, end of file is detected, or an exception is thrown.

Parameters: b The buffer into which the data is read.

```
public int read(byte[] b, int off, int len) throws
IOException
```

Reads up to len bytes of data from the input stream into an array of bytes. An attempt is made to read as many as len bytes, but a smaller number may be read, possibly zero. The number of bytes actually read is returned as an integer.

Parameters: b The buffer into which the data is read.

Off The start offset in array b at which the data is written.

Len The maximum number of bytes to read.

```
public void close() throws IOException
```

Closes this input stream, calls flush() and releases any system resources associated with the stream.

java.io.OutputStream

OutputStream is the superclass of all classes representing an output stream of bytes. It is an abstract class so it can not be instantiated on its own. Its main function is to send a byte of data to a destination. Like InputStream, an instance of OutputStream can be obtained from a Socket class.

```
public void write(int b) throws IOException
```

Writes the specified byte to this output stream.

```
public void write(byte b[]) throws IOException
```

Writes b.length bytes from the specified byte array to this output stream.

Parameters: b The data.

```
public void write(byte b[], int off, int len)
throws IOException
```

Writes len bytes from the specified byte array starting at offset off to this output stream. The general contract for write(b, off, len) is that some of the bytes in the array b are written to the output stream in order; element b[off] is the first byte written and b[off+len-1] is the last byte written by this operation.

Parameters: b The data.

off The start offset in the data.

len The maximum number of bytes to write.

```
public void flush() throws IOException
```

Flushes this output stream and forces any buffered output bytes to be written out. The general contract of flush() is that calling it is an indication that, if any bytes previously written have been buffered by the implementation of the output stream, such bytes should immediately be written to their intended destination.

WARNING: *Flush is one of the most important but often forgotten methods of streams. The non-use of this method probably accounts for most bugs when using the java.io package. Don't forget to call flush() after sending data, otherwise the data may never be sent to the destination!*

```
public void close() throws IOException
```

Closes this output stream and releases any system resources associated with this stream. A closed stream cannot perform output operations and cannot be reopened. A call to flush() is made just before the stream is closed.

java.io.DataInputStream

DataInputStream extends InputStream, so it has all the methods of InputStream (see above). This method allows data types other than bytes to be sent. This includes short, int, float, double, char, boolean and String.

```
public DataInputStream(InputStream in)
```

Returns an instance of DataInputStream. The constructor requires an InputStream object.

Parameters: in The input stream.

```
public final boolean readBoolean() throws IOException
```

Used to send a boolean value through a stream. Reads one input byte and returns true if that byte is nonzero, false if that byte is zero.

```
public final byte readByte() throws IOException
```

Reads and returns one input byte. The byte is treated as a signed value in the range -128 through 127, inclusive.

```
public final short readShort() throws IOException
```

Reads two input bytes and returns a short value.

```
public final char readChar() throws IOException
```

Reads an input char and returns the char value. (A Unicode char is made up of two bytes.)

```
public final int readInt() throws IOException
```

Reads four input bytes and returns an int value.

```
public final float readFloat() throws IOException
```

Reads four input bytes and returns a float value.

```
public final double readDouble() throws
IOException
```

Reads eight input bytes and returns a double value.

```
public final String readUTF() throws IOException
```

Reads a string using UTF-8 encoding.

java.io.DataOutputStream

If DataInputStream is the receiver then DataOutputStream is the sender. It encodes various data types into byte values and sends them across a data stream. DataOutputStream extends OutputStream, so it has all the methods described in the OutputStream API.

```
public DataOutputStream(OutputStream out)
```

Creates a new data output stream to write data to the specified underlying output stream.

Parameters: out The output stream.

```
public final void writeBoolean(boolean v) throws
IOException
```

Writes a boolean value to this output stream.

Parameters: v A boolean value.

```
public final void writeByte(int v) throws
IOException
```

Writes to the output stream the eight low-order bits of the argument v.

Parameters: v A byte value.

```
public final void writeShort(int v) throws
IOException
```

Writes two bytes to the output stream to represent the value of the argument.

Parameters: v A short value.

```
public final void writeChar(int v) throws IOException
```

Writes a char value, which is comprised of two bytes, to the output stream.

Parameters: v A char value.

```
public final void writeInt(int v) throws IOException
```

Writes an int value, which is comprised of four bytes, to the output stream.

Parameters: v An int value.

```
public final void writeFloat(float v) throws
IOException
```

Writes a float value, which is comprised of four bytes, to the output stream.

Parameters: v A float value.

```
public final void writeDouble(double v) throws
IOException
```

Writes a double value, which is comprised of eight bytes, to the output stream.

Parameters: v A double value.

```
public final void writeUTF(String str) throws
IOException
```

Writes a string to the underlying output stream using modified UTF-8 encoding in a machine-independent manner.

Parameters: str A String object.

Client-Server Creation

The architecture for creating a connection between two devices is the client-server architecture (see Figure 14-1). The client must initiate the connection and log into the server. Both client and server each contain an input and output stream of data. Normally many clients can connect to a single server, in a many to one relationship. Sometimes a server can only have one concurrent connection at a time, making essentially a one to one relationship. We will explore both of these servers in this chapter through real code examples.

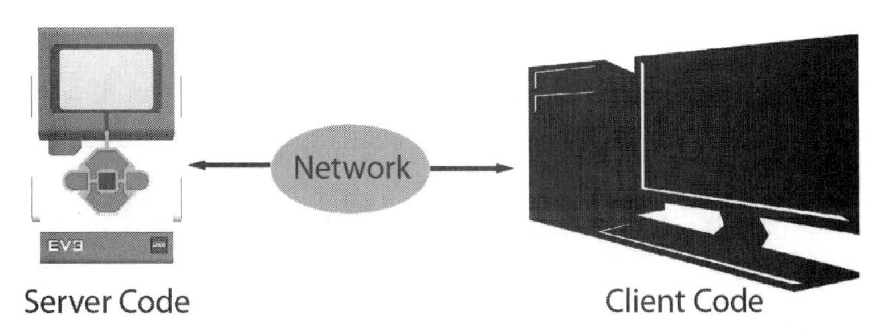

Figure 14-1: Creating a client to server connection.

Server Code

In order to establish a connection, the EV3 brick should run some code that waits for a computer to connect. Java has a class that does just this called the ServerSocket, and a method called accept() that accepts a connection from a client.

As previously mentioned, input and output streams are retrieved from a Socket class. So once we have the Socket, we can get input and output streams and begin communicating. Let's look at some actual code of the simplest possible version of a server.

```
import java.io.*;
import java.net.*;
import lejos.hardware.Battery;

public class EV3Server {

  public static final int port = 1234;

  public static void main(String[] args) throws
IOException {
    ServerSocket server = new ServerSocket(port);
    System.out.println("Awaiting client..");
    Socket client = server.accept();
    System.out.println("CONNECTED");
    OutputStream out = client.getOutputStream();
```

```
    DataOutputStream dOut = new
DataOutputStream(out);
    dOut.writeUTF("Battery: " + Battery.
getVoltage());
    dOut.flush();
    server.close();
  }
}
```

Upload this code to the EV3 brick and run it. The server will sit there waiting for a client to connect, just like the millions of servers around the world. Once the client connects, the server will send a string containing the battery voltage level. We'll now enter some client code to run on your PC.

Client Code

For this example, we will use the simplest possible code to connect to a server. We'll use a full blown GUI for another example later in this chapter. The only thing the client needs to know is what the IP address is for the server. We could hard-code this into the code, but we can also enter it at run-time when the program is run. More on this later, but first, let's enter the client code.

```
import java.io.*;
import java.net.*;

public class PCClient {

  public static void main(String[] args) throws
IOException {
    String ip = "10.0.1.1"; // BT
    if(args.length > 0)
      ip = args[0];
    Socket sock = new Socket(ip, EV3Server.port);
    System.out.println("Connected");
    InputStream in = sock.getInputStream();
    DataInputStream dIn = new DataInputStream(in);
    String str = dIn.readUTF();
    System.out.println(str);
    sock.close();
  }
}
```

Before running the program, we need to tell the client what IP address the server is on. The server is your EV3 brick, so it shows the IP address from the main menu screen. If you are using Bluetooth or USB, the address is 10.0.1.1.

We could hard code this IP address into the PCClient code, but it is better to tell it the IP at runtime. To do this, we will change a setting in Eclipse.

1. First we need to create a run configuration for PCClient. Simply run the program once in Eclipse, and when it asks if it is a Java Application or leJOS EV3 Program, select Java Application. The program will run and timeout after about 30 seconds.

2. Now that it has a run configuration we can set a runtime argument by going to Run > Run Configurations... (see Figure 14-2). On the left side, choose PCClient under the Java Application heading.

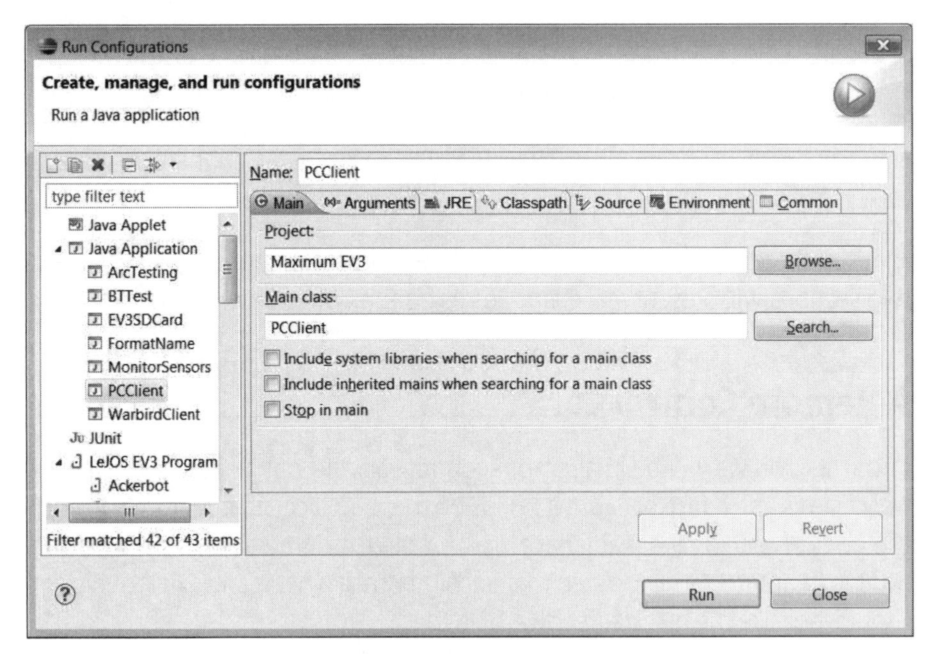

Figure 14-2: Selecting the PCClient run configuration

3. Next, click on the Arguments tab. Under program arguments, enter the IP address of your EV3 brick, such as 192.168.0.13 (see Figure 14.3).

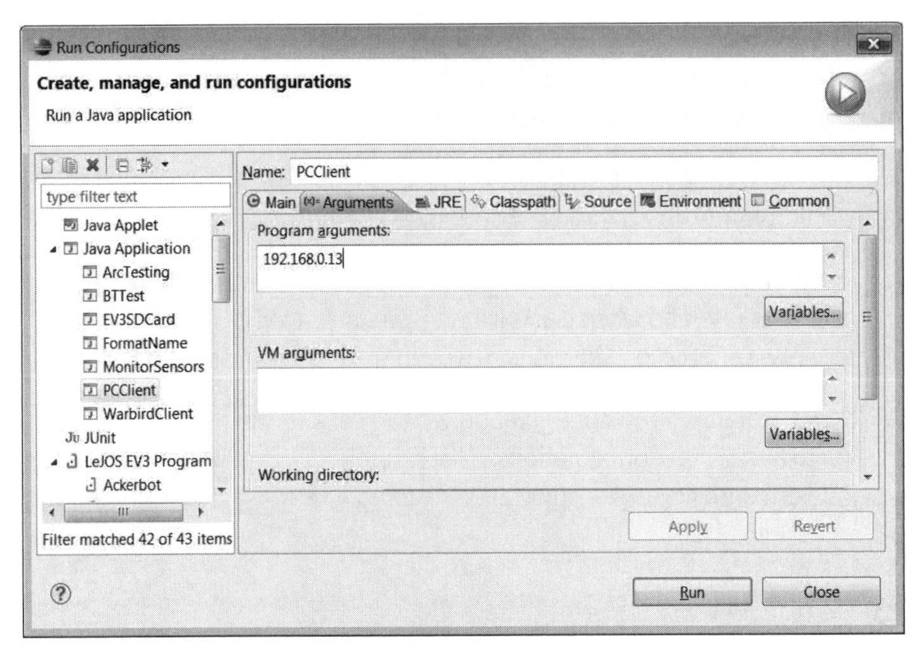

Figure 14-3: Entering the IP argument

Now click Apply and then Run. The client will immediately connect to the EV3 server and display the battery voltage in the Eclipse console area.

That's all there is to it for this very simple networking demonstration. The next two sections will present real examples using the Communications API.

A Remote Control Car

In this section we will build a remote control car. And not just any remote control car. This one is fast, because it builds on the drag racer chassis from earlier in the book. If you don't already have it assembled, it's time to revisit the instructions in Chapter 3. Omit the step that adds the small wheels and axle. Your robot should look like the first step shown below.

1

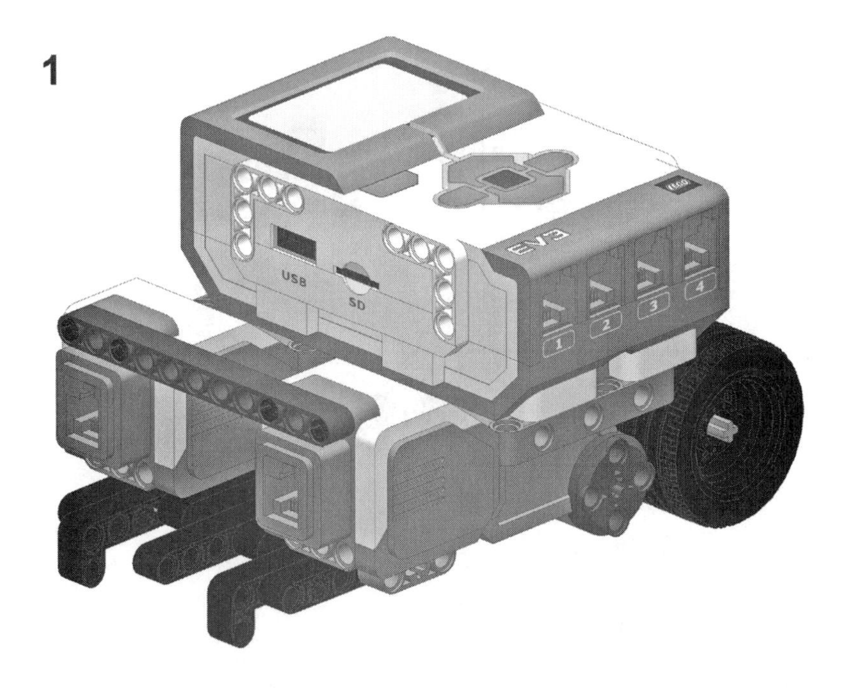

2

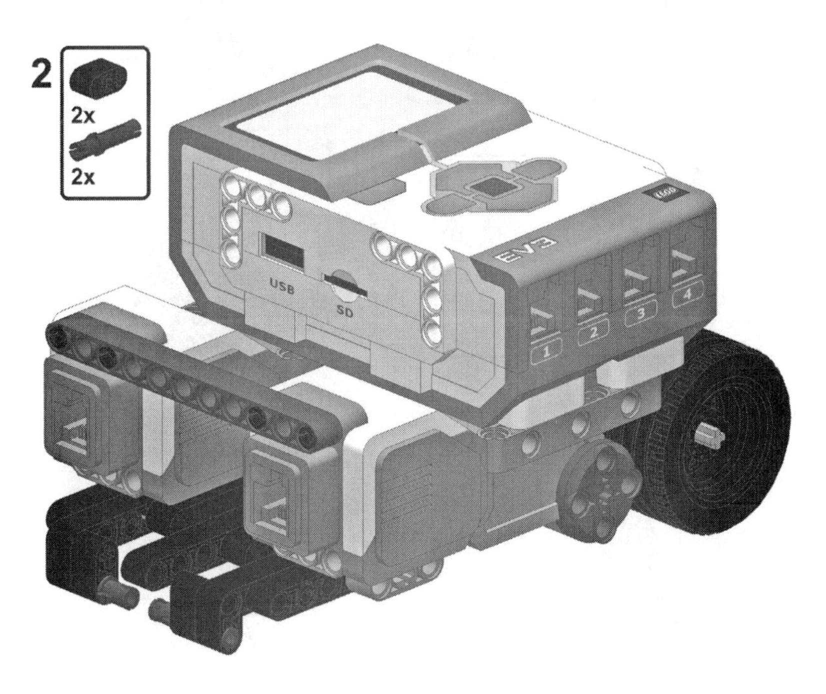

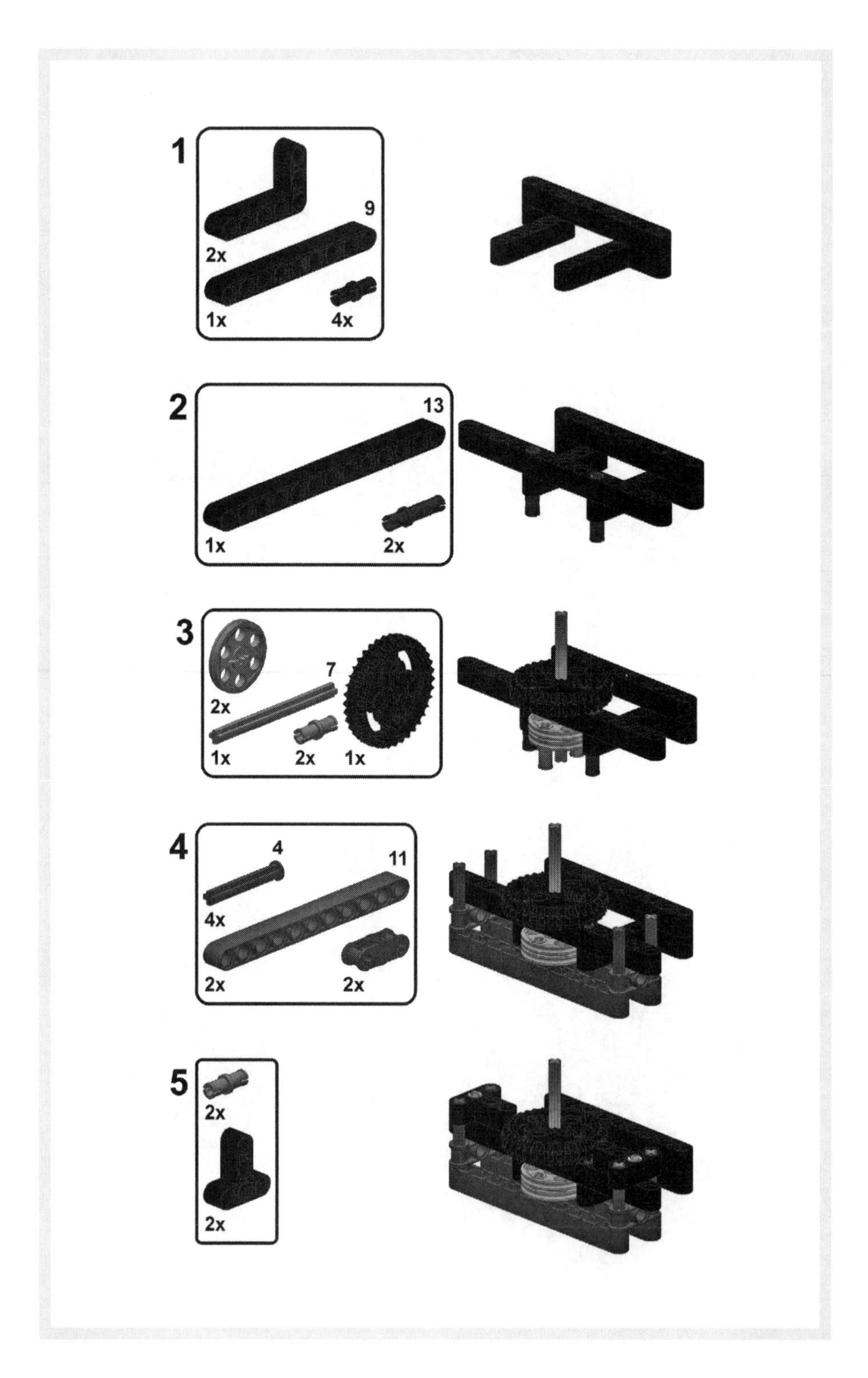
1
9
2x
1x
4x
2
13
1x
2x
3
7
2x
1x
2x
1x
4
4
11
4x
2x
2x
5
2x
2x

6

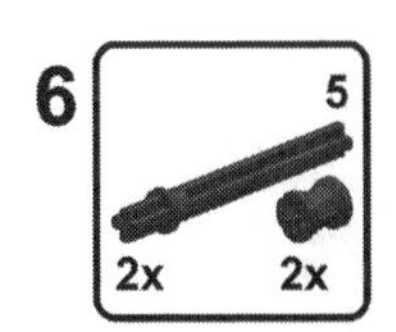

7

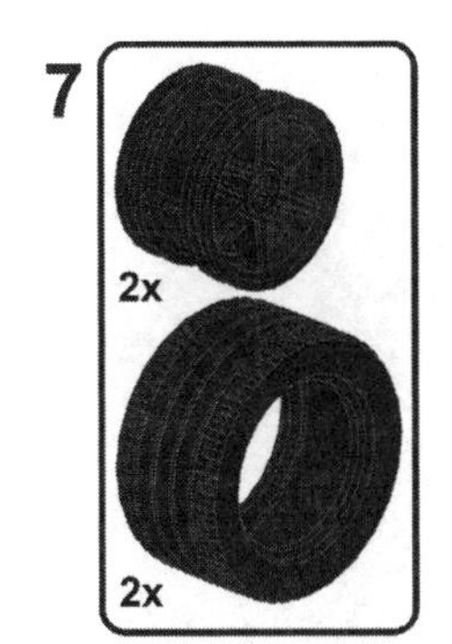

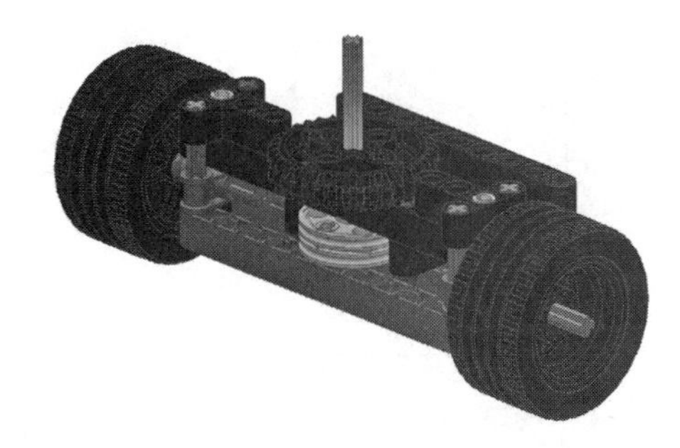

8

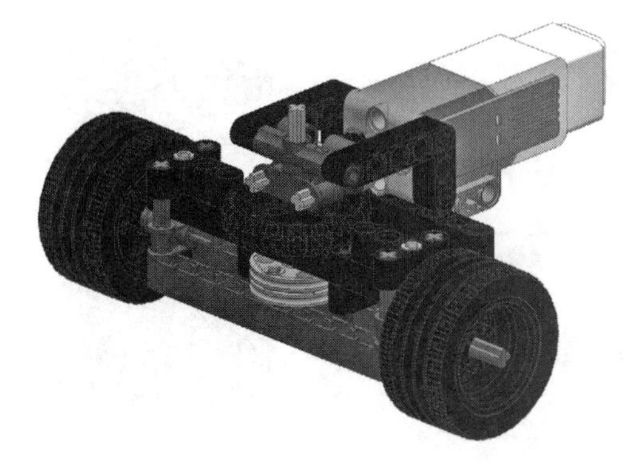

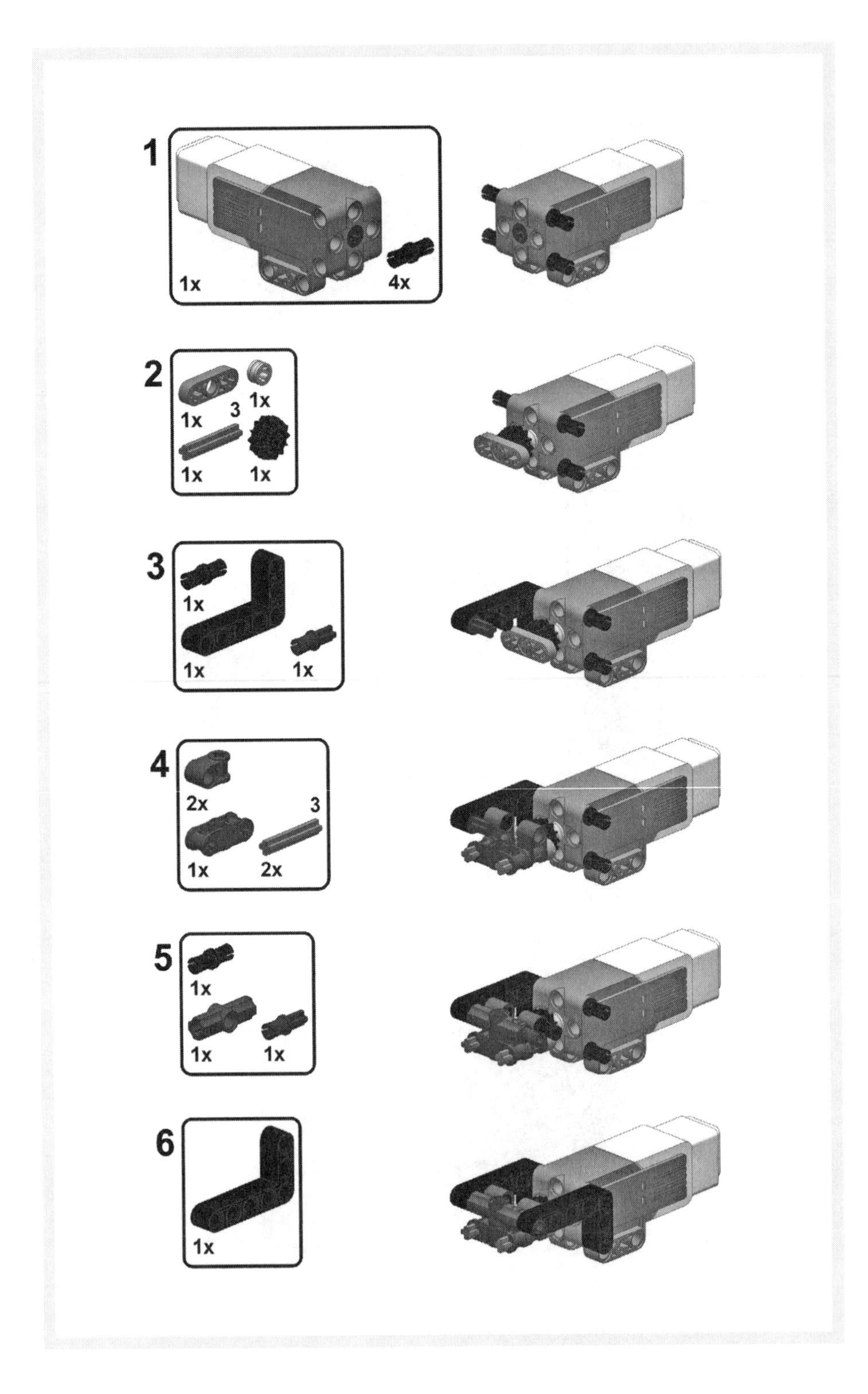
1
1x
4x
2
1x
1x
3
1x
1x
3
1x
1x
1x
4
2x
3
1x
2x
5
1x
1x
1x
6
1x

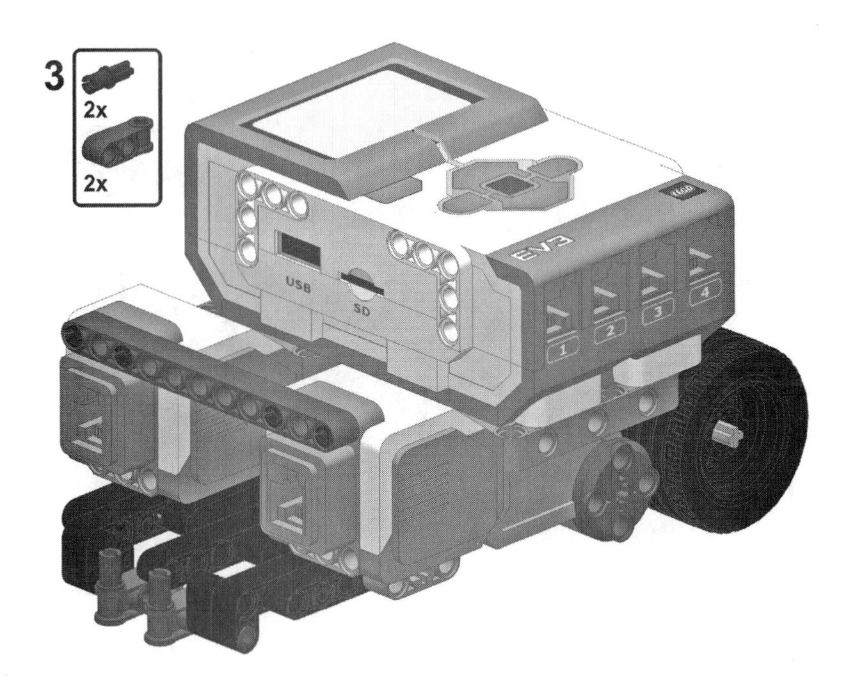
3
2x
2x
USB
SD
EV3
1
2
3
4

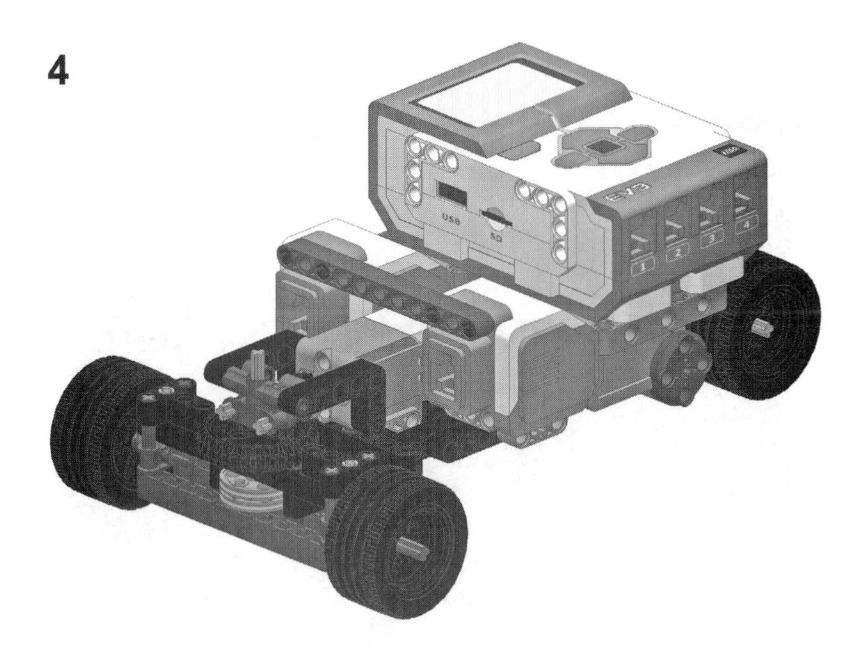
4
USB
SD
EV3
1
2
3
4

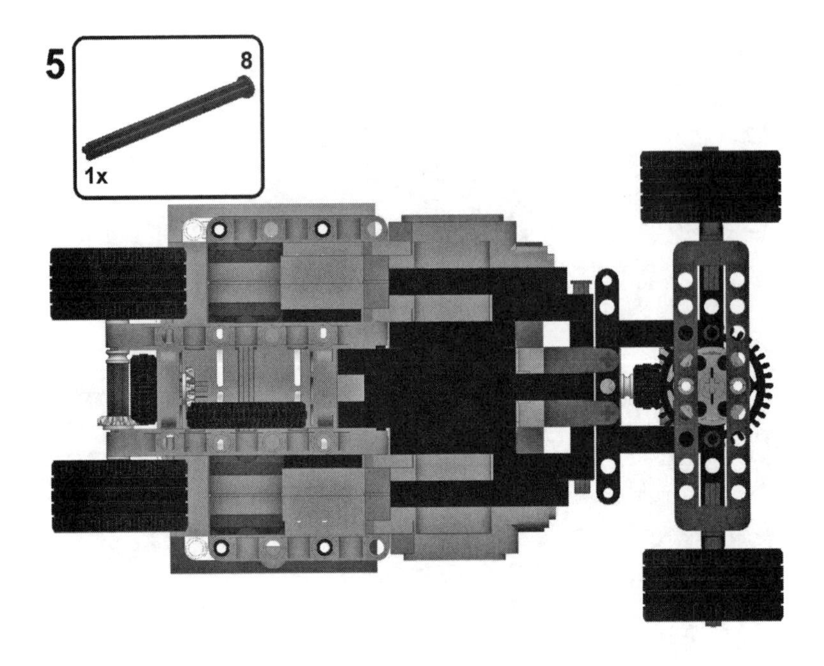
5
8
1x

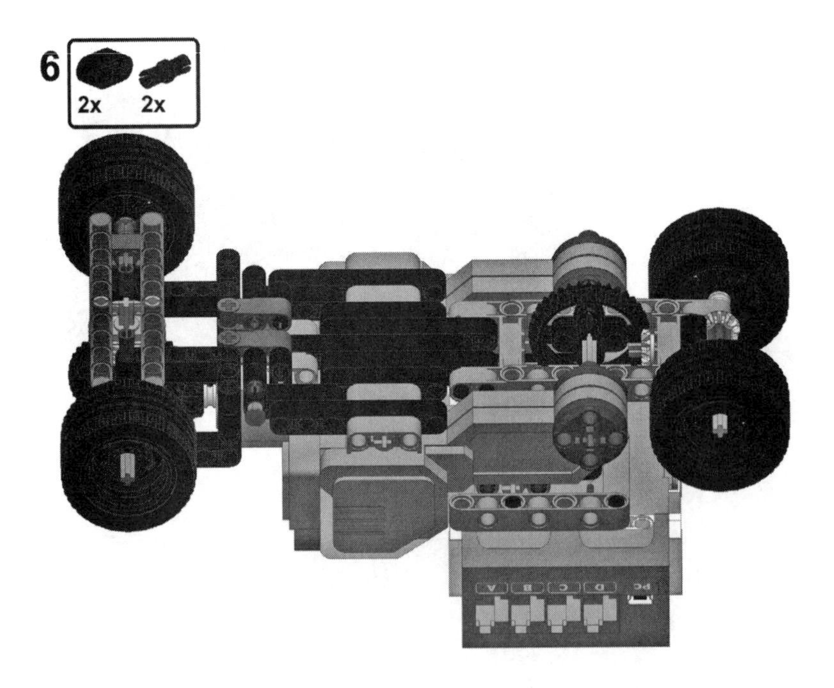
6
2x
2x

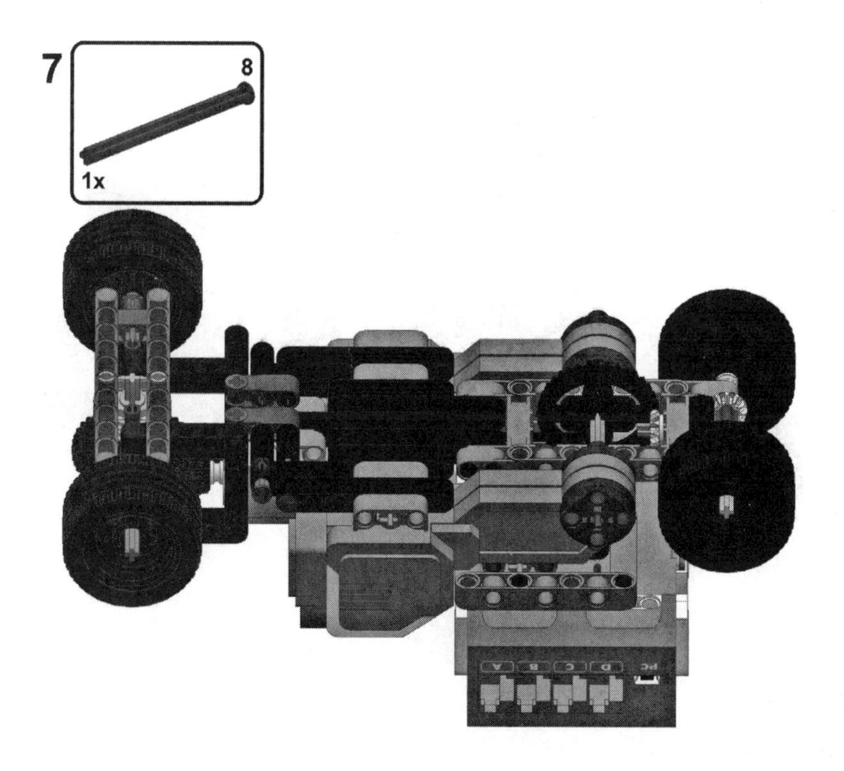
7
8
1x

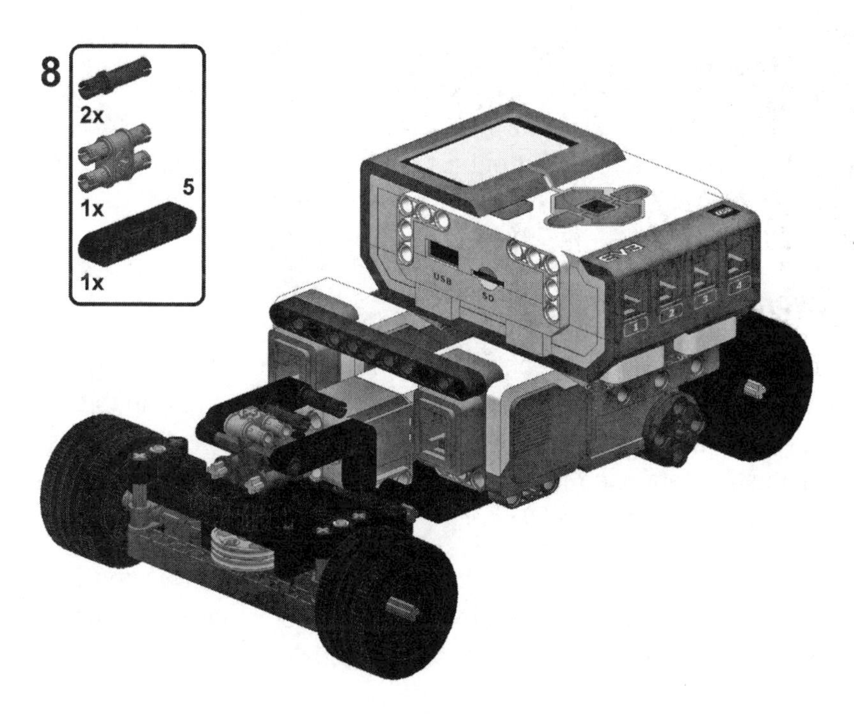
8
2x
1x
5
1x
USB
SD

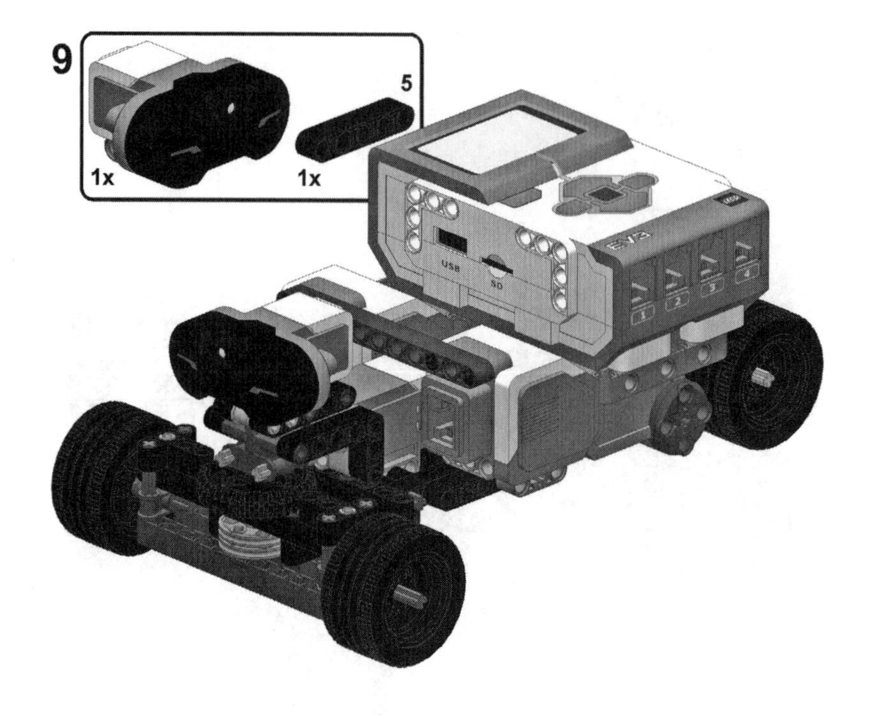

Cables

In order to connect the cable to the medium motor, I recommend temporarily removing the EV3 brick. Insert a short cable into the medium motor, and connect the other end to port A. Connect a short cable from port B to the right motor. Connect another from port C to the left motor.

Networking the Car

This project will allow you to control a car from anywhere across the globe. The architecture is a little more complex than the previous example in this chapter because it will use a GUI, it will allow multiple client connections, and it will control motors. As mentioned, it requires code to run on two separate platforms – a *client* computer and a *server* EV3 brick (Figure 14-1 above).

The client application will send user commands from the client to the server. The server acts as a sort of middleman, shuffling commands from the user to the motors and sensors. Let's enter the server code first.

```
import java.io.*;
import java.net.*;
import lejos.hardware.*;
import lejos.hardware.motor.*;
import lejos.hardware.port.*;

public class RemoteCarServer extends Thread {

  public static final int port = 7360;
  private Socket client;
  private static boolean looping = true;
  private static ServerSocket server;
  private static EV3MediumRegulatedMotor A = new
EV3MediumRegulatedMotor(MotorPort.A);
  private static EV3LargeRegulatedMotor B = new EV
3LargeRegulatedMotor(MotorPort.B);
  private static EV3LargeRegulatedMotor C = new EV
3LargeRegulatedMotor(MotorPort.C);

  public RemoteCarServer(Socket client) {
    this.client = client;

Button.ESCAPE.addKeyListener(new EscapeListener());
  }

  public static void main(String[] args) throws
IOException {
    server = new ServerSocket(port);
    while(looping) {

System.out.println("Awaiting client..");
      new RemoteCarServer(server.accept()).start();
    }
  }

  public void carAction(int command) {
    switch(command) {
    case RemoteCarClient.BACKWARD:
      B.rotate(-360, true);
      C.rotate(-360);
      break;
    case RemoteCarClient.FORWARD:
      B.rotate(360, true);
      C.rotate(360);
      break;
    case RemoteCarClient.STRAIGHT:A.rotateTo(0);
      break;
```

```
      case RemoteCarClient.RIGHT:
        A.rotateTo(-170);
        break;
      case RemoteCarClient.LEFT:
        A.rotateTo(170);
        break;
      }
    }

    public void run() {
      System.out.println("CLIENT CONNECT");
      try {
        InputStream in = client.getInputStream();
        DataInputStream dIn = new
DataInputStream(in);

        while(client != null) {
          int command = dIn.readInt();

System.out.println("REC:" + command);
          if(command == RemoteCarClient.CLOSE) {
            client.close();
            client = null;
          } else {
            carAction(command);
          }
        }

      } catch (IOException e) {
        e.printStackTrace();
      }
    }

    private class EscapeListener implements
KeyListener {

    public void keyPressed(Key k) {
      looping = false;
      System.exit(0);
    }

    public void keyReleased(Key k) {}
    }
  }
```

We could allow the robot to rotate and drive constantly when the button is held down, and stop motors when the button is released. However, this is problematic due to lag issues, making it difficult to control. In order to better control lag issues, it will rotate and drive in short increments when keyboard events are received.

The next part of our project is the client code. It generates a friendly API for the user, allowing easy connection and control of the robot (see Figure 14-4). Let's examine this code.

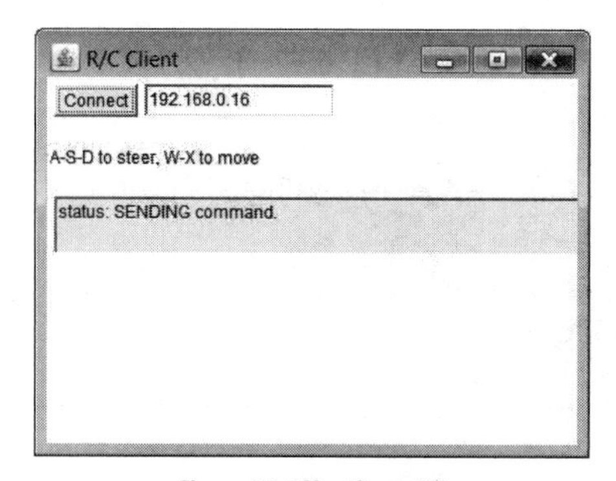

Figure 14-4 The Client GUI

```
import java.awt.*;
import java.awt.event.*;
import java.io.*;
import java.net.Socket;

public class RemoteCarClient extends Frame
implements KeyListener{
   public static final int PORT =
RemoteCarServer.port;
   public static final int CLOSE = 0;
   public static final int FORWARD = 87, //
W = main up
   STRAIGHT = 83, // S = straight
   LEFT = 65, // A = left
   RIGHT = 68, // D = right
   BACKWARD = 88; // X = main down

   Button btnConnect;
   TextField txtIPAddress;
   TextArea messages;
```

```
   private Socket socket;
   private DataOutputStream outStream;

   public RemoteCarClient(String title,
String ip) {
     super(title);
     this.setSize(400, 300);
     this.addWindowListener(new WindowAdapter() {
       public void windowClosing(WindowEvent e) {
         System.out.println("Ending
Warbird Client");
         disconnect();
         System.exit(0);
       }
     });
     buildGUI(ip);
     this.setVisible(true);
     btnConnect.addKeyListener(this);
   }

   public static void main(String args[]) {
     String ip = "10.0.1.1";
     if(args.length > 0) ip = args[0];
     System.out.println("Starting Client...");
     new RemoteCarClient("R/C Client", ip);
   }

  public void buildGUI(String ip) {
   Panel mainPanel = new Panel (new
BorderLayout());
   ControlListener cl = new ControlListener();

   btnConnect = new Button("Connect");

btnConnect.addActionListener(cl);

   txtIPAddress = new TextField(ip,16);

   messages = new TextArea("status: DISCONNECTED");
   messages.setEditable(false);

   Panel north = new Panel(new
FlowLayout(FlowLayout.LEFT));
   north.add(btnConnect);
   north.add(txtIPAddress);

   Panel center = new Panel(new GridLayout(5,1));
   center.add(new Label("A-S-D to steer, W-X
to move"));
```

```
    Panel center4 = new Panel(new
FlowLayout(FlowLayout.LEFT));
    center4.add(messages);

    center.add(center4);

    mainPanel.add(north, "North");
    mainPanel.add(center, "Center");
    this.add(mainPanel);
  }

  private void sendCommand(int command){
    // Send coordinates to Server:
    messages.setText("status: SENDING command.");
    try {
      outStream.writeInt(command);
    } catch(IOException io) {
      messages.setText("status: ERROR Problems
occurred sending data.");
    }

    messages.setText("status: Command SENT.");
  }

  /** A listener class for all the buttons of
the GUI. */
    private class ControlListener implements
ActionListener{
      public void actionPerformed(ActionEvent e) {
        String command = e.getActionCommand();
        if (command.equals("Connect")) {
          try {
            socket = new Socket(txtIPAddress.
getText(), PORT);
            outStream = new DataOutputStream(socket.
getOutputStream());
            messages.setText("status: CONNECTED");

btnConnect.setLabel("Disconnect");
          } catch (Exception exc) {
            messages.setText("status: FAILURE Error
establishing connection with server.");
            System.out.println("Error: " + exc);
          }
        }
        else if (command.equals("Disconnect")) {
          disconnect();
        }
```

```
      }
    }

    public void disconnect() {
      try {
        sendCommand(CLOSE);
        socket.close();

btnConnect.setLabel("Connect");
        messages.setText("status: DISCONNECTED");
      } catch (Exception exc) {
        messages.setText("status: FAILURE Error
closing connection with server.");
        System.out.println("Error: " + exc);
      }
    }

    public void keyPressed(KeyEvent e) {
      sendCommand(e.getKeyCode());
      System.out.println("Pressed " +
e.getKeyCode());
    }

    public void keyReleased(KeyEvent e) {}
    public void keyTyped(KeyEvent arg0) {}
  }
```

Since the client application is test code, it isn't as fleshed out as production code might be. For the server code, I decided to allow multiple computers to connect to the EV3 at the same time. This was cheap and easy to do, and only required spawning a new thread to deal with each new connection.

WARNING: *The keyboard commands are attached to the Connect/Disconnect button, since that object is likely to hold the focus. If you click on other widgets in the window and the button loses focus, the keys stop working. Tab back to the button to regain focus.*

Results

Now that we have all the code ready it's time to test it out. First, upload and run the server code on your EV3. It will sit waiting for a client to connect.

Next, run the client on your PC. Type in the IP address (or leave it as 10.0.1.1 if it is using Bluetooth) and click connect. As previously mentioned, you can see the IP address of your EV3 brick from the main menu of the EV3.

Now you can tap the A, S, and D to steer. Be sure to tap the keys, don't hold them down of they will repeat. The W and X keys drive the robot forward and backward. This should work well provided the robot is in a location that receives your WiFi signal. If you have problems connecting or staying connected, try moving closer to the WiFi router.

Across the Internet

If you want to connect to an EV3 on a different network across the Internet, you will need to do something slightly different. Most networks use a router that connects to the Internet, allowing everyone to have Internet access. You need to know the IP address that the rest of the world sees. Sometimes this changes every time the router connects to the Internet provider, depending on the service. To check the IP address, open a browser and go to your router admin page (usually http://192.168.0.1). Find the status of your connection, which also lists the IP address (see Figure 14-5).

Tip!

This robot uses iterative movement rather than continuous movement when a button is pressed. In reality, the lag for WiFi does not seem to be much of a problem as long as you are on a local network and not communicating across the Internet (see below). Try adapting the client and server code so that the robot moves when a button is pressed, and stops moving when a button is released.

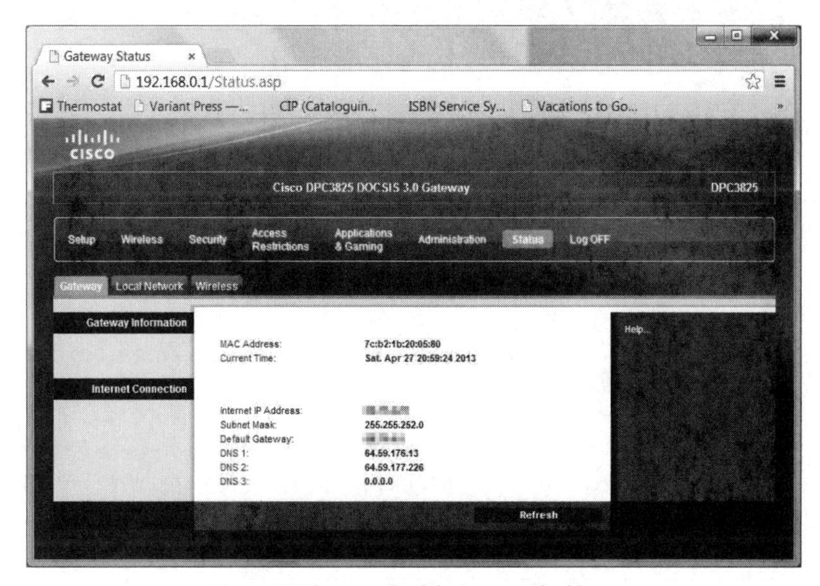

Figure 14-5 An example of the Internet IP address.

You aren't finished yet. Since you probably have more than one computer on your router, you need to set up the router to forward packets to the EV3 (which is a server with its own local IP). It does this using port numbers. Every application uses a different port number, and our robot arm application uses port 7360 (i.e. "LEGO"), as you can see in the server code. You want any data meant for port 7360 to be forwarded to your EV3. You'll have to consult your documentation for how to set up *port forwarding* on your router. You can also add the IP address of your EV3 brick to the DMZ (Demilitarized Zone) allowing outside connections directly to the EV3 brick (see Figure 14-6).

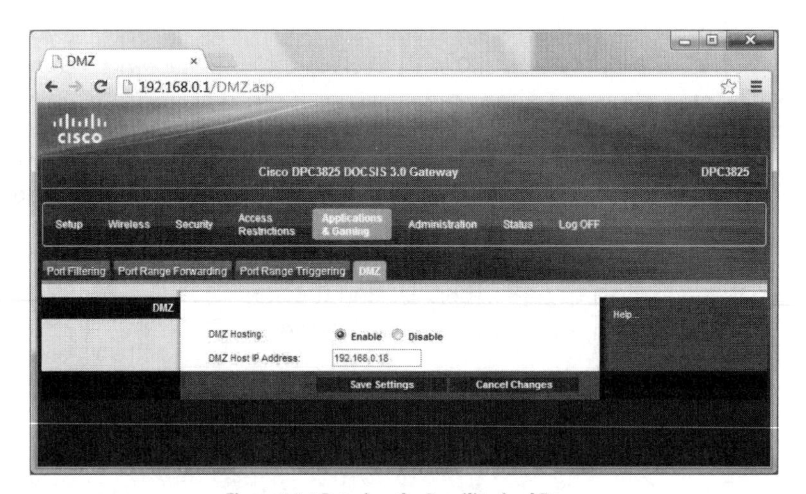

Figure 14-6 Entering the Demilitarized Zone.

Now that you are ready to go, run the client application and click connect. You can now press the different keys that control the three different motors. After each press you will see the tachometer count that shows that the motors did in fact move.

NOTE: *If you are having problems accessing a server through the Internet, chances are there is a firewall blocking some ports. Temporarily disable the firewall while testing this project, or allow firewall access through port 7360.*

If you have a webcam, I recommend pointing it at the R/C car so you can experience telepresence. You can receive a live webcam image by using one of the many free instant messaging programs, such as Skype, Google Hangouts, or one of the many open source instant messaging clients built on Jabber.

CHAPTER 15

Debugging Spray

TOPICS IN THIS CHAPTER

- Text output to PC
- Debugging within Eclipse

Cross development is when you write code on one platform (a PC) and run that code on another platform (the EV3). With cross development, sometimes it is difficult to locate bugs or monitor what is happening inside your program. Often programmers just use System.out.println() or LCD.drawString() to output values to the LCD screen. This might be good enough for you, but there is also a better way. The leJOS environment has several built-in tools to make debugging easier. You can output longer messages than the LCD screen will allow, and you can monitor the messages on your computer monitor in the comfort of an office chair rather than having to catch a (sometimes) moving robot to peek at the LCD screen. You can also use Eclipse's debugging mode to step through lines of code one at a time, set breakpoints, and monitor variables.

Basic Debugging

When programming, you sometimes want to output variables to a screen while your code is running. Normally you can output the values to the EV3 screen, but this screen is limited to eight lines of text. Also, if the program crashes, you will lose the text output before you can see what the values were.

The solution to this problem is to output text real-time to your PC. To do this, you need to run and connect to EV3 Control Center (see Chapter 8). By default, leJOS outputs text from System.out and System.err to the LCD screen. Let's generate an exception and see what it looks like on the screen. Enter the following code to generate an exception.

```
public class ThrowException {
public static void main(String[] args) {
   Thread t = null;
   t.start();
  }
}
```

Upload and run the code on the EV3. When the program starts, it will immediately throw a NullPointerException (Figure 15-1). The text flows off the screen, but you can use the arrow keys to scroll around the screen to read the full exception. Note that the information even includes line numbers in your code. Tiny, isn't it! Hit the escape button to return to the menu.

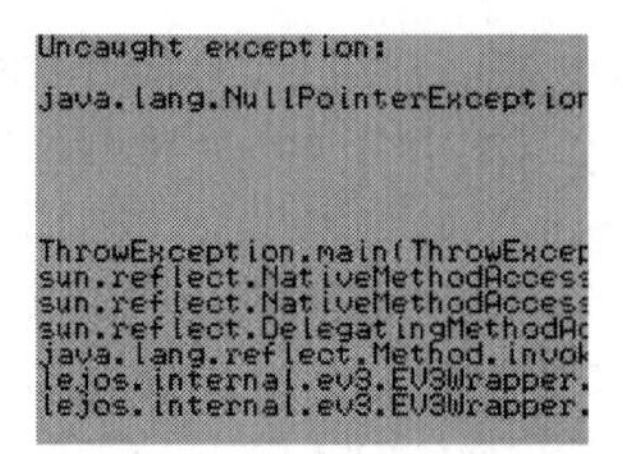

Figure 15-1: The EV3 LCD displaying an exception.

Wouldn't it be nice to view these on a real computer monitor instead? The first thing to do is reroute the output from the LCD to a console.

1. In Eclipse (with the leJOS plugin installed), click on the icon for EV3 Control Center.
2. Once the control center is up, click Search to find your brick. (Make sure your EV3 is booted up.) Once it is found, click Connect.
3. Click on the Console tab. Your EV3's output will appear here.
4. Try running the example above to throw an exception. Notice the output goes to the EV3 LCD screen and the console output (see Figure 15-2). Any output, even from System.out.println(), will appear here.

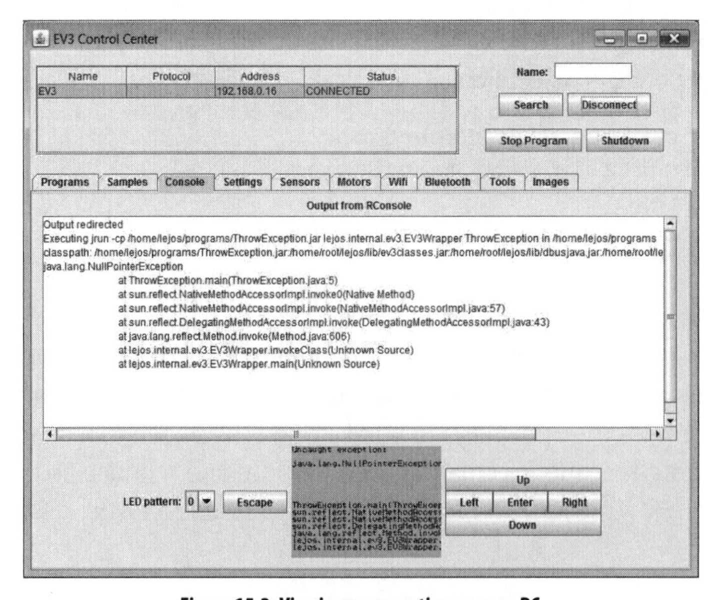

Figure 15-2: Viewing an exception on your PC.

As you can see, EV3 Control Center can be a useful tool for viewing your LCD screen and debugging code. Be sure to check out chapter 8, which explains all the features and how to use it in more detail.

Debugging with Eclipse

The above example merely allows you to output text to the console, however there is a more refined way to debug using an actual debugger module within Eclipse. A debugger allows you to stop an executing program anywhere in the code you choose. This allows you to monitor the effects of the program up to that point, as well as monitor variables to see what values they hold. This, as you might guess, is key to debugging and figuring out what is going on in your code.

Before we can use the debugger, we'll need some code. Enter the following into a new class called DebugTest.

```
import lejos.hardware.*;
import lejos.hardware.lcd.*;

public class DebugTest {

  public static void main(String[] args) {
    System.out.println("Hello std");
    System.err.println("Hello err");
    int x = 6;
    LCD.drawString("HELLO", 0, 3);
    LCD.drawString("NUMBER " + x, 0, 4);
    x = x - 4;
    LCD.drawString("I AM", 0, 5);
    LCD.drawString("NUMBER " + x, 0, 6);
    Button.waitForAnyEvent();
  }
}
```

Now that we have some code, we're going to insert a *breakpoint*. This will cause the program to pause execution at this point in the code so we can gain control of the program and execute the program line by line.

1. Move the cursor to line 10, which outputs "HELLO" to the LCD.

2. Next, select Run > Toggle Line Breakpoint. You should now see a blue dot next to the line (see Figure 15-3).

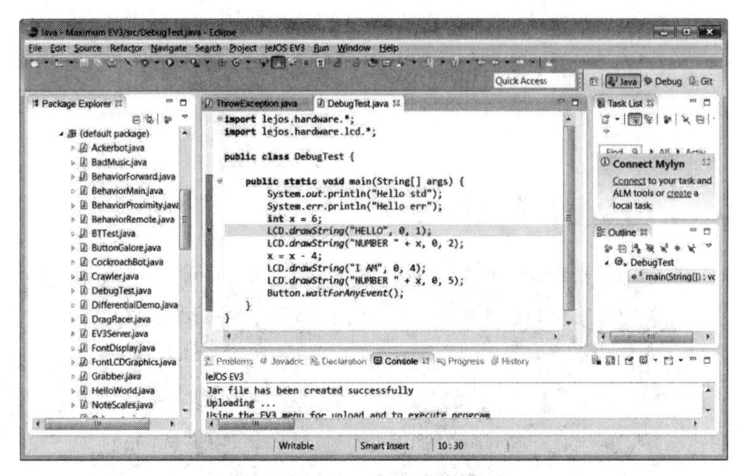

Figure 15-3: Inserting a breakpoint.

Great! Now we can begin executing the program in debugger mode. This is slightly different from running a program the normal way.

1. With your Java class selected, click the Debug icon in the toolbar (it looks like a bug). Alternately, select Run > Debug As > leJOS EV3 program.

2. This will upload your program and begin execution. However, Eclipse will display a pop-up dialog that says "Confirm Perspective Switch" and ask if you want to switch to the Eclipse Debug perspective. Say yes. This causes the screen to switch perspective to debug mode (see Figure 15-4).

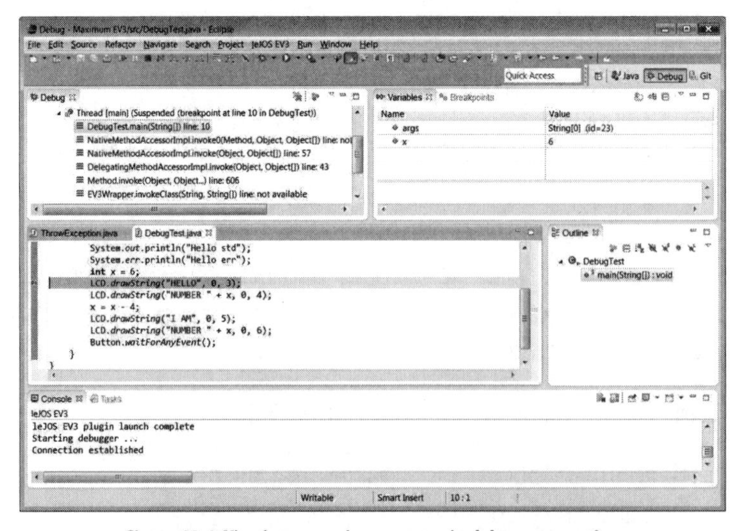

Figure 15-4: Viewing a running program in debug perspective

3. The console will inform you that the connection is established. Your program will then run until it hits the line with the breakpoint. Note that the EV3 brick displays the output from the first two lines of the program, but it has not yet executed the line with the breakpoint (see Figure 15-5). Also note that we can see the value of x in the Variables tab in the upper right window (x equals 6).

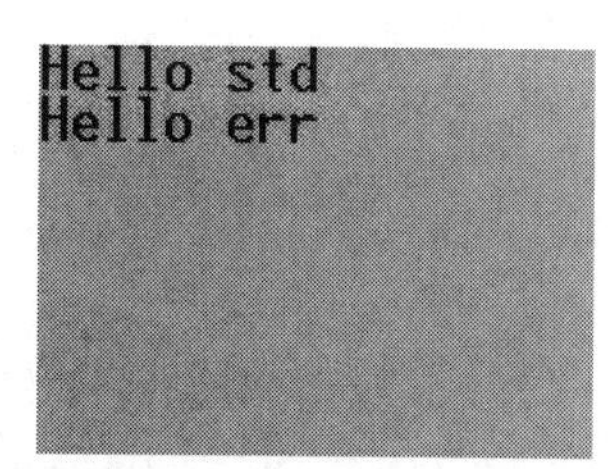

Figure 15-5: Viewing the program output up to the breakpoint

4. At this point we can *step into* (F5) or *step over* (F6) line number 10 in the code. (These are the curved yellow arrows in the toolbar. See Figure 15-4). I recommend stepping over, at least for now, which will cause the program to output "NUMBER 6" to the LCD. If you select step into, it will execute the drawString() method line by line, which is far too much detail for us and will become tedious.

5. Now line 11 is highlighted. Click step over twice and you will see the variable x change to 2, plus this line in the variables window will become highlighted in yellow (see Figure 15-6).

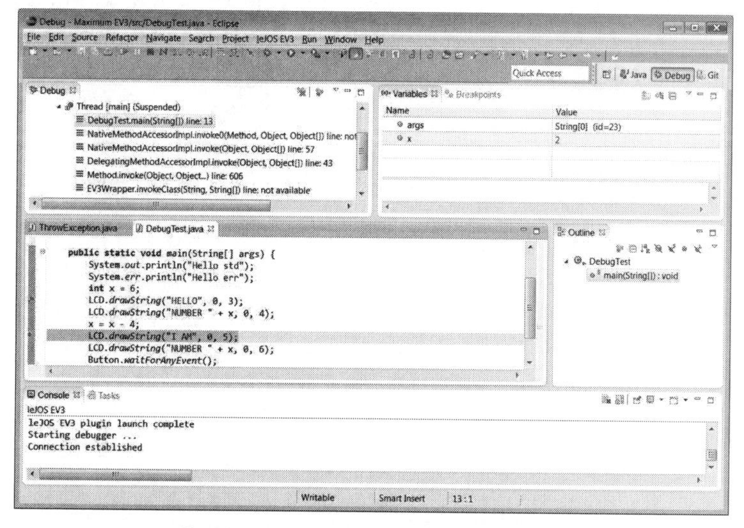

Figure 15-6: Noting a change in a variable value

6. Now we are going to change a variable as the program is running! Double click the 2 value for x and change it to 1. Now we will allow the program to continue running by pressing resume (the green play button or F8). Notice the output on the LCD says 1 (see Figure 15-7).

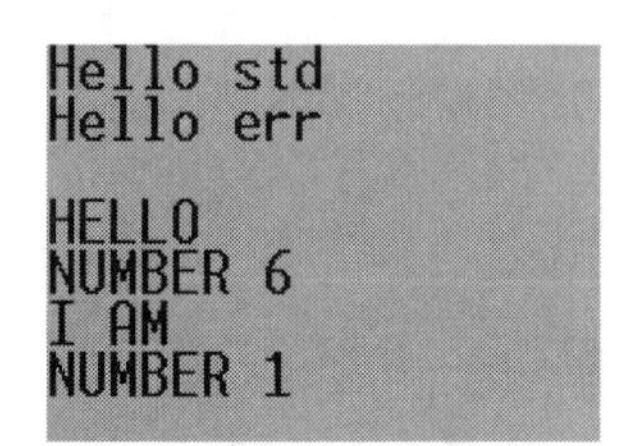

Figure 15-7: Viewing the final output of the program

> ***Website!***
>
> Programmer Lars Vogel has created a good online tutorial for beginners here.
>
> www.vogella.com/tutorials/EclipseDebugging/article.html

7. Press any key on the EV3 and the program will end, including the Eclipse debugger. You can change back to Java perspective in the upper-right corner by clicking the button that says Java.

These are the basic functions of the Eclipse debugger. If you want to learn more, there are several tutorials available. Search the Internet for "Eclipse debugger" and you will see several good online tutorials and videos to help you go deeper.

CHAPTER 16

Sensor Framework

TOPICS IN THIS CHAPTER

- Introducing the Sensor Framework
- Generic Sensor Interfaces
- Adapters
- Warbird
- Filters

When leJOS was first released for the RCX, there were very few sensors available for the kit. Other than writing a simple sensor driver, we put very little thought into a cohesive plan for sensors. When the LEGO NXT was released, we started to notice that there were many different sensors available, and many different versions of the same sensor. For example, accelerometers and gyros were produced by multiple vendors. As expected, we developed a gyro class for each gyro on the market. However, if another class (such as a navigation class) needed a gyro in a constructor, we didn't want to write the class three separate times for each gyro on the market. Instead, we developed interfaces that produced uniform results. For example, we created a gyroscope interface that was implemented by each of the three gyro sensors.

There were still a few shortcomings with our architecture. One thing we noticed was that some sensors reported multiple types of data. For example, one gyro might only supply angular velocity for two axes, whereas another might support three axes. In the case of one sensor, it supplied both gyro and accelerometer data. Many of these sensors had to switch into different modes to supply different types of data.

Other things also needed more thought, such as coming up with a consistent naming scheme and using standard units of measurement. For these and other reasons, when the EV3 was released, the leJOS developers did an overhaul of the sensor classes, eventually coming up with a whole new sensor framework. Let's examine this framework and how to use it.

The leJOS Sensor Framework

There are dozens of sensor classes in the package lejos.hardware.sensor. These sensors have a specific naming structure:

Vendor + Sensor Type + "Sensor" (+ model number/version)

The LEGO sensors forego the LEGO name and merely indicate which kit they belong to. Thus, the sensors in the EV3 kit are called EV3ColorSensor, EV3TouchSensor, and EV3IRSensor.

Each of these sensors implements the interface SensorModes. Normally you can't just retrieve a sensor value (called a *sample*) directly from a sensor class. Instead, you must first get a SensorMode instance from the Sensor class. Why? Because many times a sensor has more than one mode of operation.

You can use a method from SensorModes called getMode(String mode) to get a SensorMode. And how can you tell which modes a sensor has? By using a method called getSensorModes() which returns an array of Strings identifying the available modes. The sample code below shows how to retrieve and output the modes for a color sensor to the LCD screen.

```
EV3ColorSensor cs = new
EV3ColorSensor(SensorPort.S3);
ArrayList<String> modes = cs.getAvailableModes();
  for(int i=0;i<modes.size();i++) {
  System.out.println(modes.get(i));
}
Button.waitForAnyPress();
```

When you run this code snipped in a main() method, it will output all four modes of the color sensor, as shown in Figure 16-1.

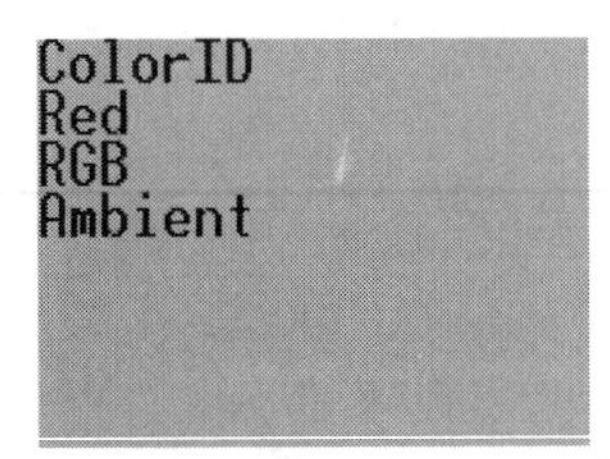

Figure 16-1: Outputting the sensor modes to the EV3

At this point, you might be wondering why you would have to poke around with sample code in order to find out what modes a sensor is capable of producing. Well, fear not! The Javadocs for each sensor explicitly state which modes you can retrieve from a sensor.

OK, so now you have a SensorMode. How do you get an actual data value from the sensor? As it turns out, the SensorMode implements the SampleProvider interface. In other words, all SensorMode objects are also SampleProviders. (In fact, the instructions for how to retrieve samples are located in the Java docs for SampleProvider.) So how do you get a sample?

As the following code demonstrates, you request an array of float values. Why an array? Because some sensors return more than one value, such as the EV3 ultrasonic sensor. To find out how many samples it is returning, use the method SampleProvider.sampleSize(). Let's examine all of this in code.

```
  EV3IRSensor sensor = new
EV3IRSensor(SensorPort.S4);
  SampleProvider distance= sensor.
getMode("Distance");
  float[] sample = new float[distance.
sampleSize()];
  for(int i=0;i<20;i++) {
    distance.fetchSample(sample, 0);
    System.out.println(i + ". " + sample[0]);
    Delay.msDelay(250);
  }
```

The above code outputs values from the IR sensor. It does this for 20 iterations then ends. This is the official version of accessing data from sensors, but there are easier ways to access data using adapters.

A Word about Generic Sensor Interfaces

The LEGO EV3 has a mini-economy of sensors manufactured by third parties. For example, compass sensors are produced by both HiTechnic, Mindsensors, and Dexter Industries. Although the individual compass sensors are designed and programmed differently, they all return direction readings.

On the leJOS project, we wanted our API to be compatible with sensors, even if they came from different manufacturers. To accomplish this, we created a series of interfaces that describe the main functions of each family of sensors. For example, we created an interface called DirectionFinder which can return a standard value.

So what does this accomplish? Now a class can accept either of these sensors in a method or constructor. The following code sample shows a sample class with a constructor that accepts a DirectionFinder.

```
import lejos.robotics.DirectionFinder;

public class MyRobot {
  private DirectionFinder df;

  public MyRobot(DirectionFinder compass) {
    this.df = compass;
  }
}
```

The sample code above is not functional, but as you can see, the class will now accept any direction finder that exists or comes along in the future. This effectively future-proofs our API to allow people to add new sensors without us having to reprogram the other classes in leJOS.

The lejos.robotics package also includes other sensor types, such as RangeFinder, which is used by the EV3UltrasonicSensor class (more on that in the Adapters section below). You can make your own sensors that implement these interfaces. For example, my (very ancient) book Core LEGO Mindstorms Programming contained projects to build a compass sensor and a range sensor. If you have an RCX to NXT/EV3 adapter, you can program your own class to implement the DirectionFinder and RangeFinder interfaces, and thus allow your homebrew sensor to work with the leJOS API.

In fact, the DirectionFinder does not have to be a compass sensor. Other sensors can determine direction in other ways, such as the GyroDirectionFinder located in the lejos.utility package.

You can even experiment with exotic methods for determining direction. Imagine a sensor that estimates direction based on the position of the sun, given that it knows the latitude, longitude, and current time. Or more simply, it could estimate direction based on spotting a fixed light source in a room by using a light sensor mounted on a motor.

Adapters

You'll notice that using the sensor framework takes quite a few lines of code to access a simple value from a sensor. You can use a less robust but simpler class to get data with fewer line calls, and also to provide some backward compatibility with other parts of the leJOS API. These classes are called adapters and they reside in the lejos.robotics package.

There are currently six adapters:

- AccelerometerAdapter
- ColorAdapter
- DirectionFinderAdapter
- GyroscopeAdapter
- RangeFinderAdapter
- TouchAdapter

Each adapter implements an interface corresponding to the name of the adapter. For example, RangeFinderAdapter implements the RangeFinder interface, which is also located in the lejos.robotics package.

Each adapter requires a SampleProvider in the constructor of the appropriate sensor. This requires three lines of code to create a RangeFinder object. For example, you can create a RangeFinder for the EV3IRSensor as follows:

```
 SensorModes sensor = new
EV3IRSensor(SensorPort.S1);
 SampleProvider distance= sensor.
getMode("Distance");
 RangeFinderAdaptor rf = new
RangeFinderAdaptor(distance);
 float range = rf.getRange();
```

In fact, because most sensors implement SampleProvider, you can create an instance of an adapter with only two lines of code. Let's look at an example using the EV3 touch sensor this time:

```
 EV3TouchSensor sensor = new
EV3TouchSensor(SensorPort.S1);
 TouchAdapter touch = new TouchAdapter(sensor);
```

Follow this with a call to the isPressed() method of the TouchAdapter in order to read the state of the touch sensor switch:

```
boolean pressed = touch.isPressed();
```

Filters

Filters are used to alter a sample or to alter the flow of samples, or both. For example, a filter can alter the flow by returning values less frequently than the actual sensor would. Filters can do almost anything with the data, as the examples below will show. They take a sample from a SampleProvider, modify it, and then pass the sample on. They are in fact SampleProviders themselves and can basically be treated as sensors.

The leJOS API comes with some ready made filters found in the lejos.robotics.filter package. The example below shows how to use a filter to get the running average of the last five samples from an infrared sensor.

```
  SensorModes sensor = new
EV3IRSensor(SensorPort.S4);
  SampleProvider distance = sensor.
getMode("Distance");
  SampleProvider averager = new
MeanFilter(distance, 5);
  float[] sample = new float[averager.
sampleSize()];
  averager.fetchSample(sample, 0);
```

The code shows that a filter constructor takes a SampleProvider as its first parameter followed by some configuration parameters. In this case, the MeanFilter is set to average five samples. Instead of fetching a sample from the sensor class, it is now fetched from the filter.

Filters can be stacked on top of each other to allow for more complex manipulations. The second filter takes the identifier of the first filter in its constructor and so on. One always fetches the sample from the last filter in the stack.

As of this writing there are 16 filters present in the lejos.robotics.filter package. They do everything from collect mean and median values, to calibrating readings. Take a look through the package to get an idea of what they are capable of. If you don't find what you are looking for, perhaps you could program the filter yourself and submit it to the leJOS project so that others can benefit from your creation.

Warbird

Now that we know a little more about the sensor framework, let's try a full code example using a robot and the EV3 color sensor. We'll use an adapter and a filter in this example. First we need to build a new robot named Warbird.

Warbird is, as his name suggests, a mighty winged bird of war. He has powerful tank treads and a shoulder mounted cannon, plus some ferocious tail-feather swords to strike fear in his opponent's heart. We'll use Warbird to try out some concepts with the sensor framework.

NOTE: *Warbird will be used in a few more chapters after this one. You probably will want to leave him assembled after you are done this chapter.*

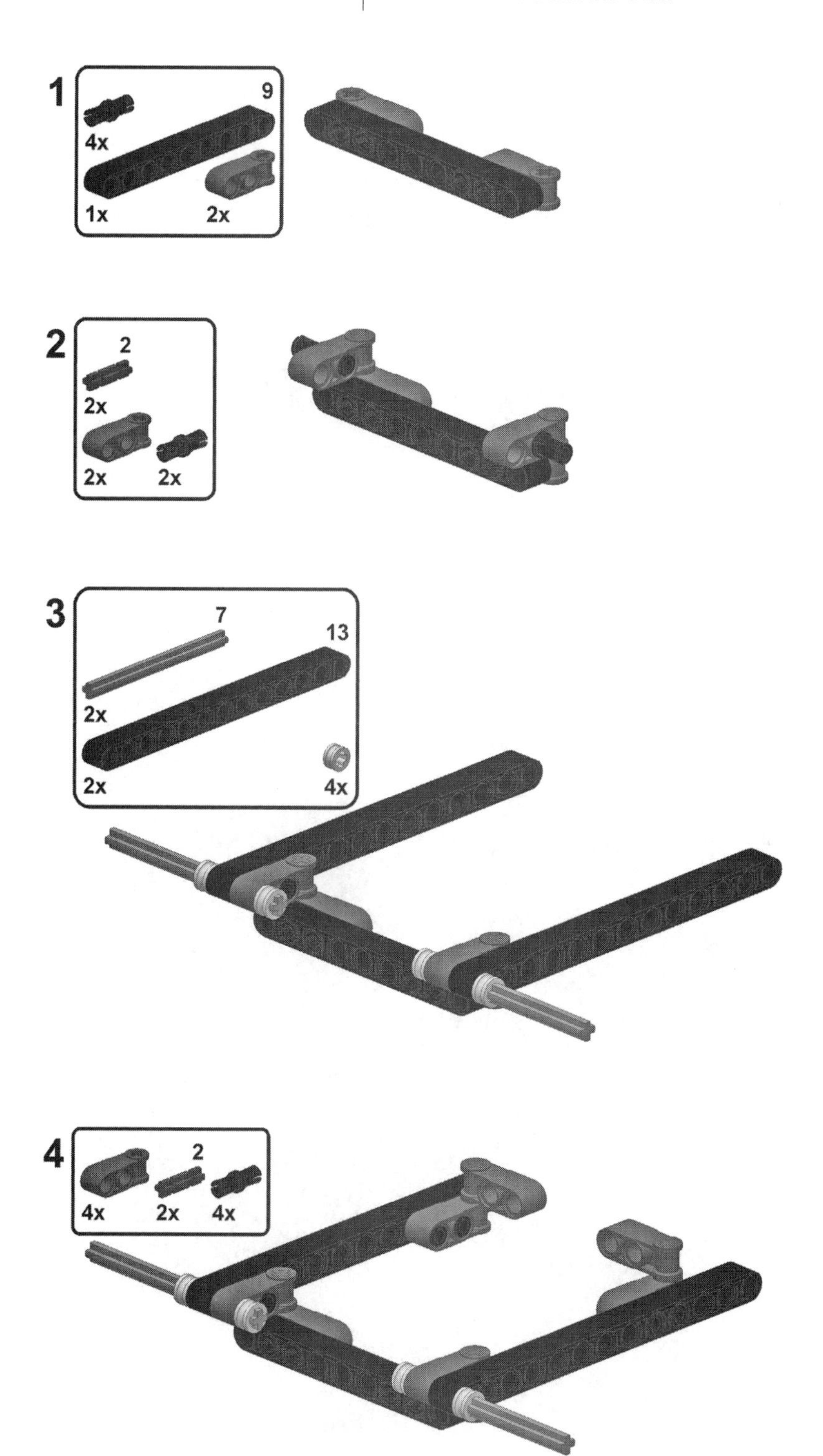
1
9
4x
1x
2x
2
2
2x
2x
2x
3
7
13
2x
2x
4x
4
2
4x
2x
4x

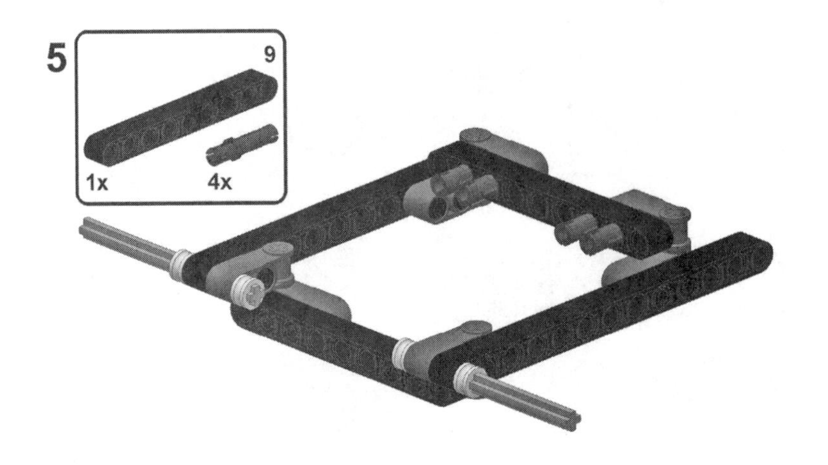
5
9
1x
4x

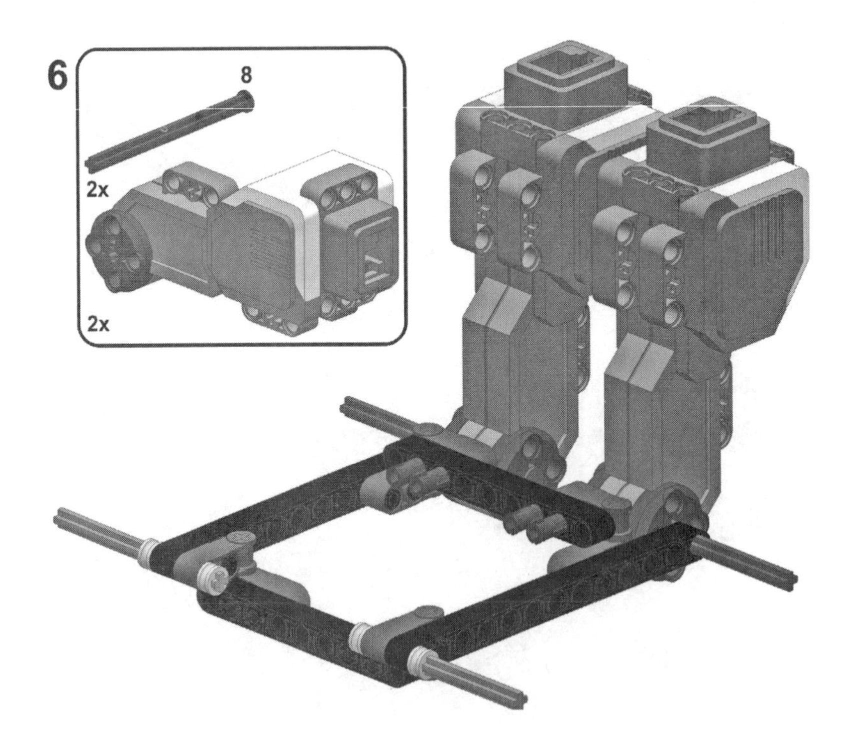
6
8
2x
2x

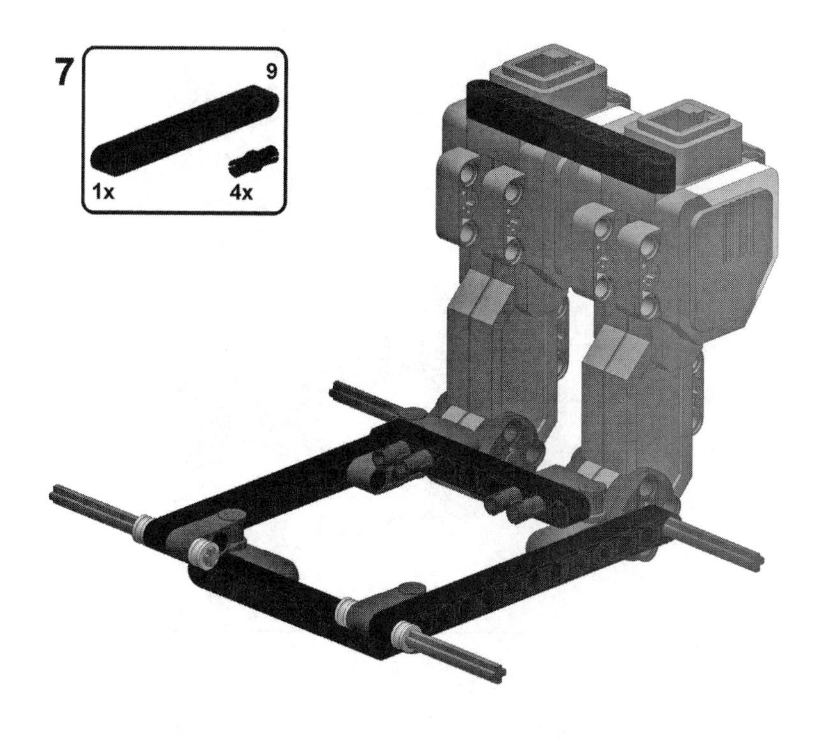
7
9
1x
4x

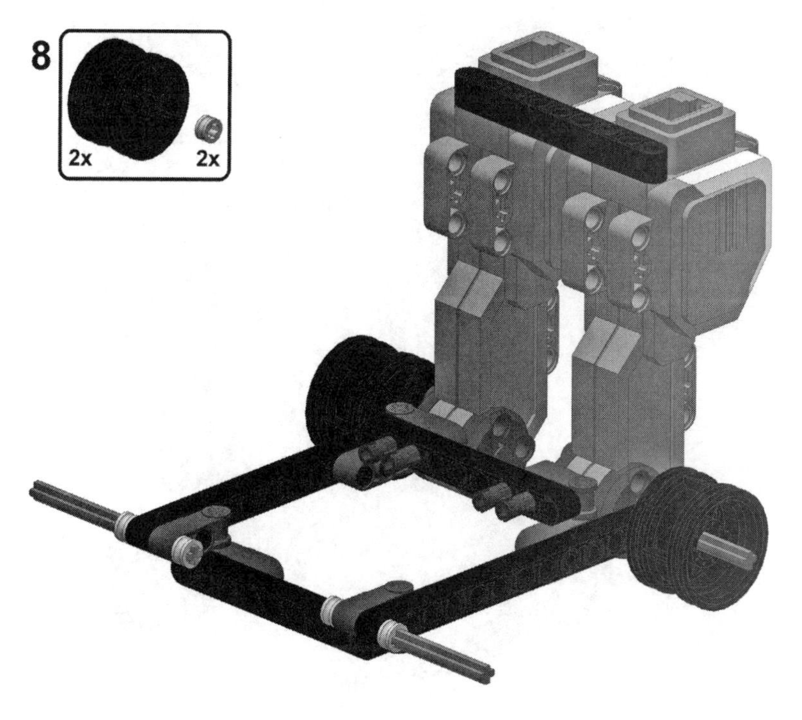
8
2x
2x

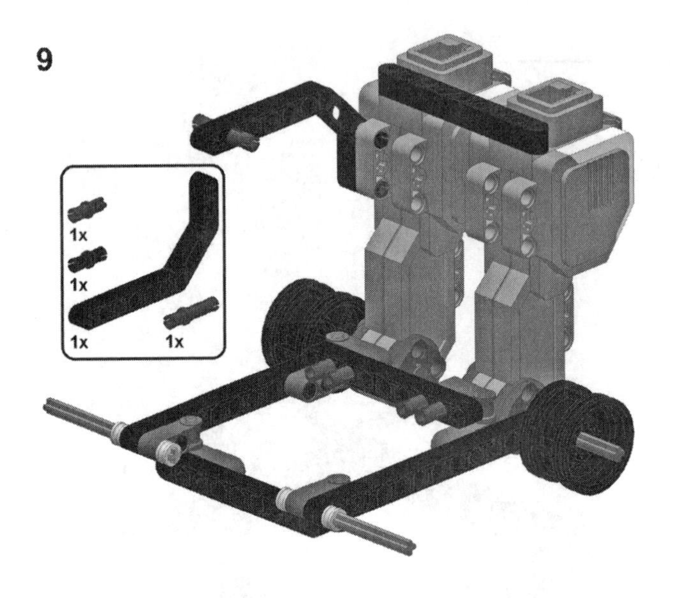
9
1x
1x
1x
1x

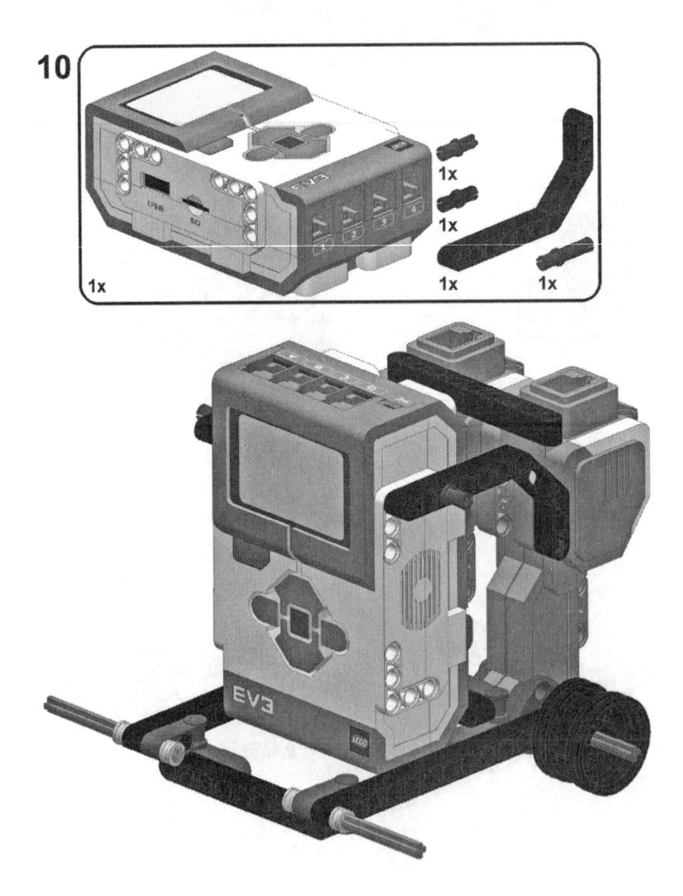
10
USB
SD
EV3
1x
1x
1x
1x
1x
EV3

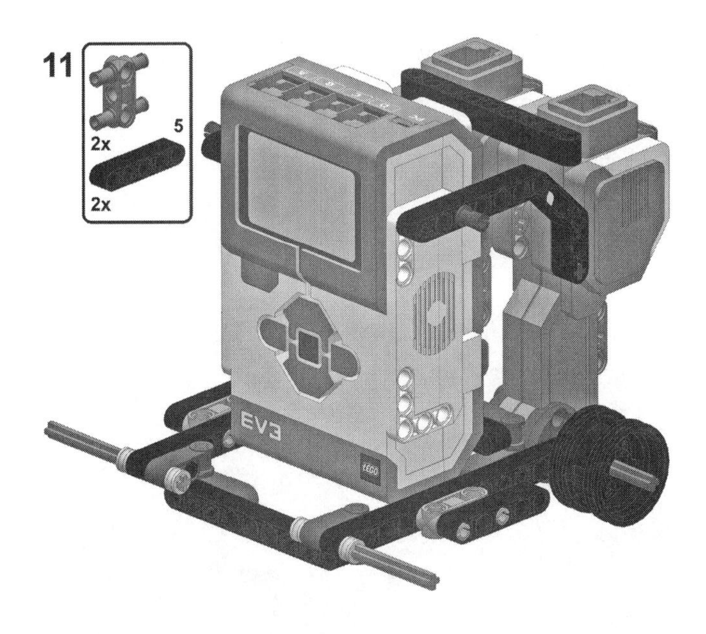
11
5
2x
2x
EV3

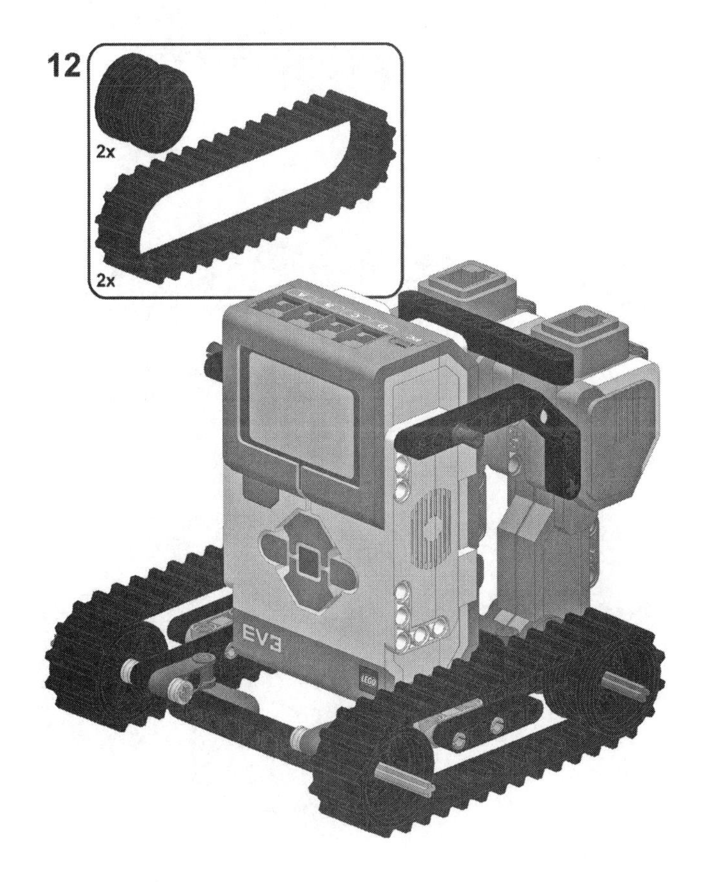
12
2x
2x
EV3

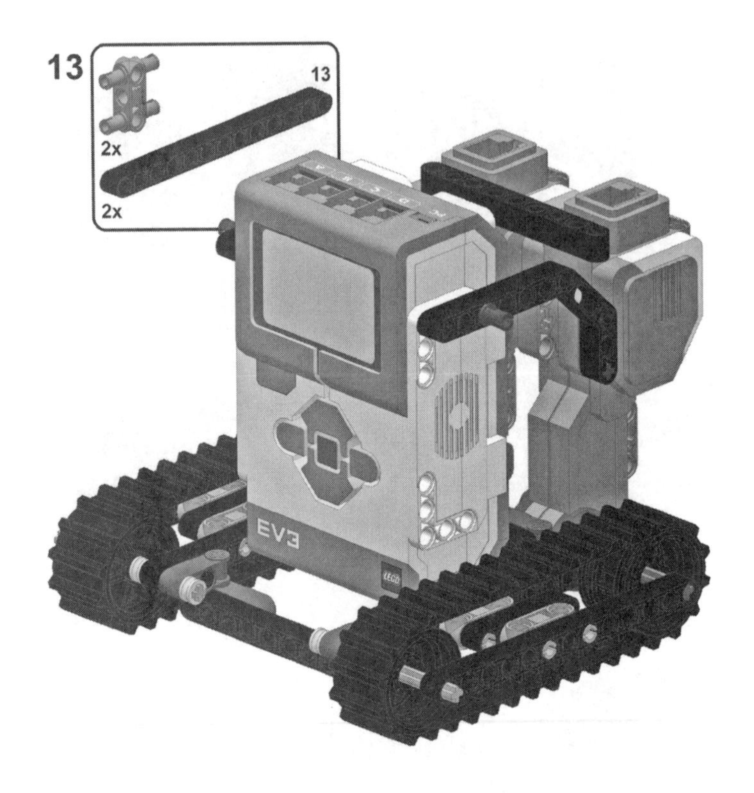
13
2x
13
2x
EV3

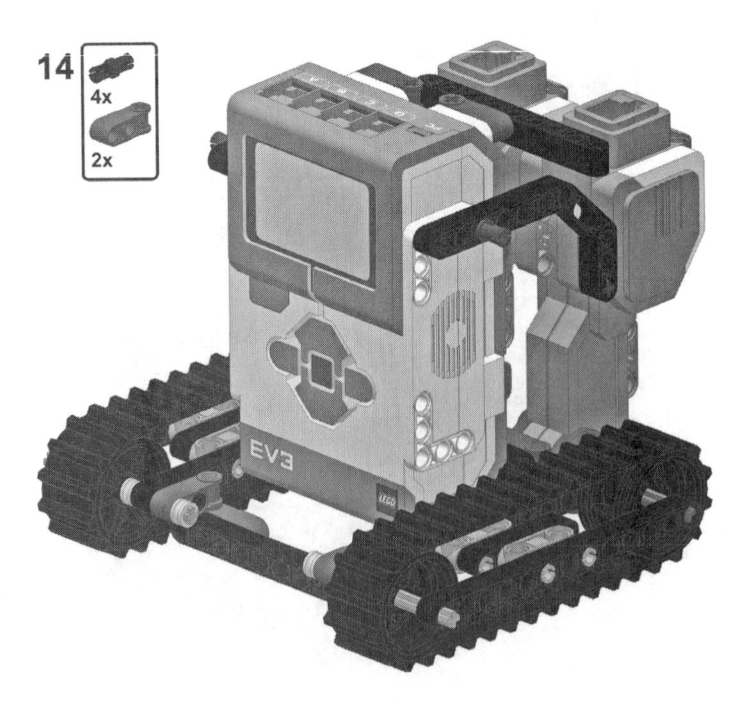
14
4x
2x
EV3

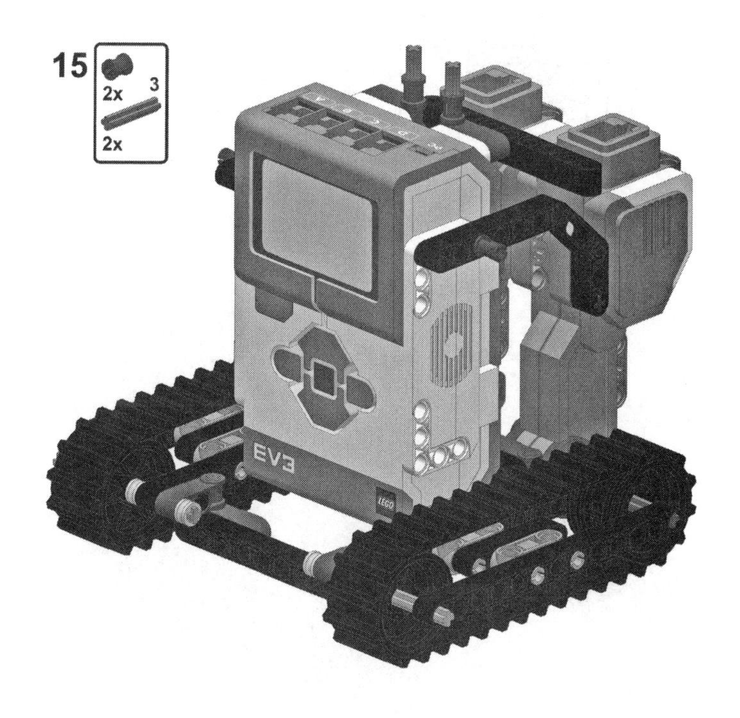
15
2x
3
2x

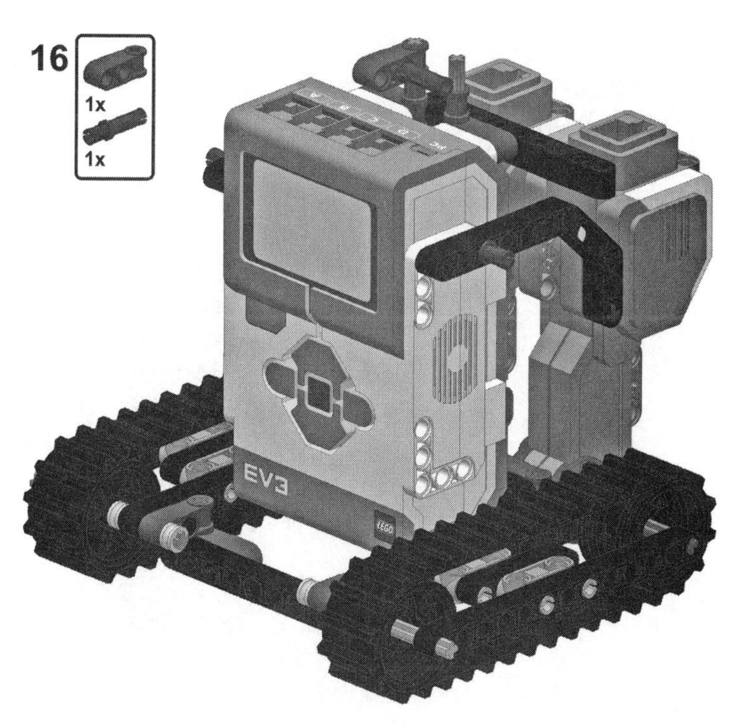
16
1x
1x

17
1x
1x
3
1x
EV3

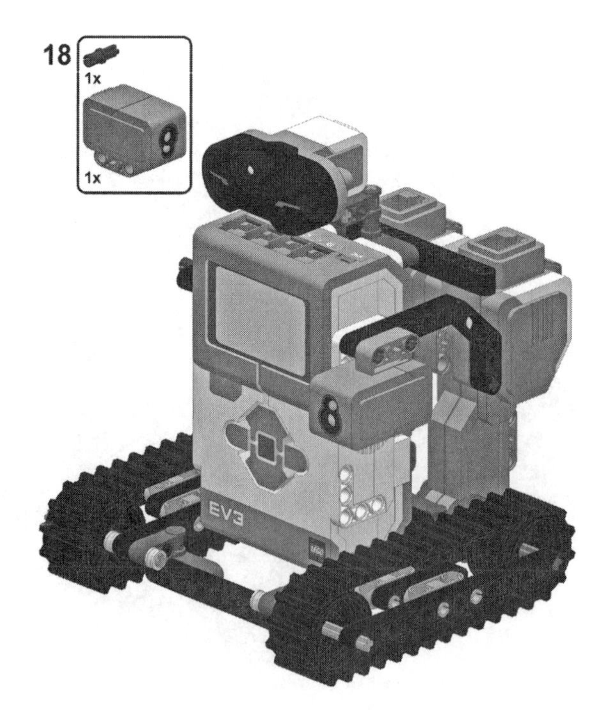
18
1x
1x
EV3

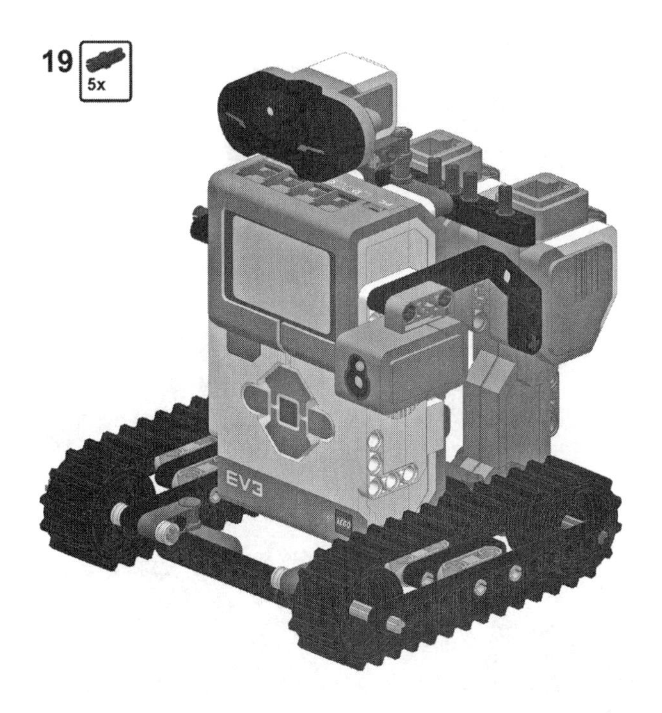
19
5x

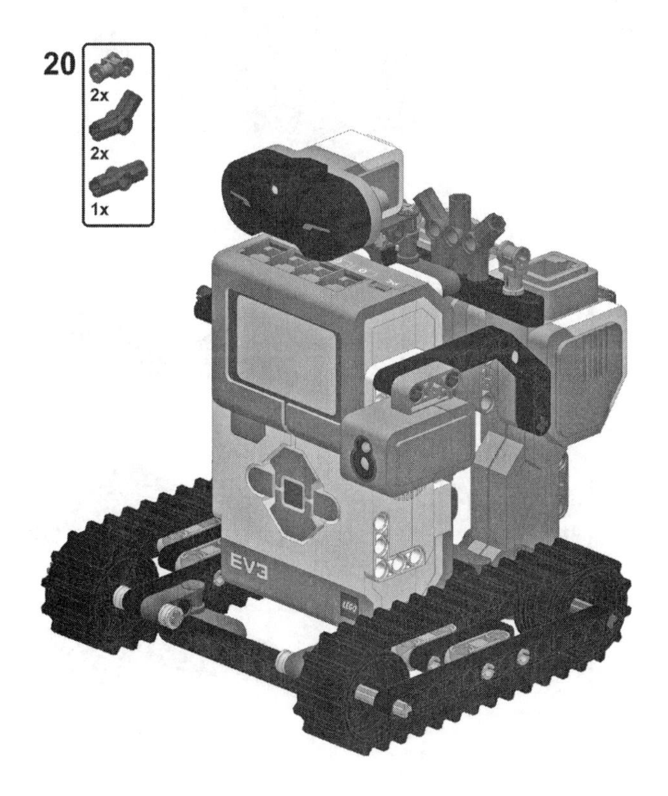
20
2x
2x
1x

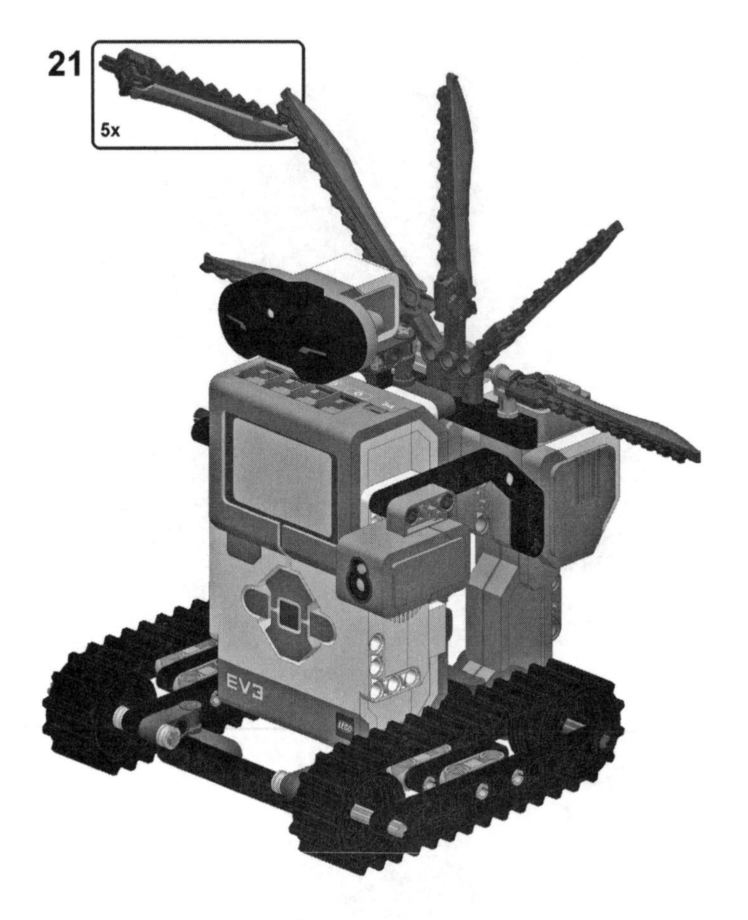
21
5x
EV3

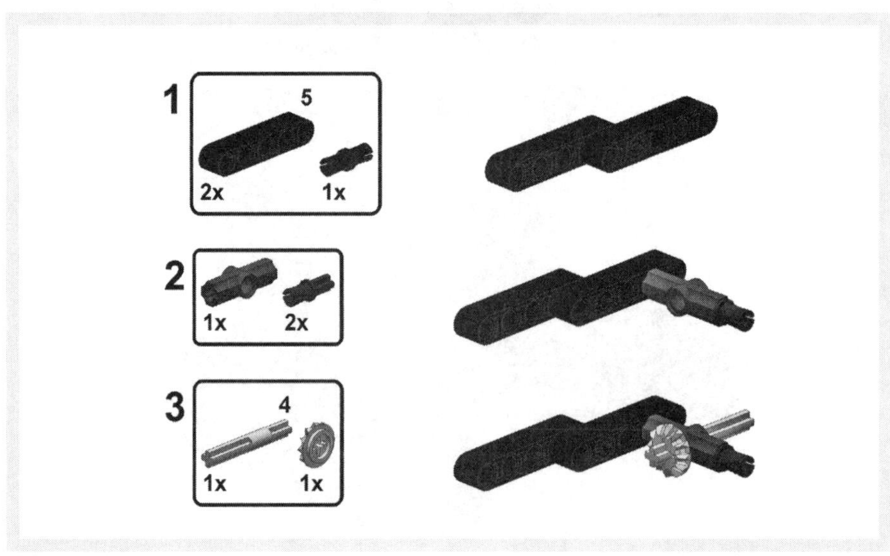
1
5
2x
1x
2
1x
2x
3
4
1x
1x

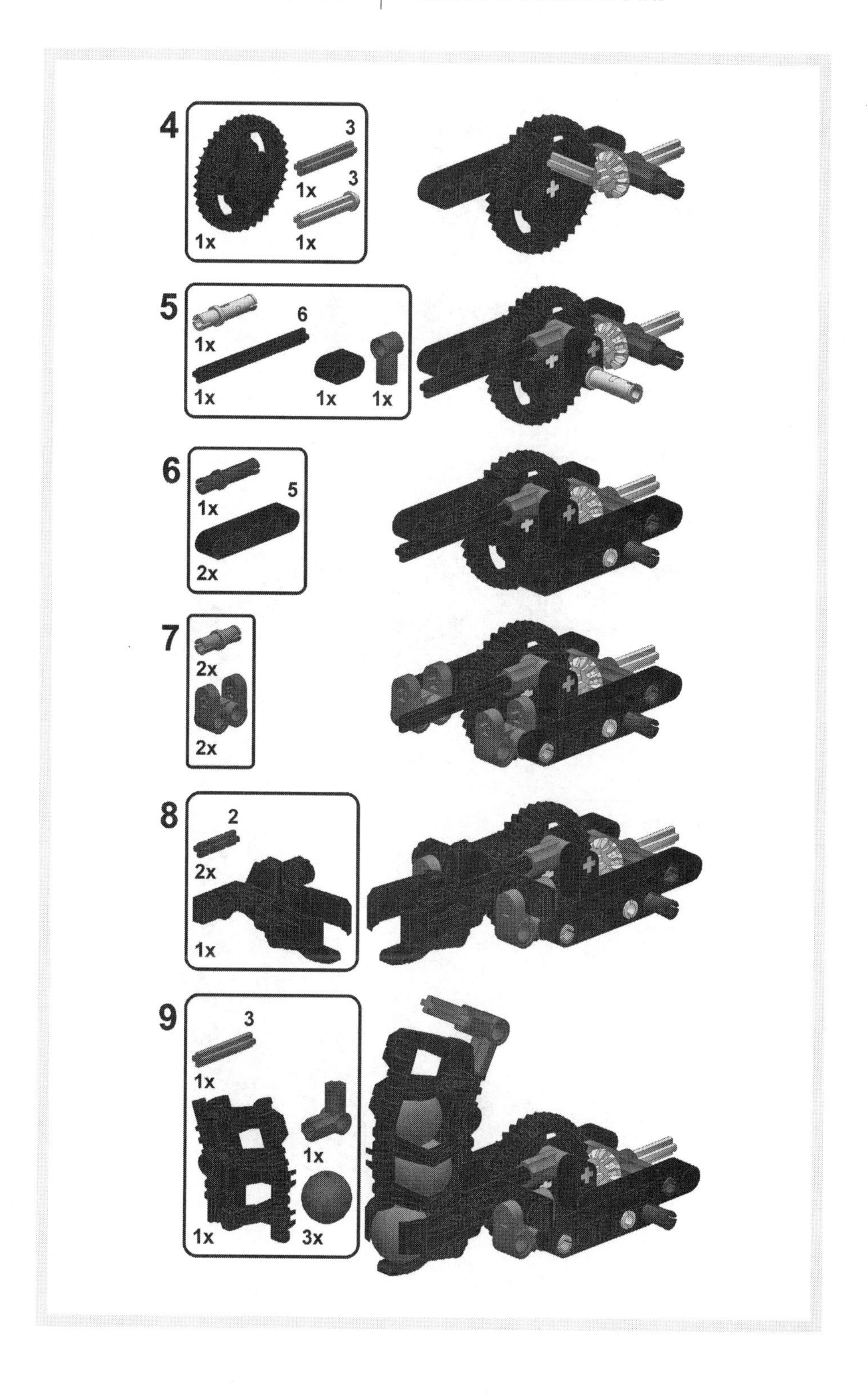
4
3
1x
3
1x
1x
5
1x
6
1x
1x
1x
6
1x
5
2x
7
2x
2x
8
2
2x
1x
9
3
1x
1x
1x
3x

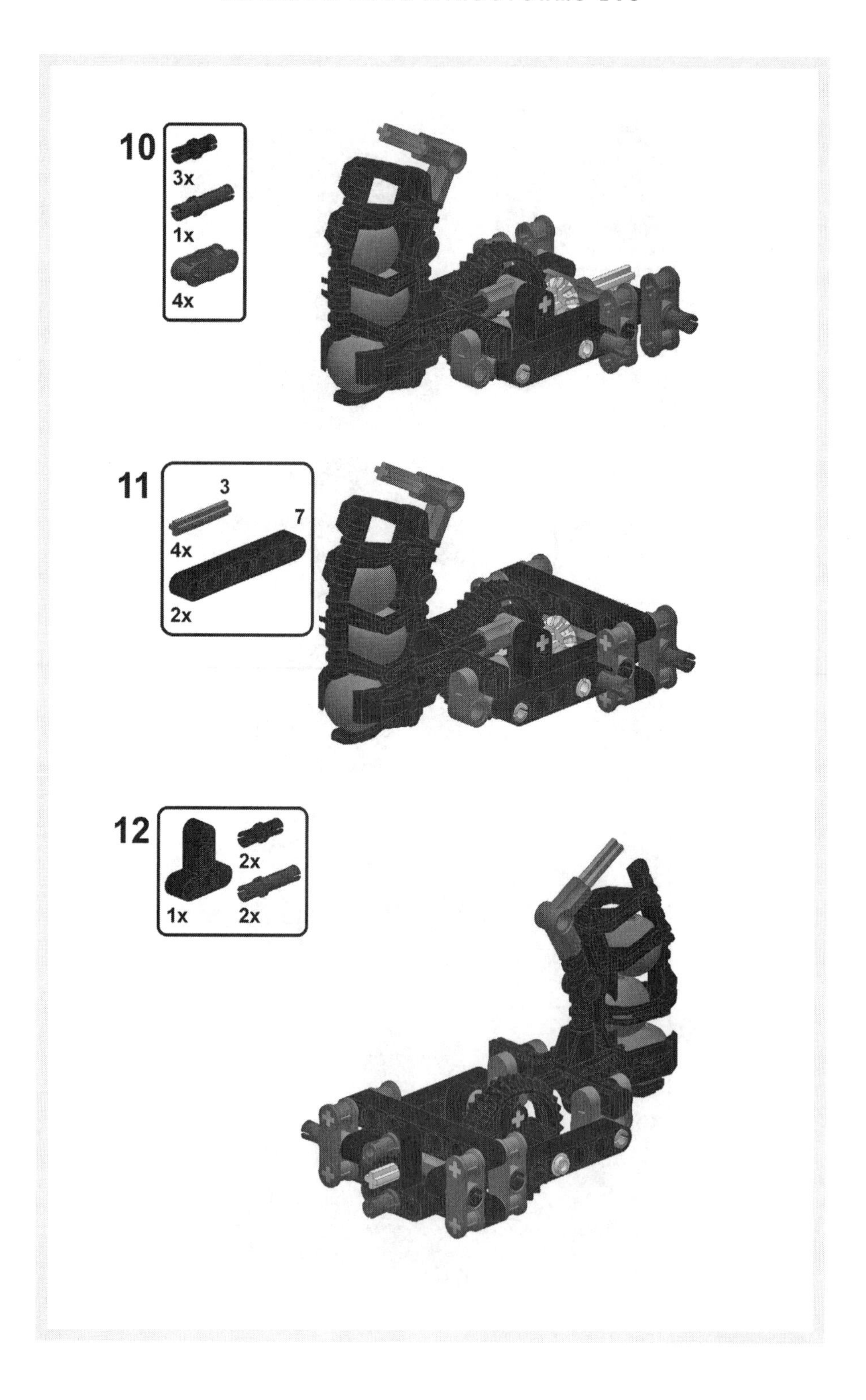
10
3x
1x
4x
11
3
7
4x
2x
12
2x
1x
2x

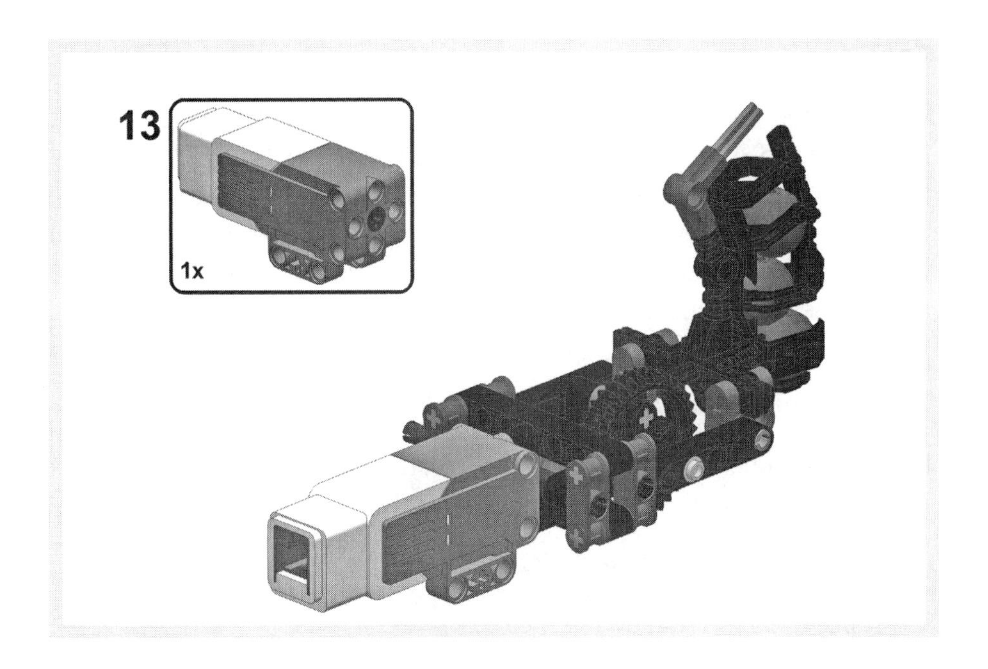
13
1x

22
EV3

Cables

You may need to temporarily take off Warbird's head to insert the motor cables. Remove the head, then insert a short cable from port B to the right large motor. Insert another short cable in port D to the left motor. Finally, insert a medium cable from port A to the medium motor. Connect a medium cable to port 3, but wind the cable forward in front of the EV3 brick, then up and around back to the color sensor. Connect another medium cable from port 4 to the IR sensor.

Programming Warbird

Let's code a program to fire a shot from Warbird's cannon every time a red color is held up in front of the color sensor.

```
import lejos.hardware.*;
import lejos.hardware.motor.*;
import lejos.hardware.port.*;
import lejos.hardware.sensor.*;
import lejos.robotics.*;
import lejos.utility.Delay;

public class ColorAttack implements KeyListener {

  public static boolean keepGoing = true;

  public static void main(String[] args) {
    Button.ESCAPE.addKeyListener(new ColorAttack());

    EV3MediumRegulatedMotor m = new EV3Medium
RegulatedMotor(MotorPort.A);

    EV3ColorSensor cs = new
EV3ColorSensor(SensorPort.S3);
    ColorAdapter ca = new ColorAdapter(cs);
    while(keepGoing) {
      Color c = ca.getColor();
      // calculate red percentage:
      double total = c.getRed() + c.getBlue() +
c.getGreen();
      double red_ratio = c.getRed() / total;

      if(red_ratio > 0.33) {
        Sound.playTone(1000, 100);
        m.rotate((36/12)*-360, false);
```

```
      } else
        Sound.playTone(500, 100);
      Delay.msDelay(1000);
    }
    cs.close();
    m.close();

  }

  public void keyPressed(Key k) {}

  public void keyReleased(Key k) {
    keepGoing = false;
  }
}
```

The code basically instantiates a ColorAdapter as shown earlier in the chapter. It then gets a color reading every second and uses all three RGB values to determine the red ratio in the color mix. If the red ratio is the highest of the three colors, then it launches a ball. There is also a key listener for the escape key in order to exit the program.

Using Warbird

Upload the code and run the program. You should hear a beep at one second intervals. A higher frequency beep indicates it has detected red. Try placing a red card about 5 cm in front of the color sensor and it should launch a ball. You can also hold up your hand in front of the sensor, since it is reddish. Now try holding a blue of green card in front of the sensor, and nothing should happen.

> **Tip!**
>
> You can adjust the sensitivity to red by playing with the line if(red_ratio > 0.33). Simply raise the value from 0.33 to 0.5 and it will be less sensitive to the color red (in other words, more discriminating of red objects).

CHAPTER 17

IR Sensor and Beacon

TOPICS IN THIS CHAPTER

- Remote Control Tool
- Remote Commands
- Proximity Detection
- IR Beacon

One of the most interesting developments of the EV3 kit is the addition of an IR sensor and IR beacon (see Figure 17-1). Unlike other components of the EV3 kit, these two devices are tightly bound to each other. Together, they allow three unique functions: detecting proximity to other objects, remote control, and seeking an IR beacon. This chapter shows examples for all three uses.

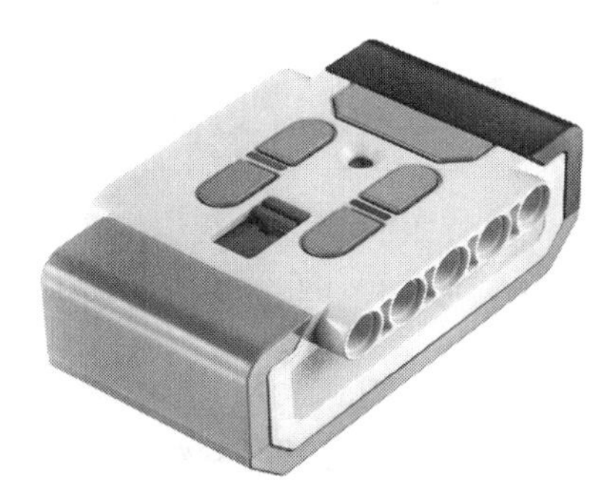

Figure 17-1: Discovering the amazing IR Beacon

Remote Control (without Code)

Let's examine one of the best parts of the EV3 kit, the ability to remote control a robot without any additional hardware. The EV3 remote control has five buttons and a four-way switch. In this section, we will concern ourselves with the four up-down buttons and the four-way switch.

You can use the IR beacon right now by using it as a remote control from the leJOS menu. You will need to connect some motors to the EV3 in order for it to do anything. If you still have Warbird assembled from the last chapter, it has motors connected to ports B and D. It also has an IR sensor connected to port 4.

1. From the menu select Tools (Figure 17-2), then Remote Control.

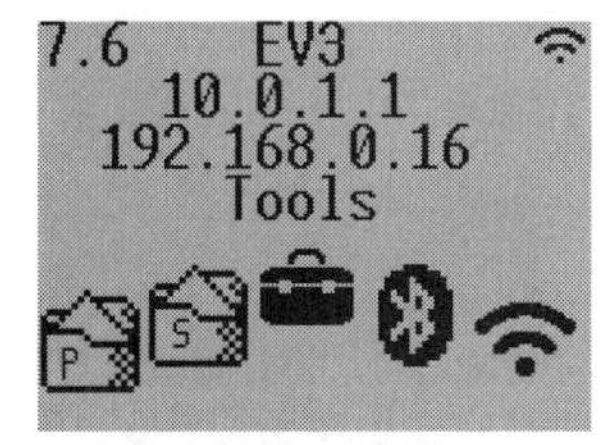

Figure 17-2: Selecting the Tools menu from the leJOS main menu

2. The Remote Control tool will automatically detect the port of the IR sensor, but it might take a moment. Once the main screen appears, it indicates which buttons on the IR beacon perform which actions (see Figure 17-3).

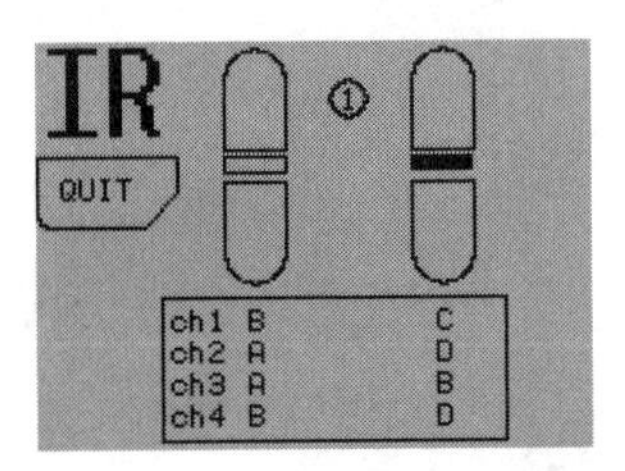

Figure 17-3: Controlling motors with the Remote Control menu tool

3. Switch the physical slider on the IR beacon to channel 4. This will allow you to control motors B and D. Now press the up and down buttons to drive Warbird around.

4. If you want to control the ball launcher, you can switch to channel 2 and hold down the left buttons to activate motor A.

IR Remote Commands

Now let's see how to code something that interacts with the EV3 IR beacon commands. This section will show a very basic example of how to receive button presses from the IR beacon. It doesn't do anything fancy, but it will allow you to build on the code to control your own robots.

The four-way switch selects which channel you want to use. Each channel effectively gives you four more buttons to use to control a robot. You can also hold down buttons at the same time to produce more commands. The code sample below shows how to figure out which channel the command came from.

The commands can be read from the EV3IRSensor class. To read the command for a particular channel, you can use the getRemoteCommand(int chan) method. The channel ranges from 0 to 3 and are visible in the circular opening on the face of the IR beacon.

```
import lejos.hardware.lcd.LCD;
import lejos.hardware.port.SensorPort;
import lejos.hardware.sensor.EV3IRSensor;
import lejos.utility.Delay;
```

```
public class RemoteControl {

  public static void main(String [] agrs) {
    EV3IRSensor ir = new EV3IRSensor(SensorPort.S4);
    boolean keep_looping = true;
    int channel = 0;

    while(keep_looping) {
      Delay.msDelay(25);

      // Get the IR commands
      byte [] cmds = new byte[4];
      ir.getRemoteCommands(cmds, 0, cmds.length);

      // Figure out which channel is active:
      int command = 0;
      for(int i=0;i<4;i++) {
        if(cmds[i] > 0) {
          channel = i+1;
          command = cmds[i];
        }
      }

      LCD.drawString("COM:" + command + "  ", 0, 2);
      LCD.drawString("CHAN:" + channel, 0, 4);

      if(command == 8) keep_looping = false;
    }
    ir.close();
  }
}
```

As you can see, you can press all four buttons and recognize which channel is selected on the remote. You can also hold two buttons down at the same time to produce even more commands. This allows up to 11 combinations of buttons. In order to exit the program, press both down buttons at the same time.

IR Proximity Detection

As you've probably gathered already, the IR sensor is also able to detect distance to objects. In fact, that's probably the biggest single feature of the IR sensor, considering how useful that is for robotics. In this section we will examine the simplest method possible for detecting distances.

The easiest way to get range (distance) readings from the IR sensor is to use an adapter. In this case, we will select the RangeFinderAdapter located in the lejos.robotics package. This class has a simple method for retrieving range values called getRange(). Let's look at some code.

```
import lejos.hardware.Button;
import lejos.hardware.lcd.LCD;
import lejos.hardware.port.SensorPort;
import lejos.hardware.sensor.*;
import lejos.robotics.*;
import lejos.utility.Delay;

public class Proximity {

  public static void main(String [] agrs) {
    EV3IRSensor ir = new EV3IRSensor(SensorPort.S4);
    SensorMode distMode = ir.getMode("Distance");
    RangeFinderAdapter ranger = new
    RangeFinderAdapter(distMode);
    boolean keep_looping = true;

    while(keep_looping) {
     Delay.msDelay(50);

     LCD.drawString("D: " + ranger.getRange() + "
cm ", 0, 2);

     if(Button.ESCAPE.isDown()) keep_
looping = false;
    }
    ir.close();
  }
}
```

The code above is quite simple. First it creates an EV3IRSensor object. It then gets a SensorMode, which is what the aforementioned RangeFinderAdapter requires in the next line. The code then loops endlessly outputting the range to the LCD screen. You can press the escape button to end the program.

NOTE: *The display will read NaN when there is nothing detected by the IR sensor. (This stands for* Not a Number.*)*

Remote Infrared Beacon

Of the five buttons on the remote control, four of them are simple up-down buttons, but the fifth button near the top puts the IR remote into seek mode. In this mode, the remote control acts like a beacon by sending out a continuous stream of IR pings. This allows the IR sensor to detect the range and bearing to the beacon (see Figure 17-4).

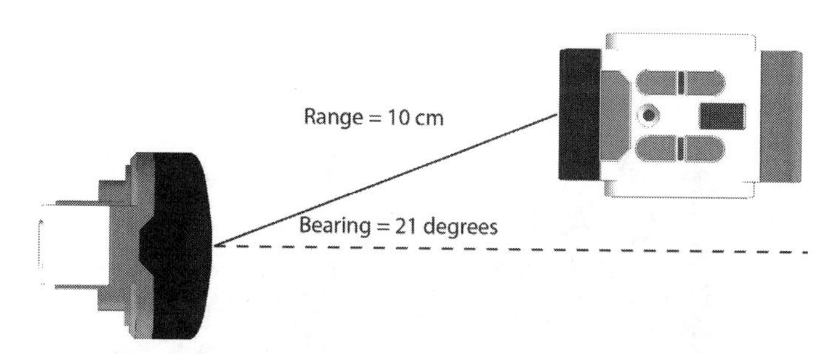

Figure 17-4: Identifying the range and bearing to the beacon.

The IR sensor can detect up to four different IR remotes (when they are in seek mode), as long as they are tuned to different channels. For example, if you have two IR remotes, you could place one on channel 1 and the other on channel 2 to allow the EV3 to distinguish between them.

Range values are reported in centimeters, while bearing (direction) is in degrees. The beacon is only visible to the IR sensor within a 50 degree range (see Figure 17-5). This means the values will generally range between -25 and 25 degrees.

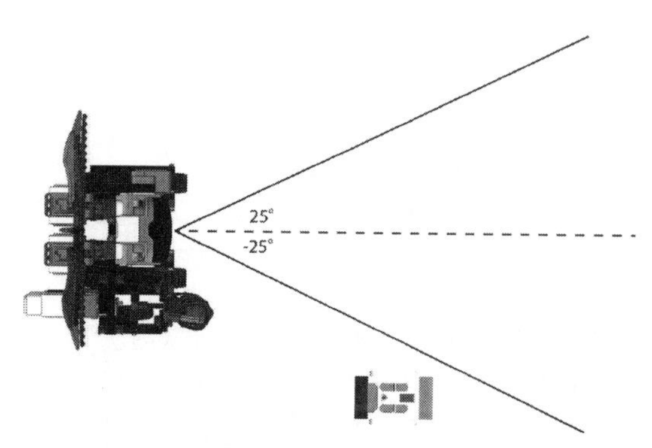

Figure 17-5: Failing to acquire the beacon when out of visual range

Let's try a practical example using some functional code. For the following example, you will need the Warbird (see Figure 17-6) from Chapter 16. The purpose of this project is to allow the robot to follow the beacon. When it senses it is within 25 cm of the beacon it will stop and fire its shoulder cannon. We also want it to be able to chase the beacon if it is within range. If it ever loses visual range of the IR beacon it will attempt to reestablish it by rotating in a circle.

Figure 17-6: Admiring Warbird from a respectful distance

```
import lejos.hardware.Button;
import lejos.hardware.lcd.LCD;
import lejos.hardware.motor.*;
import lejos.hardware.port.*;
import lejos.hardware.sensor.*;
import lejos.utility.Delay;

public class Seeker {

  static EV3IRSensor ir = new
EV3IRSensor(SensorPort.S4);
  static EV3MediumRegulatedMotor a = new EV3Medium
RegulatedMotor(MotorPort.A);
```

```
  static EV3LargeRegulatedMotor b = new EV3Large
RegulatedMotor(MotorPort.B);
  static EV3LargeRegulatedMotor d = new EV3Large
RegulatedMotor(MotorPort.D);

  public static void main(String [] agrs) {
    boolean keep_looping = true;
    boolean fired = false;

    SensorMode seek = ir.getMode("Seek");
    int speed = 100;
    float [] vals = new float[8];

    while(keep_looping) {
      Delay.msDelay(50);

      seek.fetchSample(vals, 0);
      int bearing = (int)vals[0];
      int range = (int)vals[1];
      LCD.drawString("B: " + bearing + " R: " +
range + "cm ", 0, 2);

      if(range==0&(bearing==128|bearing==0)) {
        // Acquire IR beacon:
        b.setSpeed(speed);
        d.setSpeed(speed);
        b.forward();
        d.backward();
        fired = false;
      } else if(range<25){
        // Firing range!
        b.flt();
        d.flt();
        if(!fired) {
          a.rotate((36/12)*-360);
          fired = true;
        }
      } else if(bearing<0) {
        // veer left B faster
        b.setSpeed(2*speed);
        d.setSpeed(speed);
        b.backward();
        d.backward();
        fired = false;
      } else if(bearing>0){
        // veer right D faster
```

```
          b.setSpeed(speed);
          d.setSpeed(2*speed);
          b.backward();
          d.backward();
          fired = false;
        } else {
          // Forward march!
          b.setSpeed(2*speed);
          d.setSpeed(2*speed);
          b.backward();
          d.backward();
          fired = false;
        }

        if(Button.ESCAPE.isDown()) keep_
looping = false;
      }
      ir.close();
      a.close();
      b.close();
      d.close();
    }
}
```

Using the IR Seeker

Make sure the IR beacon is switched to channel 1, then begin the program. It can detect the beacon up to about 100 cm away. Place the beacon on the floor, then hit the seek mode button. You can tell it is in seek mode when the green LED is lit on the beacon. Now watch the robot home in on the beacon and fire when it is within 25 cm. It will only fire once each time it is in firing range. You can then move the beacon around to somewhere else and it will follow it and fire again when it is within range.

CHAPTER 18

Behavior-Based Robotics

TOPICS IN THIS CHAPTER

- Rodney Brooks' Behavior-Based Robotics
- Subsumption Architecture
- The leJOS Behavior API

Behavior-Based Robotics is a branch of programming that uses very little memory and produces insect-level intelligence. The EV3 brick has more than enough memory to add interesting behaviors like navigating towards a beacon, avoiding objects, finding objects, and moving objects. The more behaviors added to your robot, the more interesting it becomes. Let's examine Behavior-Based Robotics.

Behavior-Based Robotics

Behavior-Based Robotics was first defined by Professor Rodney Brooks while at the MIT Artificial Intelligence Laboratory. The strategy of Behavior-Based Robotics is different than most AI programming styles developed before it. The traditional models tend to rely on large data models of the world. For this reason, they can also be slow to react to changes in the environment. Rodney Brooks took his strategy from the insect world. He noticed that insects are able to perform in the real world with excellent success, despite having very little in the way of memory or intelligence.

If we were to compare an insect to a computer, we would conclude that the insect possessed only a small amount of working memory. After all, it has been shown that insects do not remember things from the past, and can not be trained with Pavlovian methods. So it appears insects rely on a strategy of many simple behaviors that, when alternated with one another, appear as complex behavior. These strategies are effectively hard-wired into the insect and are not learned in the way mammals learn their behavior.

Out of this strategy came the practical implementation of Behavior-Based Robotics, called *subsumption architecture*. This architecture decomposes complex behaviors into several smaller, simple behaviors.

There are essentially two discrete structures that build up a behavior: sensors (inputs) and actuators (outputs). This is true of any organism, not just robots. For example, a mosquito monitors the environment using antennae and compound eyes (sensors), and then reacts using its muscular system. Similarly, if certain sensors are stimulated in a robot, it triggers a reaction by the motors. These pairs of conditions and actions are called behaviors (Figure 18-1).

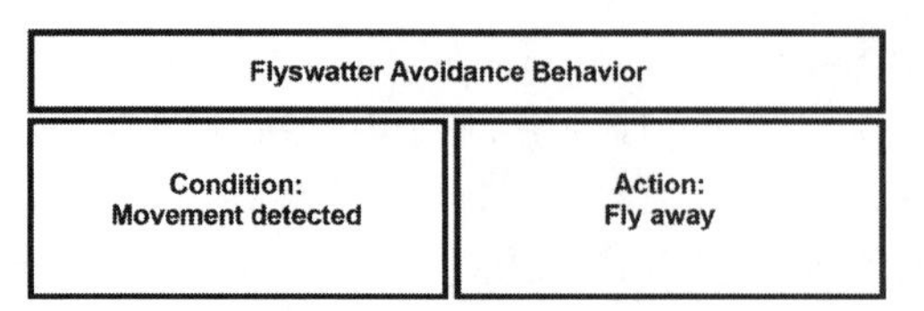

Figure 18-1 Diagram for a Mosquito behavior.

Behaviors

Behavior actions can be very simple, such as moving forward, or they can be complex, such as mapping the data in a room. A behavior is just a program the robot (or organism) follows for a period of time. For example, when you want to communicate with someone and are writing an e-mail, you are in letter writing mode, but when your stomach starts to grumble you go to the kitchen and switch over to eating mode. By building up several stimulus/behavior pairs you can theoretically achieve a complex logical model.

Sensors are not restricted to determining when to switch from one mode to another. In fact, they can be used within the action portion of the code as well. If the behavior is to map the boundaries of a room using an ultrasonic sensor, then the action method uses this sensor.

Let's look at a specific example. A firefighting robot could share a single temperature sensor to control more than one reaction. The robot actively searches for areas with greater heat to help it locate fire. If it moves forward and the temperature sensor indicates it is warmer, then it tries another step in that direction. If it gets cooler, then it tries another direction. But assuming there is actually a fire, there will come a time when the temperature gets so high that the robot would melt if it got any closer. To counter this, we could use the same sensor with a higher level behavior to back off if the temperature gets close to the melting point of plastic (Table 18-1).

Behavior	Condition	Action	Priority
Seek heat	Temperature hotter in another area.	Move towards heat.	low
Extinguish flames	Ultraviolet light sensor detects flame	Spray CO2	medium
Avoid flames	Temperature > 160 degrees	Move away from heat	highest

Table 18-1: Firefighting Robot Behavior

A behavior takes over when a condition becomes true. Sometimes a behavior does not have to depend on an external sensor to become true. For example, the robot could monitor an internal condition such as time. When two minutes have elapsed, the condition becomes true. In a way, the timer is a time sensor. Other factors, such as counting the number of bricks a robot has picked up, can also be monitored. Once a robot has collected 10 bricks, a higher level behavior of seeking home and dumping the bricks could become active. Any condition, internal or external to the robot, can activate a behavior.

This doesn't have to be for practical behaviors only. I have seen robots that switch from one emotional mode to another. For example, there is an R2-D2 type of robot that switches from happy mode to bored mode, and sad mode. The bored mode kicks in if the robot has been wandering around for a while without encountering a stimulus. Each of the emotional states is dependent upon what the robot senses in its surroundings, which is really not that different from complex organisms such as humans.

Managing Priorities

There are times when more than one behavior could become activated. In order for a robot to determine which mode gains control, the different behaviors need to have priorities. For example, an animal has several basic goals, such as eating, mating, defending itself, exploring, and protecting its young. Some of these behaviors are more important than others, but all are necessary for the overall survival of the species.

One of the most important behaviors is eating, since starvation prevents an organism from completing any of the other important goals. So we would say eating is a high level priority for an animal. But what if another ferocious animal is attacking it while it is eating? Defending itself (attack mode/retreat mode) would be the highest level behavior for an animal in this situation. Likewise a robot is useless without power, so when the battery level is too low it could seek the recharging station. It is important that the code interrupts current action if a higher level action needs to take over. When this happens, we say that the lower priority behavior has been *suppressed*.

Rodney Brooks developed a standard diagram for representing hierarchies of behaviors (Figure 18-2). A typical diagram shows the behavior, and the order of priorities. When a behavior suppresses other behaviors it is indicated with an S. The rule for Behavior-Based Robotics is that all lower level behaviors are suppressed when a higher level behavior takes over. This means only one behavior can be running at any given time. It might seem that this limits a robot, but most organisms really do only one thing at a time. In order to do more than one behavior at a time, a single behavior must include more than one function within it.

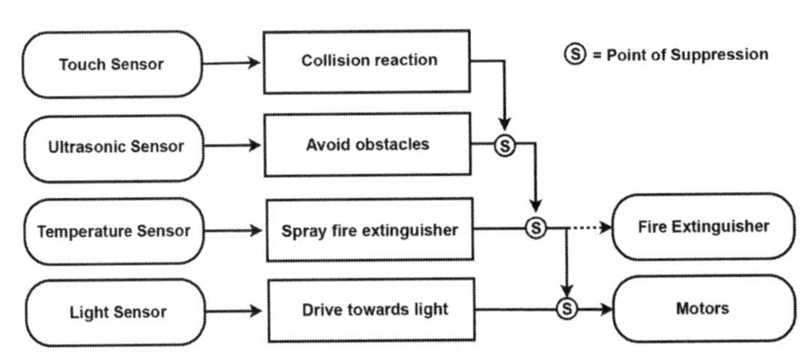

Figure 18-2: Hierarchy of firefighting behavior.

The Subsumption API

Many new programmers, when they begin programming robots, think of program flow as a series of if-then statements, which is reminiscent of structured programming. This type of programming is very easy to start but can end up as spaghetti code; all tangled up and difficult to expand.

The Behavior-Based Robotics model, in contrast, requires a little more planning before coding begins, but the payoff is that each behavior is nicely encapsulated within an easily-understood structure. This will theoretically make your code easier for other programmers to understand, and more importantly, make it very easy to add or remove specific behaviors. Let's examine how to do this in leJOS.

The Behavior API is very simple and is composed of only one interface and one class. The Behavior interface is used to define behaviors. It is very general, since the individual implementations of behaviors vary widely. Once all the Behaviors are defined, they are given to an Arbitrator to regulate. All classes and interfaces for behavior robotics are located in the lejos.robotics.subsumption package. The API for the Behavior interface is as follows.

lejos.robotics.subsumption.Behavior

- `boolean takeControl()`

  Returns a boolean value to indicate if this behavior should become active. For example, if a touch sensor indicates the robot has bumped into an object, this method should return true.

- `void action()`

  The code in this method initiates an action when the behavior becomes active. For example, if takeControl() detects that the robot has collided with an object, the action() code could make the robot back up and turn away from the object.

- `void suppress()`

  The code in the suppress() method should immediately terminate the code running in the action() method. The suppress() method can also be used to update any data before this behavior completes.

As you can see, the three methods in the Behavior interface are quite simple. If a robot has two discreet behaviors, then the programmer needs to create two classes, with each class implementing the Behavior interface. Once these classes are complete, the code should hand the Behavior objects to the Arbitrator.

lejos.robotics.subsumption.Arbitrator

- `public Arbitrator(Behavior [] behaviors)`

  Creates an Arbitrator object that regulates when each of the behaviors will become active. The higher the index array number for a Behavior, the higher the priority level.

Parameters: behaviors An array of Behaviors.

- `public void start()`

  Starts the arbitration system.

The Arbitrator class is even easier to understand than Behavior. When an Arbitrator object is instantiated, it is given an array of Behavior objects. Once it has these, the start() method is called and it begins arbitrating–deciding which behaviors should become active. The Arbitrator calls the takeControl() method on each Behavior object, starting with the object with the highest index number in the array.

It works its way through each of the behavior objects until it encounters a behavior that wants to take control. When it encounters one, it executes the action() method of that behavior once and only once. If two behaviors both want to take control, then only the higher level behavior will be allowed (Figure 18-3).

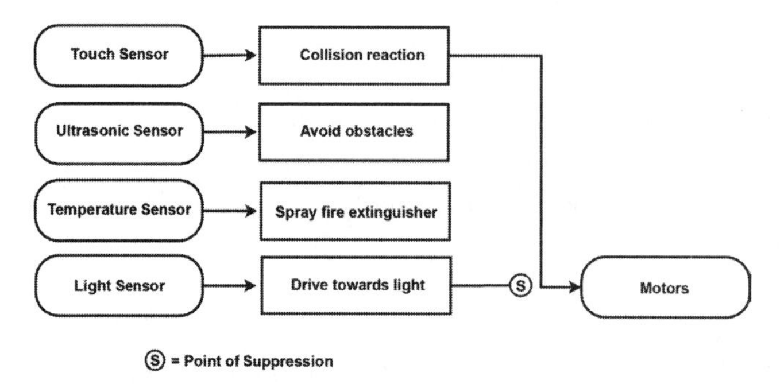

Figure 18-3: Higher level behaviors suppress lower level behaviors.

Programming Behavior-Based Robotics

Now that you are familiar with the Behavior API under leJOS, let's look at a simple example using three behaviors. We'll use Warbird from the previous two chapters to demonstrate the subsumption architecture. For this example, we will program some behavior for Warbird. We'll give it three simple behaviors:

1. Drive forward.
2. If too close to an object, find a new direction.
3. Allow user to take control with IR remote.

The behavior hierarchy is displayed in Figure 18-4.

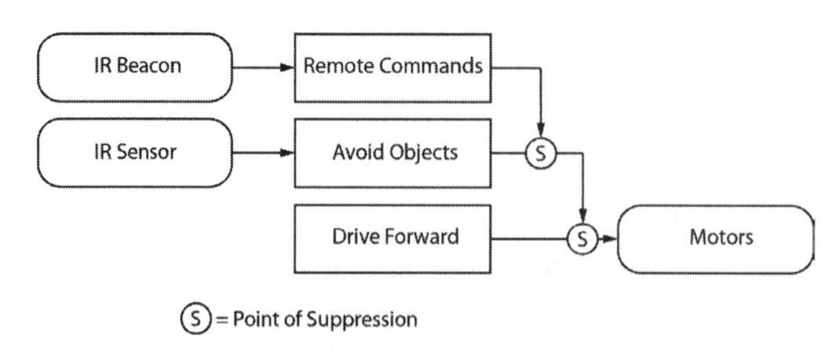

Figure 18-4: Examining Warbird's behavior

Let's start with the first behavior. As we saw in the Behavior interface, we must implement the methods action(), suppress(), and takeControl(). The behavior for driving forward will take place in the action() method. It simply needs to make the motors rotate. Next, the suppress() method will need to stop this action when it is called.

```
import lejos.robotics.*;
import lejos.robotics.subsumption.*;

public class BehaviorForward implements Behavior {

  RegulatedMotor left;
  RegulatedMotor right;

  public BehaviorForward(RegulatedMotor left,
RegulatedMotor right) {
    this.left = left;
    this.right = right;
  }

  public boolean takeControl() {
    return true;
  }

  public void action() {
    left.backward();
    right.backward();
  }

  public void suppress() {
    // nothing to suppress
  }
}
```

That's all it takes to define our first Behavior to drive the robot forward. This robot will always drive forward, unless something suppresses the action, so this Behavior should always want to take control (it's a bit of a control freak). The takeControl() method should return true, no matter what is happening. This may seem counter intuitive, but rest assured that higher level behaviors will be able to cut in on this behavior when the need arises.

Also, suppress() is empty because as soon as the motors start rotating in action(), the method returns. Therefore it doesn't need suppress code to stop the action() method.

The second behavior is a little more complicated than the first, but similar. The main action of this behavior is to reverse and turn when the robot comes within 40 centimeters of an object. The complete listing for this behavior is as follows:

```
import lejos.robotics.RegulatedMotor;
import lejos.robotics.subsumption.Behavior;

public class BehaviorProximity implements
Behavior {

  RegulatedMotor left;
  RegulatedMotor right;
  SharedIRSensor ir;
  boolean backing_up = false;

  public BehaviorProximity(RegulatedMotor left,
RegulatedMotor right, SharedIRSensor ir) {
    this.left = left;
    this.right = right;
    this.ir = ir;
  }

  public boolean takeControl() {
    return (ir.distance < 40);
  }

  public void action() {
    backing_up = true;

    left.rotate(600, true);
    right.rotate(600);

    left.rotate(-450, true);
    right.rotate(450);

    backing_up = false;
  }

  public void suppress() {
    // Wait until backup done
    while(backing_up) {Thread.yield();}
  }
}
```

The code in the action() method here only takes a moment to execute. Therefore, if suppress() is called, it merely waits for the action() code to complete.

Now that we have two of the behaviors defined, we need to create the main method using an Arbitrator. There are actually two separate classes in this file:

```
import lejos.robotics.*;
import lejos.robotics.subsumption.*;
import lejos.hardware.lcd.LCD;
import lejos.hardware.motor.*;
import lejos.hardware.port.*;
import lejos.hardware.sensor.*;

public class BehaviorMain {

  static Arbitrator arby;

  public static void main(String[] args) {
    RegulatedMotor left = new EV3LargeRegulatedMotor
(MotorPort.B);
    RegulatedMotor right = new EV3LargeRegulatedMotor
(MotorPort.D);
    SharedIRSensor ir = new SharedIRSensor();

    Behavior b1 = new BehaviorForward(left, right);
    Behavior b2 = new BehaviorProximity(left,
right, ir);
    Behavior [] behave = {b1, b2};
    arby = new Arbitrator(behave);
    arby.start();
  }
}

class SharedIRSensor extends Thread {

  EV3IRSensor ir = new EV3IRSensor(SensorPort.S4);
  SampleProvider sp = ir.getDistanceMode();
  public int control = 0;
  public int distance = 255;

  SharedIRSensor() {
    this.setDaemon(true);
    this.start();
  }

  public void run() {
    while (true) {
      float [] sample = new float[sp.sampleSize()];
      control = ir.getRemoteCommand(0);
      sp.fetchSample(sample, 0);
```

```
        if((int)sample[0] == 0)
          distance = 255;
        else
          distance = (int)sample[0];
        LCD.drawString("Control: " + control, 0, 0);
        LCD.drawString("Distance: " + distance +
" ", 0, 1);
        Thread.yield();
      }
    }
  }
```

The second class allows us to share the IR sensor among several different classes without any conflicts. It also outputs the control command and distances to the LCD screen continuously.

The code in BehaviorMain is easy to understand. The first few lines in the main() method create the motors and ir sensor objects. The next lines instantiate the Behaviors. Note that the Behaviors share the motors. This is perfectly legal in subsumption architecture. The next line places the Behaviors into an array, with the lowest priority Behavior taking the lowest array index. The next line creates the Arbitrator, and the final line starts the Arbitration process.

Give it a try!

NOTE: *To end this program, press the square enter button and the down button at the same time. Next, we will add a behavior that allows you to end the program.*

Adding a Third Behavior

This is a lot of work for two simple behaviors, but now let's see how easy it is to insert a third behavior without altering any code in the other classes. This is the part that makes subsumption architecture very appealing for robotics programming.

The third behavior is more complicated because it will monitor the IR remote for input. However, when the user is done with controlling the robot it will wait 3 seconds before relinquishing control in case the user wants to continue controlling:

```
import lejos.robotics.*;
import lejos.robotics.subsumption.*;

public class BehaviorRemote implements Behavior {

  RegulatedMotor left;
  RegulatedMotor right;
  SharedIRSensor ir;
  boolean _continue = false;

  public BehaviorRemote(RegulatedMotor left,
RegulatedMotor right, SharedIRSensor ir) {
  this.left = left;
  this.right = right;
  this.ir = ir;
 }

 public boolean takeControl() {
  return (ir.control != 0);
 }

 public void action() {

  _continue = true;
  long old_time = 0;

  while (_continue) {

   switch(ir.control) {
   case 0:
    left.stop();
    right.stop();
    break;
   case 1:
    left.backward();
    old_time = System.currentTimeMillis();
    break;
   case 2:
    left.forward();
    old_time = System.currentTimeMillis();
    break;
   case 3:
    right.backward();
    old_time = System.currentTimeMillis();
    break;
   case 4:
    right.forward();
```

```
        old_time = System.currentTimeMillis();
        break;
      case 5:
        right.backward();
        left.backward();
        old_time = System.currentTimeMillis();
        break;
      case 6:
        right.forward();
        left.backward();
        old_time = System.currentTimeMillis();
        break;
      case 7:
        left.forward();
        right.backward();
        old_time = System.currentTimeMillis();
        break;
      case 8:
        right.forward();
        left.forward();
        old_time = System.currentTimeMillis();
        break;
      case 9:
        System.exit(1);
      }

  if(System.currentTimeMillis() - old_time > 3000)
  _continue = false;
    }
  }

  public void suppress() {
  _continue = false;
  }
}
```

The action(), takeControl() and suppress() methods above are easy to understand. You'll need to modify the BehaviorMain class in order to add this behavior:

```
Behavior b3 = new BehaviorRemote(left, right, ir);
Behavior [] behave = {b1, b2, b3};
```

This beautifully demonstrates the real benefit of Behavior-Based Robotics coding. Inserting a new behavior is simple and it is grounded in object-oriented design. Each behavior is a self contained, independent object.

Results

Subsumption architecture is an interesting programming architecture that has potential, but also some recognized problems. The main disadvantage of this model is that the more layers you add to a program, the more likely it becomes that the behaviors will interfere with one another.

The promise is that you will get interesting behavior from your robot, but you are limited by two physical considerations: sensors and motors. Most robots just roll around, so how much interesting behavior can be produced? You can go forward, backward, or turn.

The sensors need to be varied too. A touch sensor can do one thing—indicate pressed or not. A light sensor is better because it has variation, but the reactions to it are limited. The more sensors you have, the more varied reactions can become.

> ***Tip!***
>
> When creating a Behavior-Based Robotics system, it is best to program the behaviors one at a time and test them individually. If you code all the behaviors and then upload them all at once to the EV3 brick, there is a good chance a bug will exist somewhere in the behaviors that will be difficult to find. By programming and testing them one at a time, it will be easier to identify where a problem was introduced.

It would be nice if all behaviors were as simple as the examples given above, but in more complex coding unexpected results can sometimes be introduced. Threads, for example, can sometimes be difficult to halt from the suppress() method. This can lead to two different threads fighting over the same resources—often the same motor! Another potential problem that can occur in multi-threaded programs is events such as touch sensor hits going undetected.

Since behaviors are totally suppressed when a higher level takes over, robots tend to do one thing at a time rather than multitasking. For example, if both motors are moving forward and a higher level behavior takes command, it is not clear if all lower level motor movements should be stopped. What if the higher level behavior only uses one of the motors? Should the other keep moving forward? And will this lead to odd behavior?

So why use the Behavior API? The best reason is because in programming we strive to create the simplest, most powerful solution possible, even if it takes slightly more time. The importance of reusable, maintainable code has been demonstrated repeatedly in the workplace, especially on projects involving more than one person. If you leave your code and come back to it several months later, things that looked clear suddenly are no longer so obvious. Behavior-Based Robotics allows you to add and remove behaviors without even looking at the rest of the code.

CHAPTER 19

SCARA Arm

TOPICS IN THIS CHAPTER

- SCARA intro
- Building Scarab
- Programming
- Using the arm

In this chapter we will construct a robot arm capable of moving objects from one location to another across a small area. The arm is a well known design, called SCARA, which stands for Selective Compliant Assembly Robot Arm (see Figure 19-1). The unique trait of SCARA arms is that they only move along the x and y axes, and cannot move up and down along the Z axis. In other words, they are adept at moving along a two dimensional (planar) coordinate system. Lucky for us, this type of arm is ideal for the game of chess, allowing the robot to move chess pieces around on a chessboard.

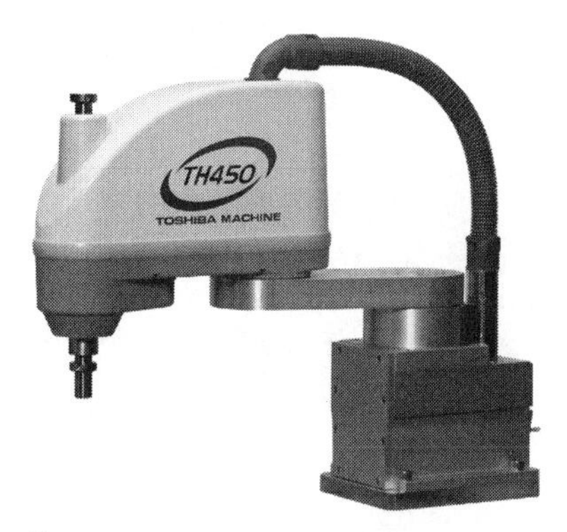

Figure 19-1 Assembling products with a SCARA

In this chapter we will build a SCARA robot. Even though it is built from LEGO, it is quite large and functional. The arm is about as long as a child's arm and capable of reaching to the edge of a medium sized chessboard.

SCARA Robots

The SCARA concept was invented in Japan in 1981 by Hiroshi Makino, a university professor. His goal was to create a robot that could assemble electronics by drilling holes in a Printed Circuit Board (PCB) and inserting electronic components like resistors and transistors.

The tool at the end of the SCARA arm is called an *end effector*. This end effector can be almost anything, such as an electromagnet, a drill, or a claw.

Previous to SCARA, most assembly robots were Cartesian coordinate robots, which used motors for each axis to move the end effector along each coordinate. A CNC milling machine is a popular example of one of these robots.

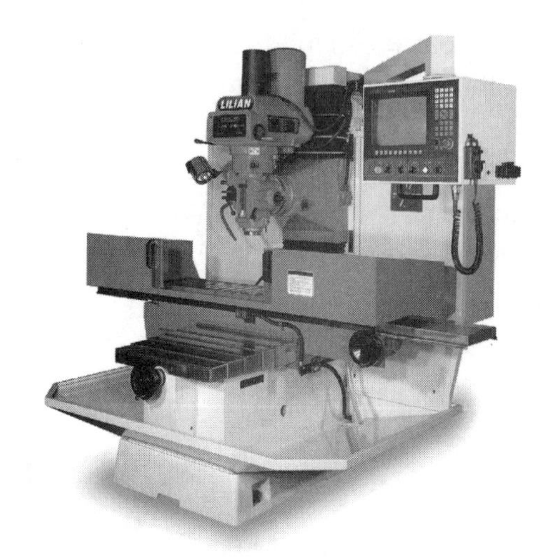

Figure 19-2 A CNC milling machine

In contrast, SCARA uses articulated movement—two joints that rotate, much like a human arm. The advantage of SCARA is that it can reach into confined spaces and then retract. It also has a smaller footprint, and it is capable of folding itself up, like your own arm.

Building a SCARA

Robot arms are both easily built and accurately controlled with LEGO EV3. The servo motors allow the EV3 to precisely control and keep track of the position of the robot arm. The SCARA has the following parts: base, main arm, forearm, and claw (see Figure 19-3). Each of these parts is designed as a modular component.

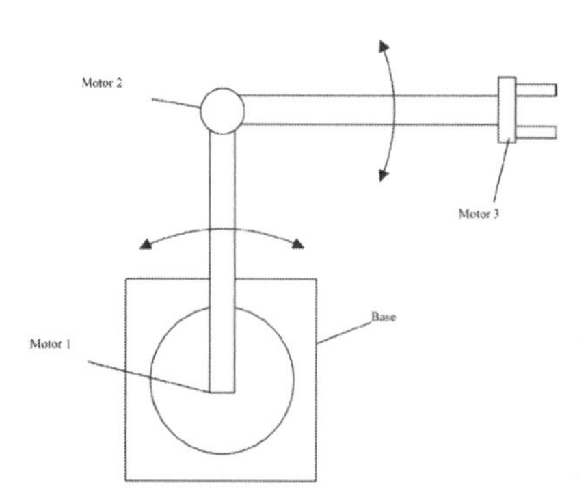

Figure 19-3: Viewing an abstract diagram of the SCARA.

Playing chess seems like a good challenge for our SCARA, so our goal will be to move chess pieces on a full sized chess board. Let's review the list of requirements for moving chess pieces:

- The claw must be able to pick up a chess piece without interfering with the surrounding chess pieces.
- The claw must be able to pick up pieces varying in height from 3 cm to 7 cm, with different shapes and random orientation.
- The arm must be able to move over the entire surface of a chess board without interfering with other pieces, even while holding a piece.
- The claw must be able to drop a chess piece without interfering with the surrounding chess pieces.
- The arm must only use three EV3 motors.

It's time to build the arm. We will call this robot SCARA Buddy, or Scarab for short.

Tip!

Prepare youself for one of the most difficult builds in the book. Although this model is complicated, it should be manageable because you are adding only a few parts per step. Remember to gather the parts for each step. If you can't find where a part goes, compare the current step with the last step and try to spot the difference. Also, the model is added to from all sides and rotates frequently. Note the rotate icon which indicates you need to pay special attention to the direction it is facing.

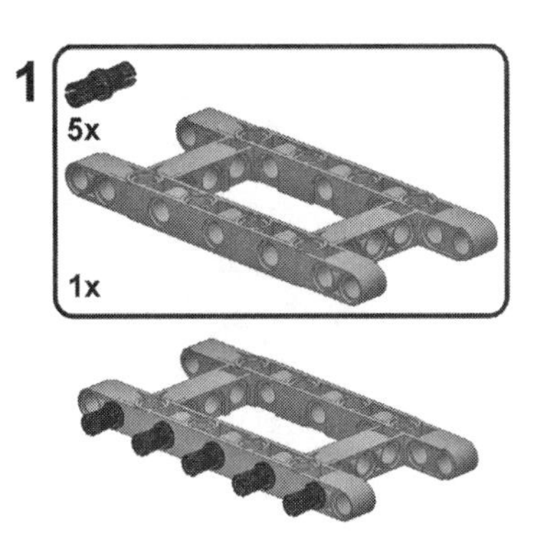

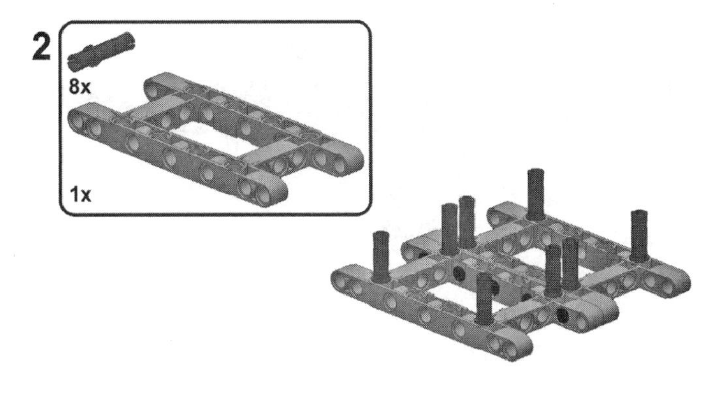

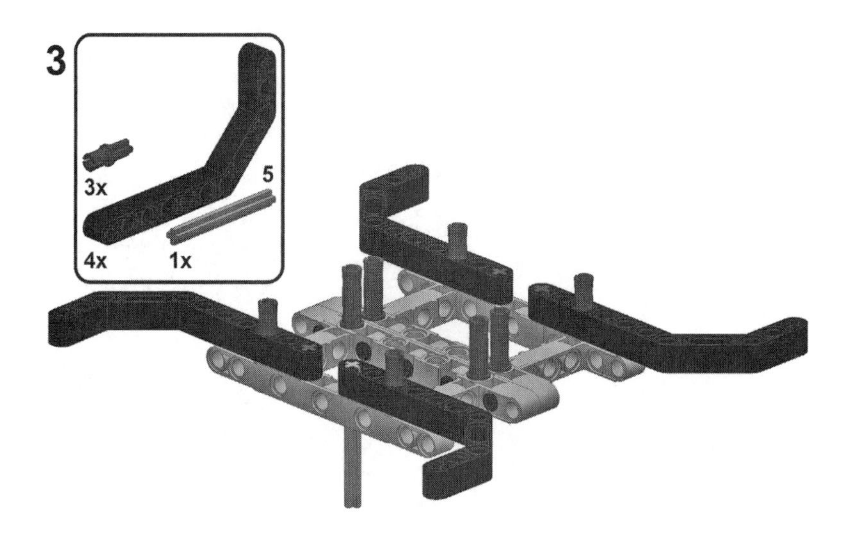
3
3x
5
4x
1x

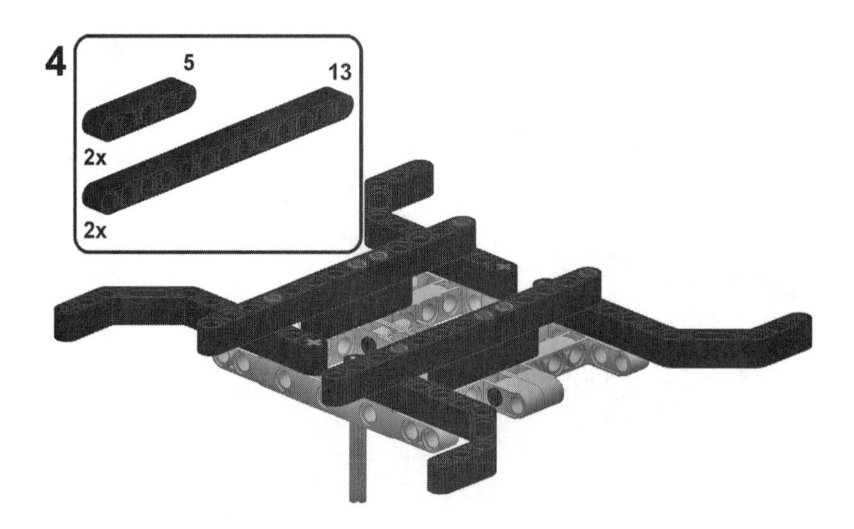
4
5
13
2x
2x

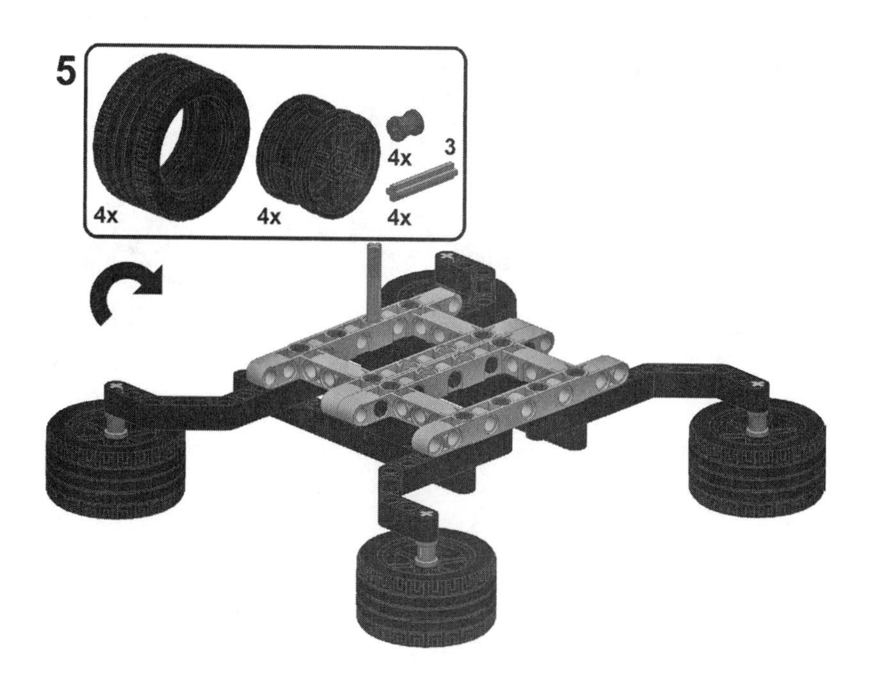
5
4x
4x
4x
3
4x

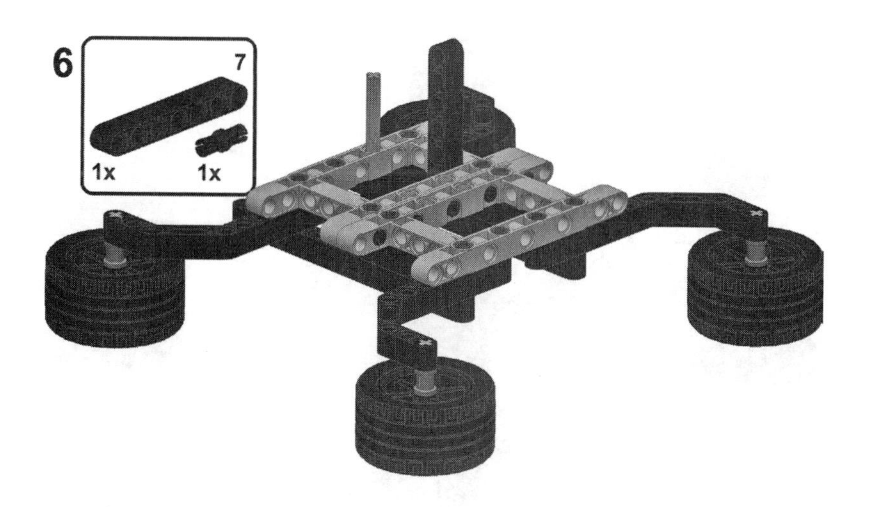
6
7
1x
1x

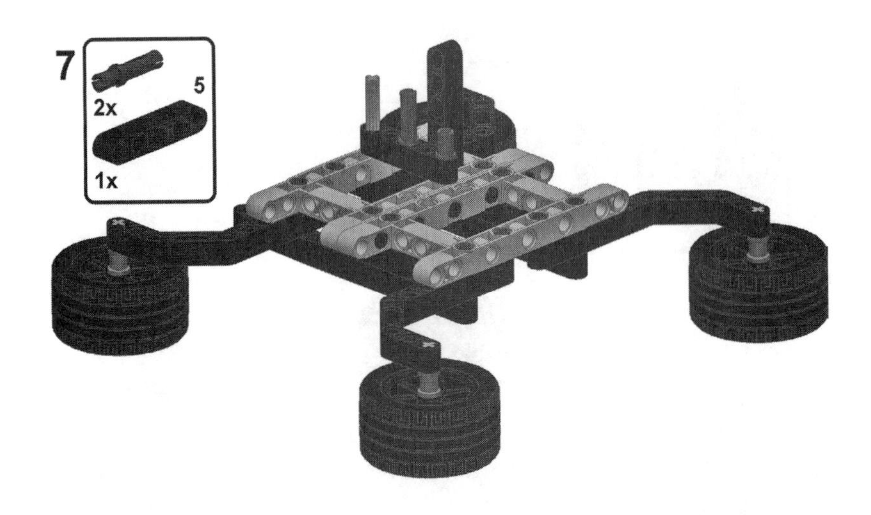
7
2x
5
1x

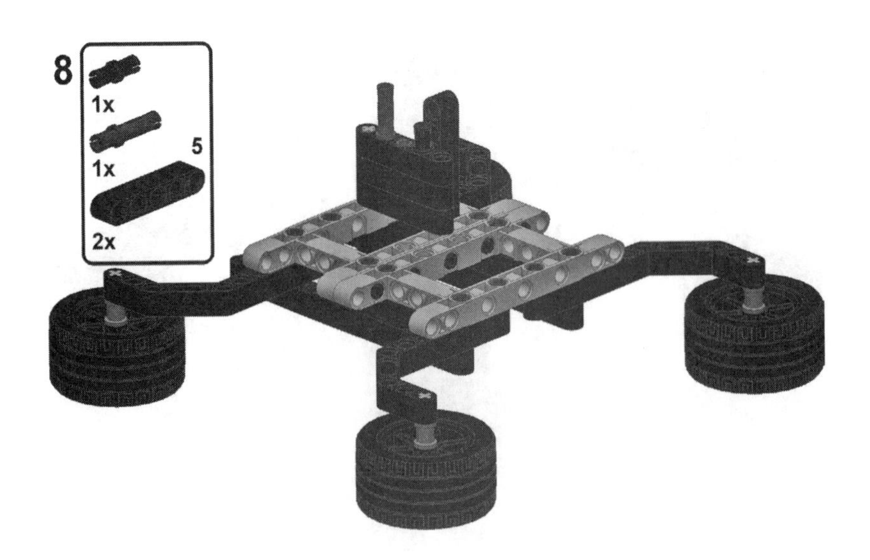
8
1x
1x
5
2x

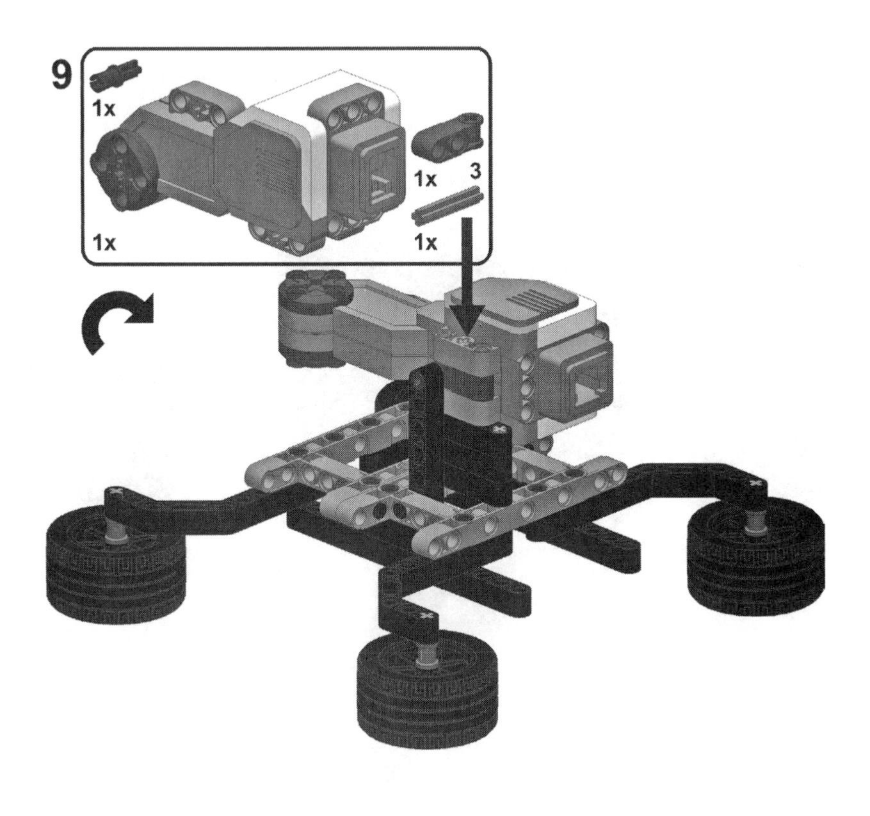
9
1x
1x
1x
3
1x

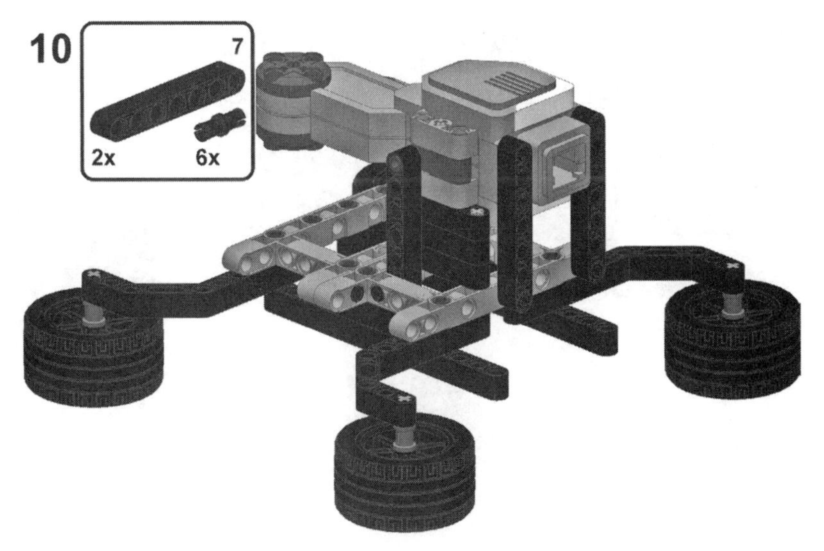
10
7
2x
6x

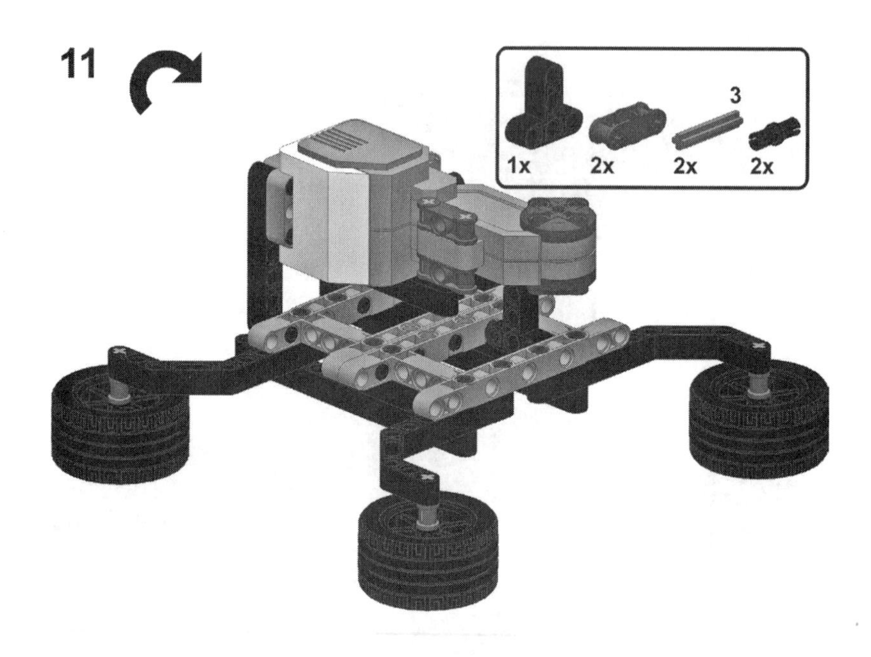
11
3
1x
2x
2x
2x

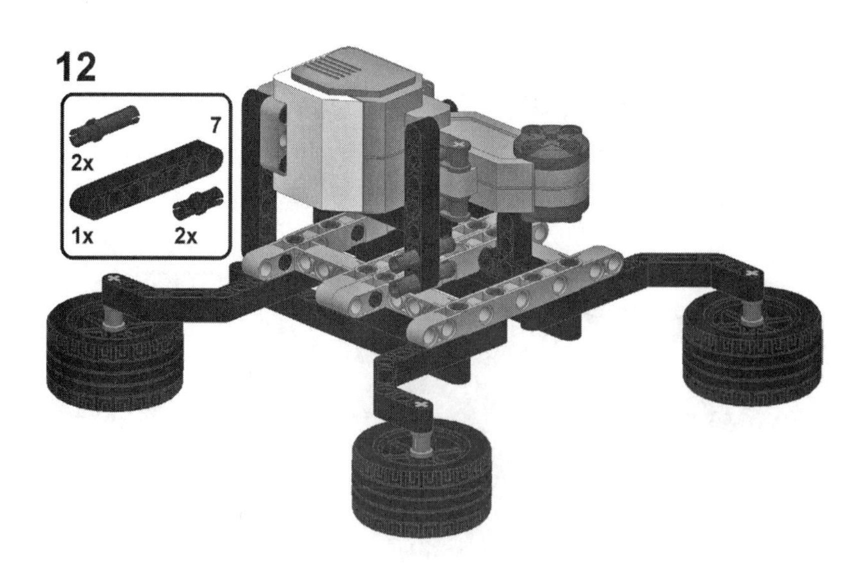
12
7
2x
1x
2x

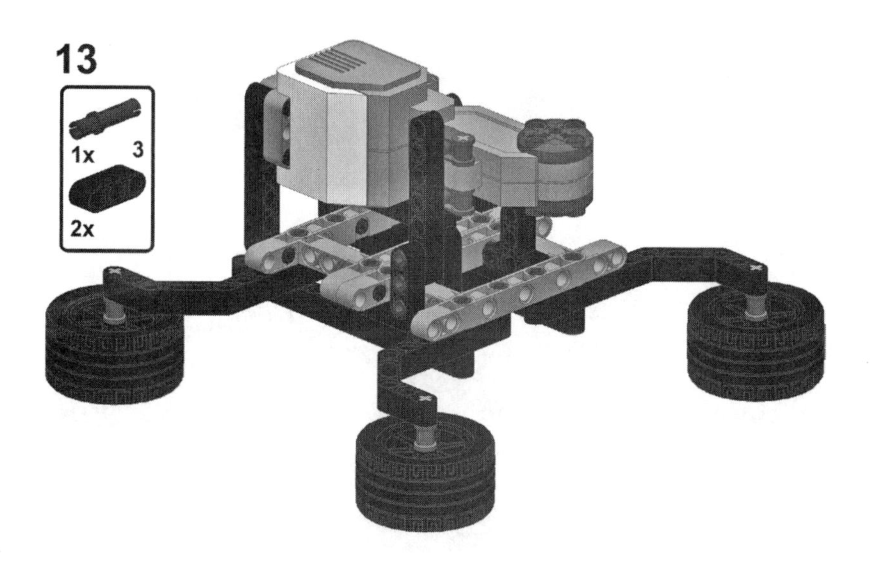
13
1x
3
2x

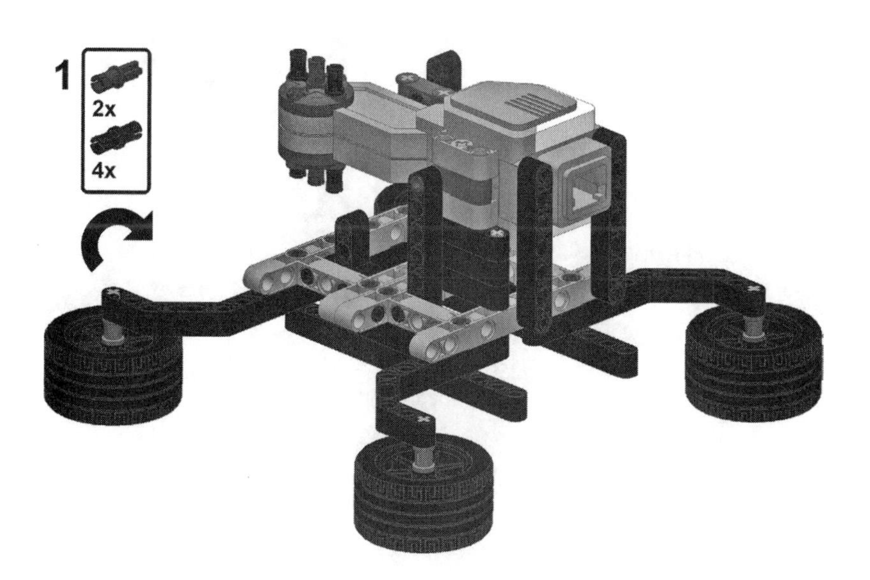
1
2x
4x

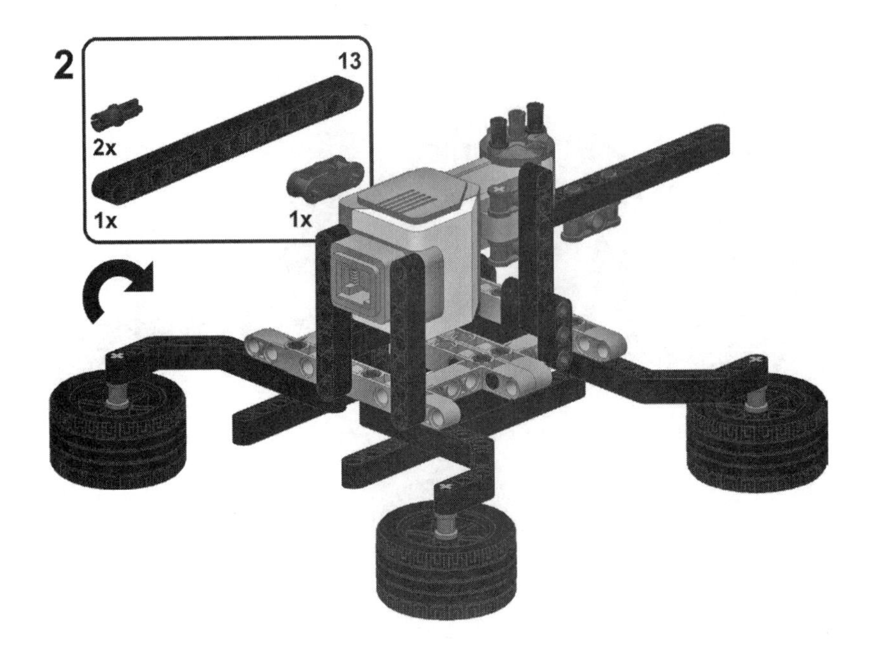
2
13
2x
1x
1x

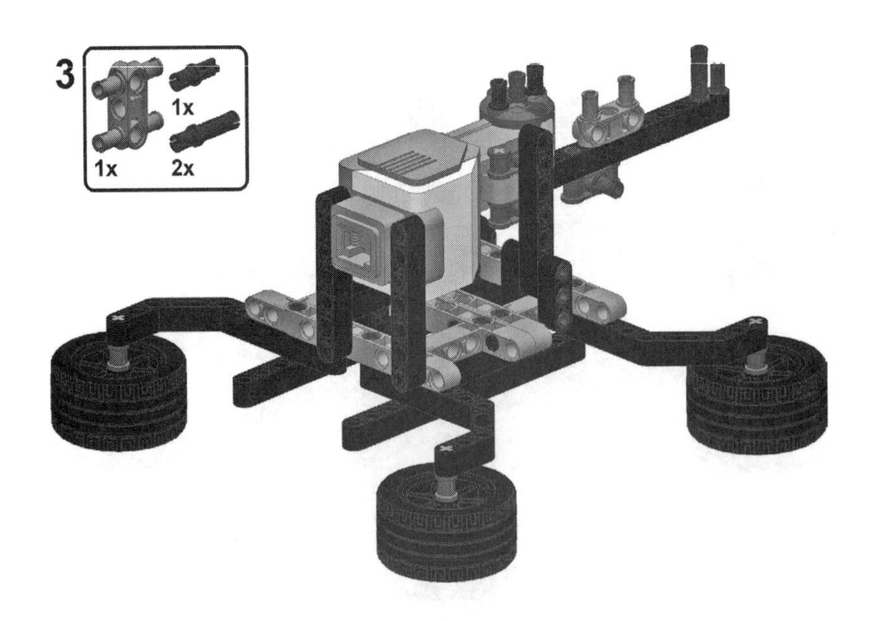
3
1x
1x
2x

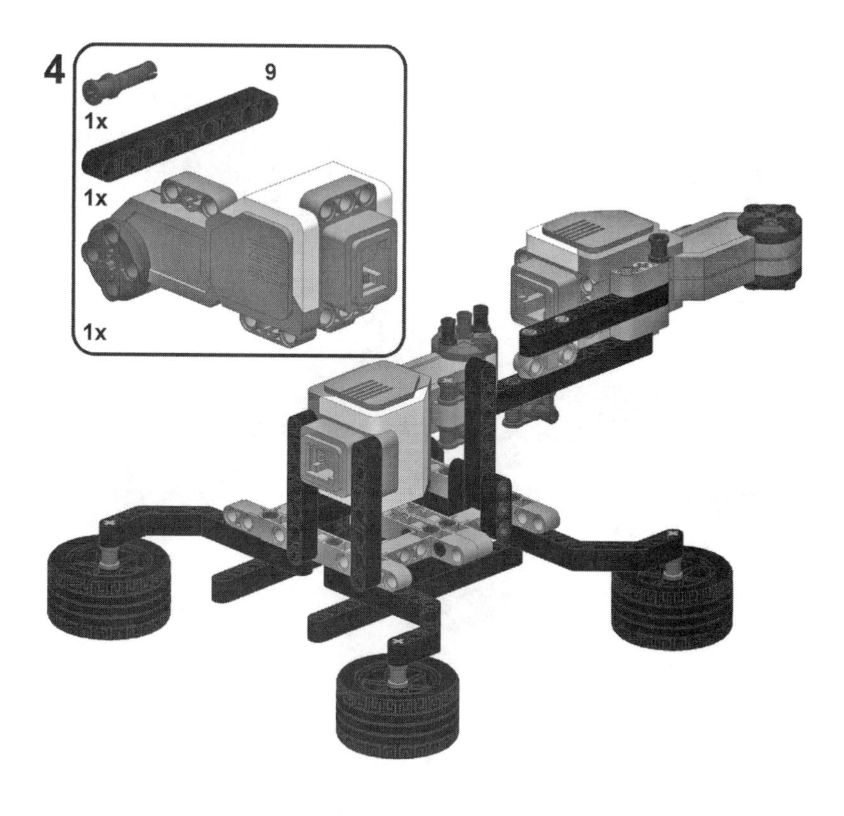
4
9
1x
1x
1x

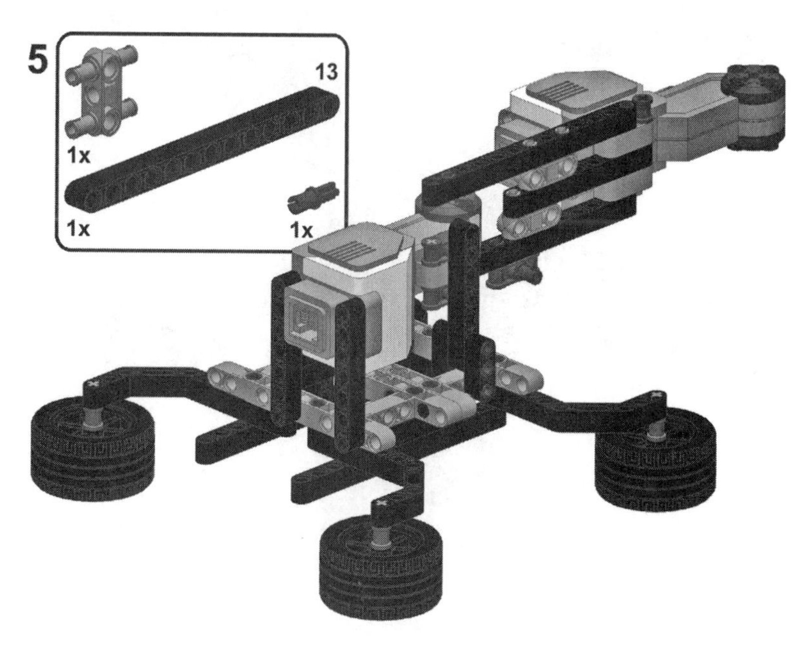
5
13
1x
1x
1x

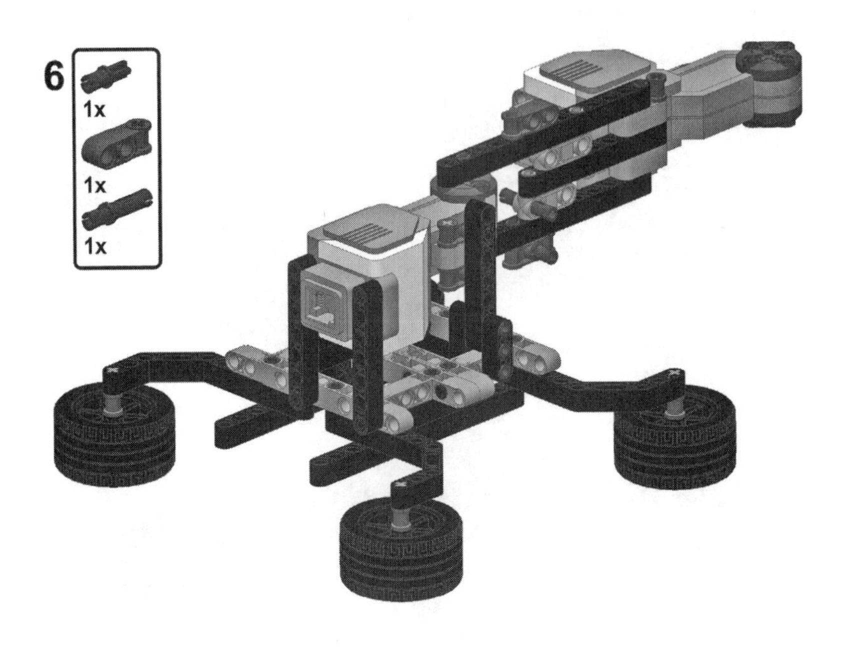
6
1x
1x
1x

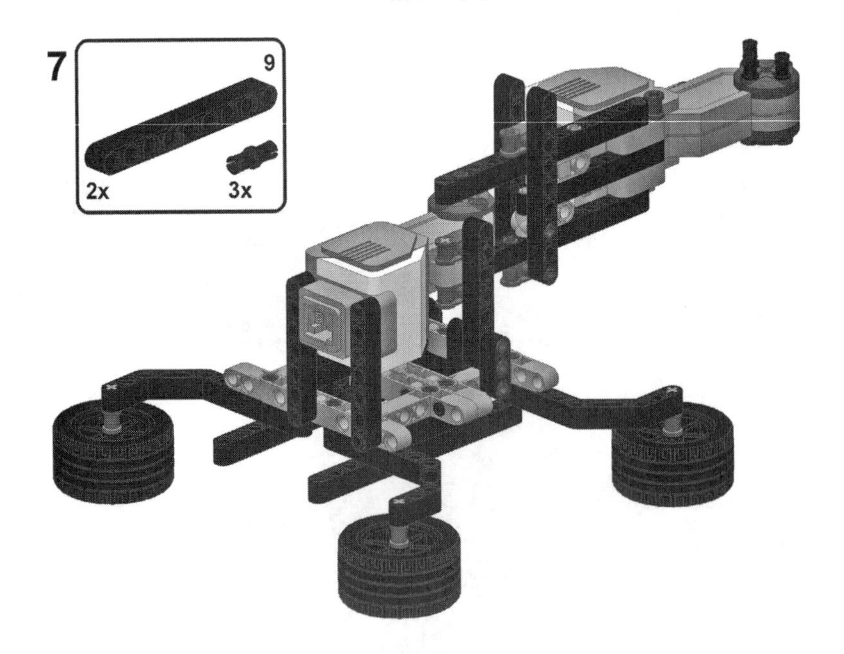
7
9
2x
3x

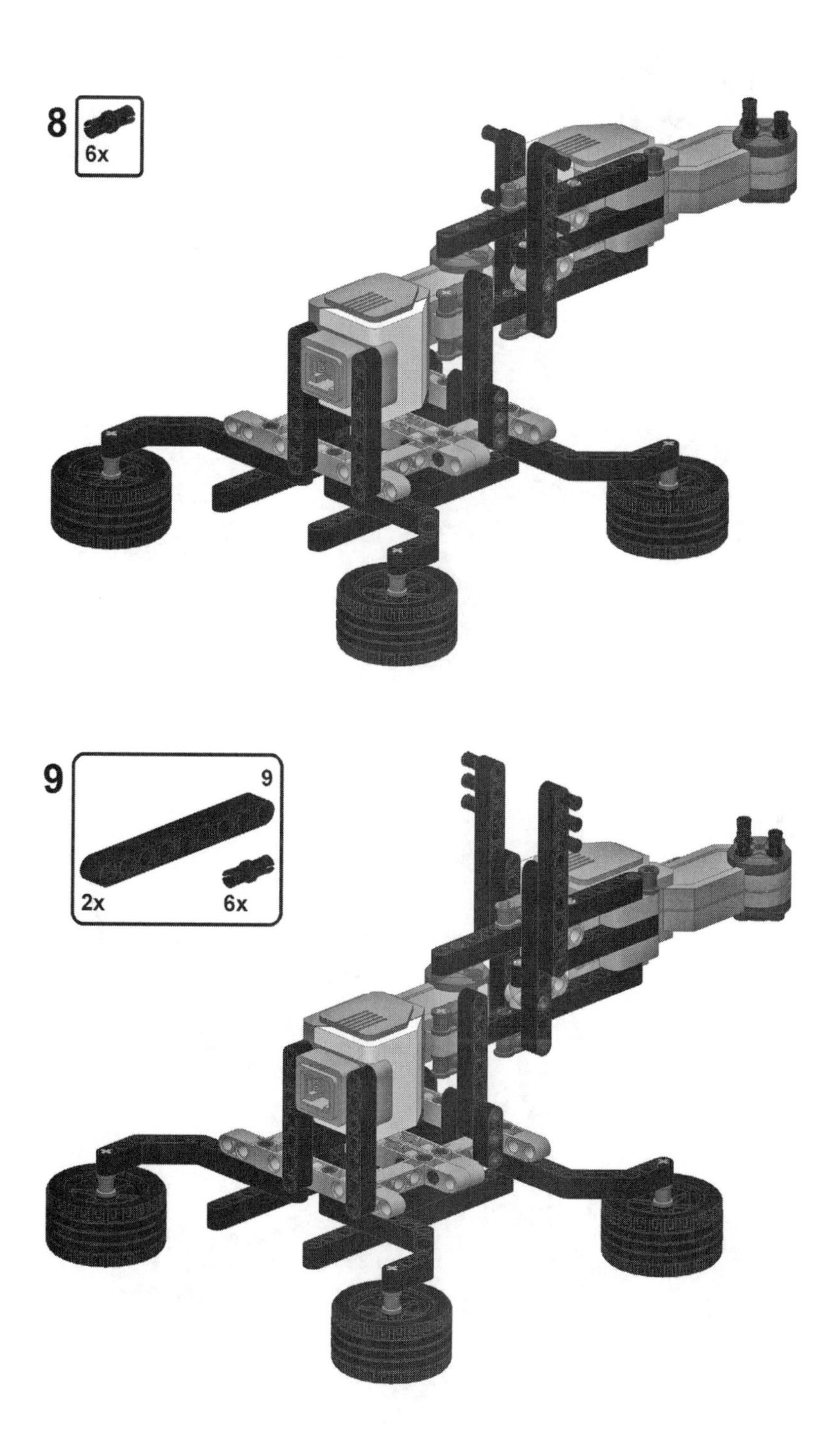
8
6x
9
9
2x
6x

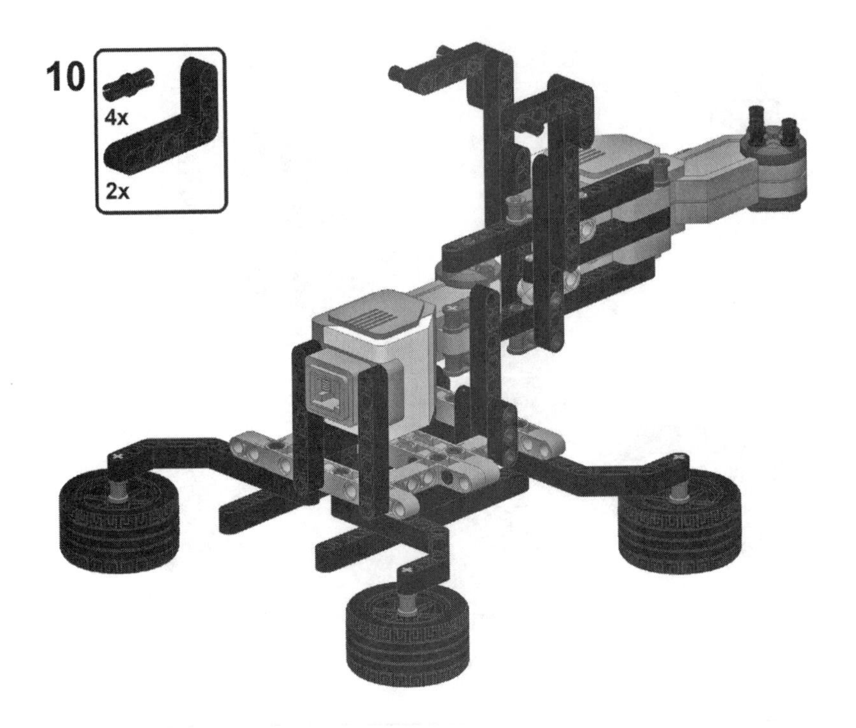
10
4x
2x

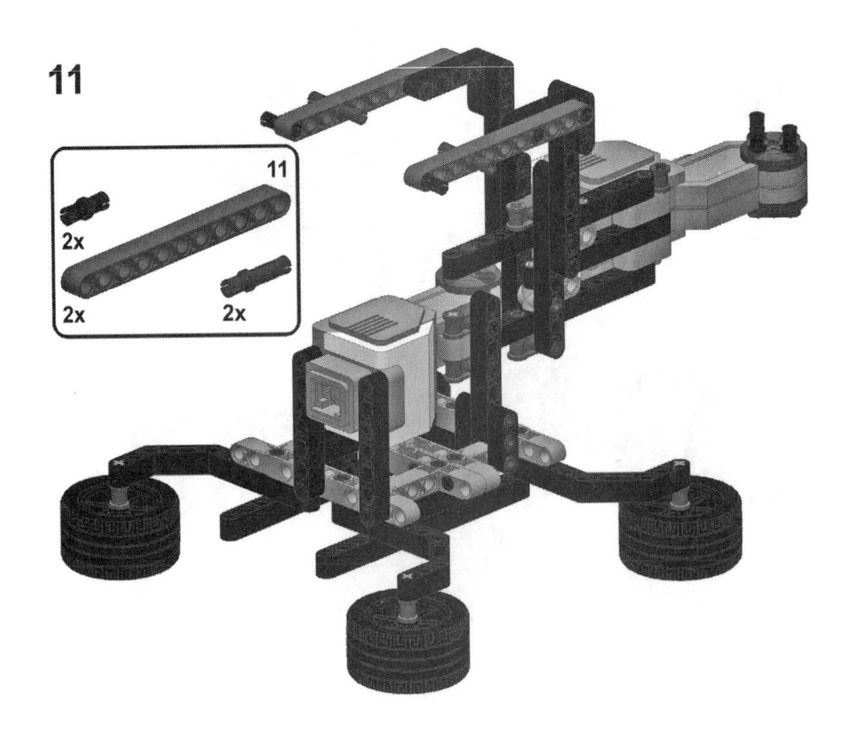
11
11
2x
2x
2x

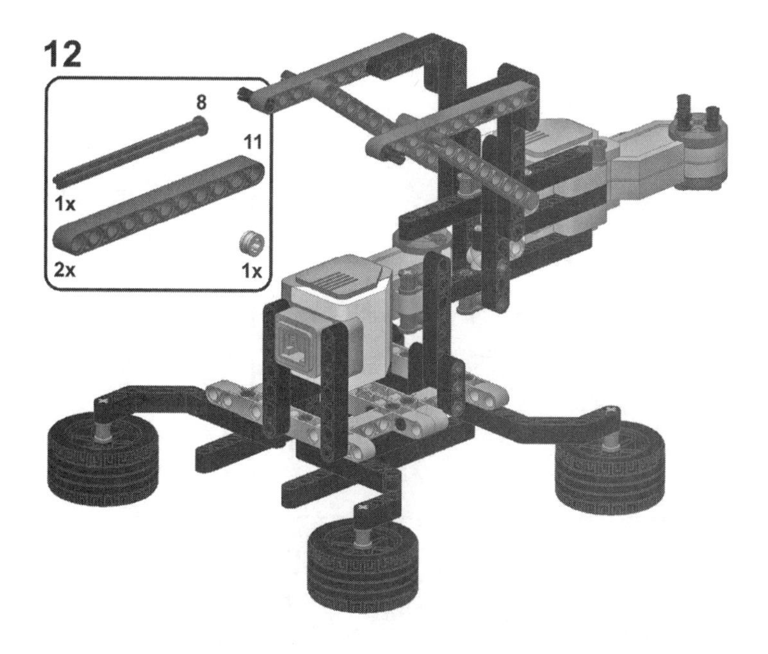
12
8
11
1x
2x
1x

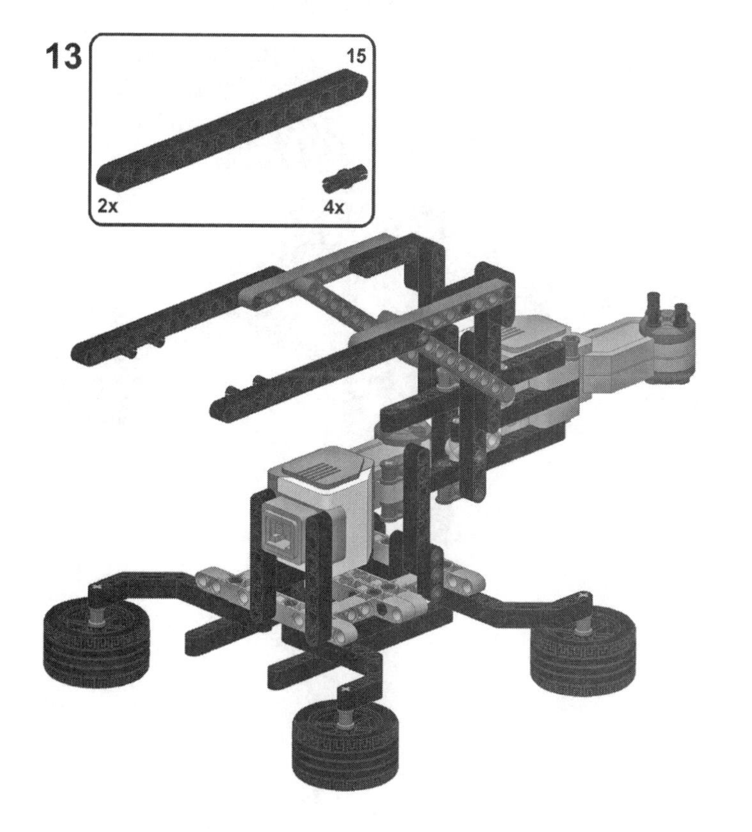
13
15
2x
4x

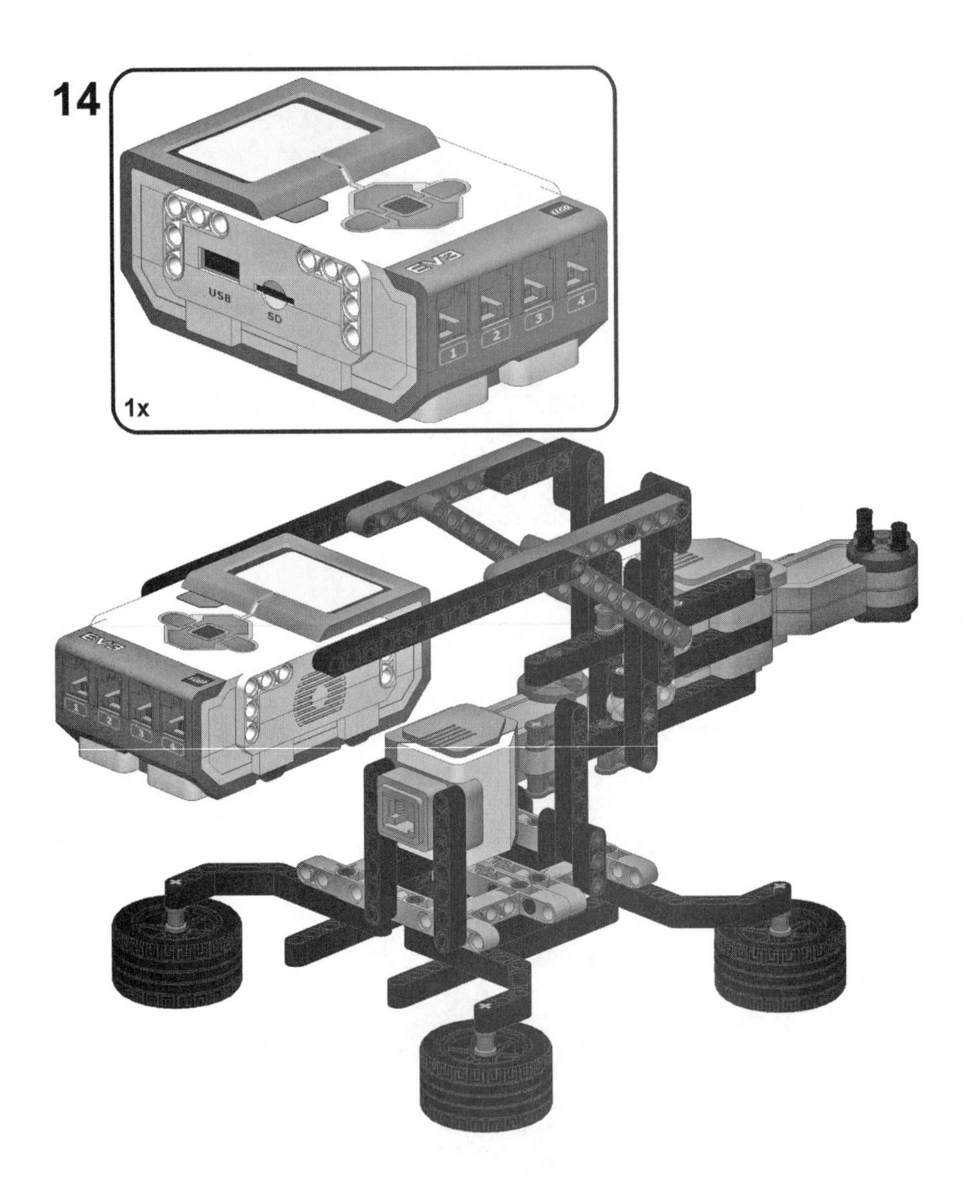
14
USB
SD
EV3
1
2
3
4
1x

1

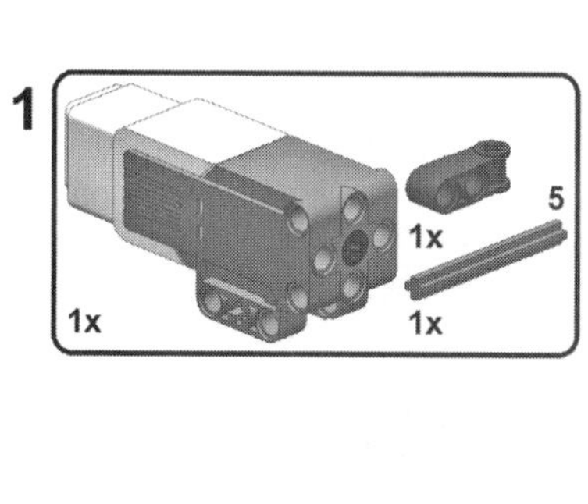

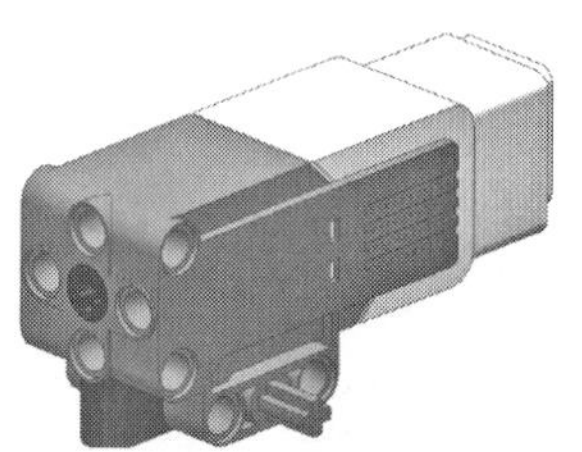

2

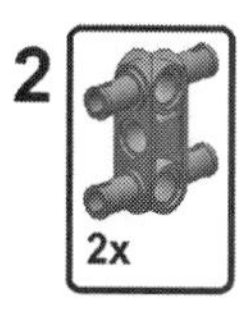

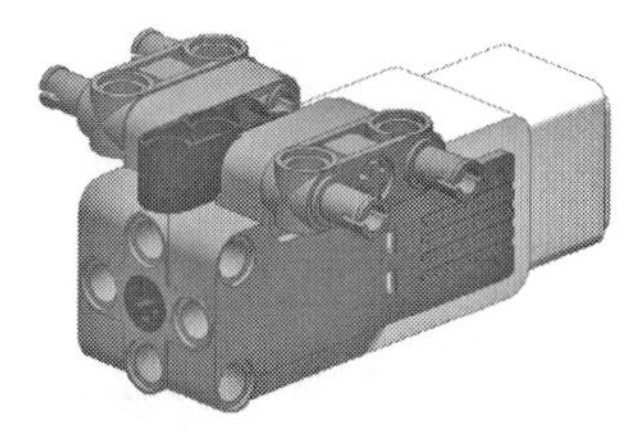

3

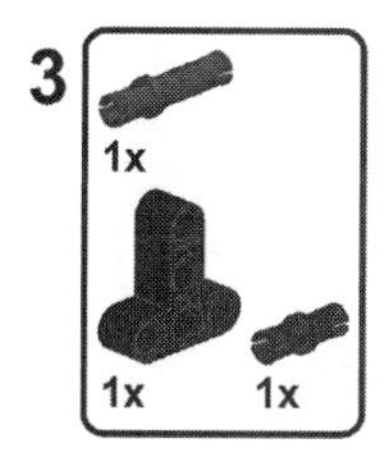

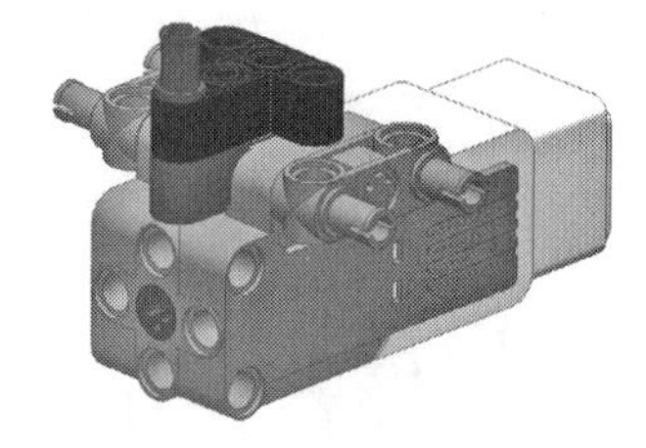

4

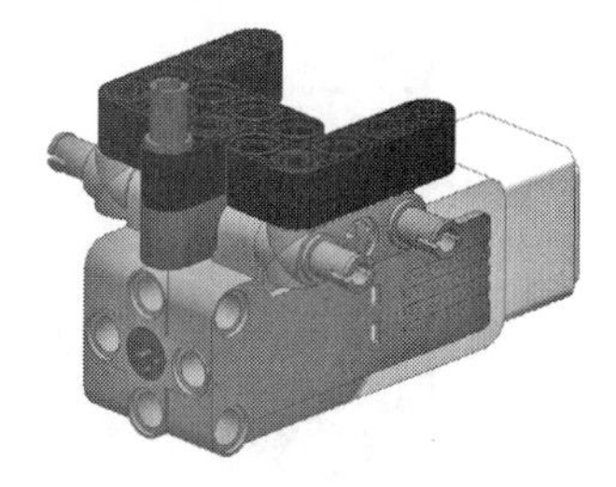

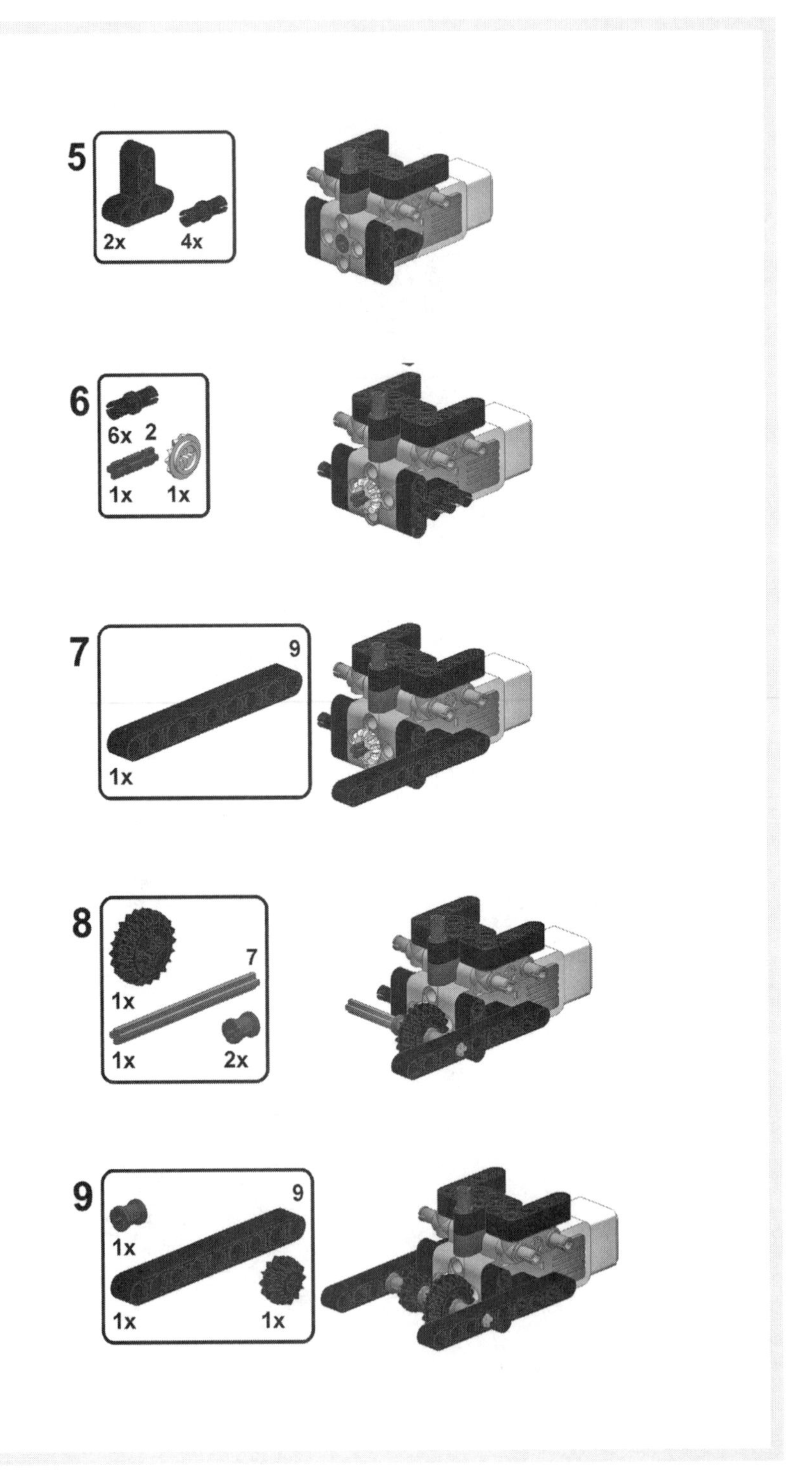
5
2x
4x
6
6x 2
1x
1x
7
9
1x
8
7
1x
1x
2x
9
9
1x
1x
1x

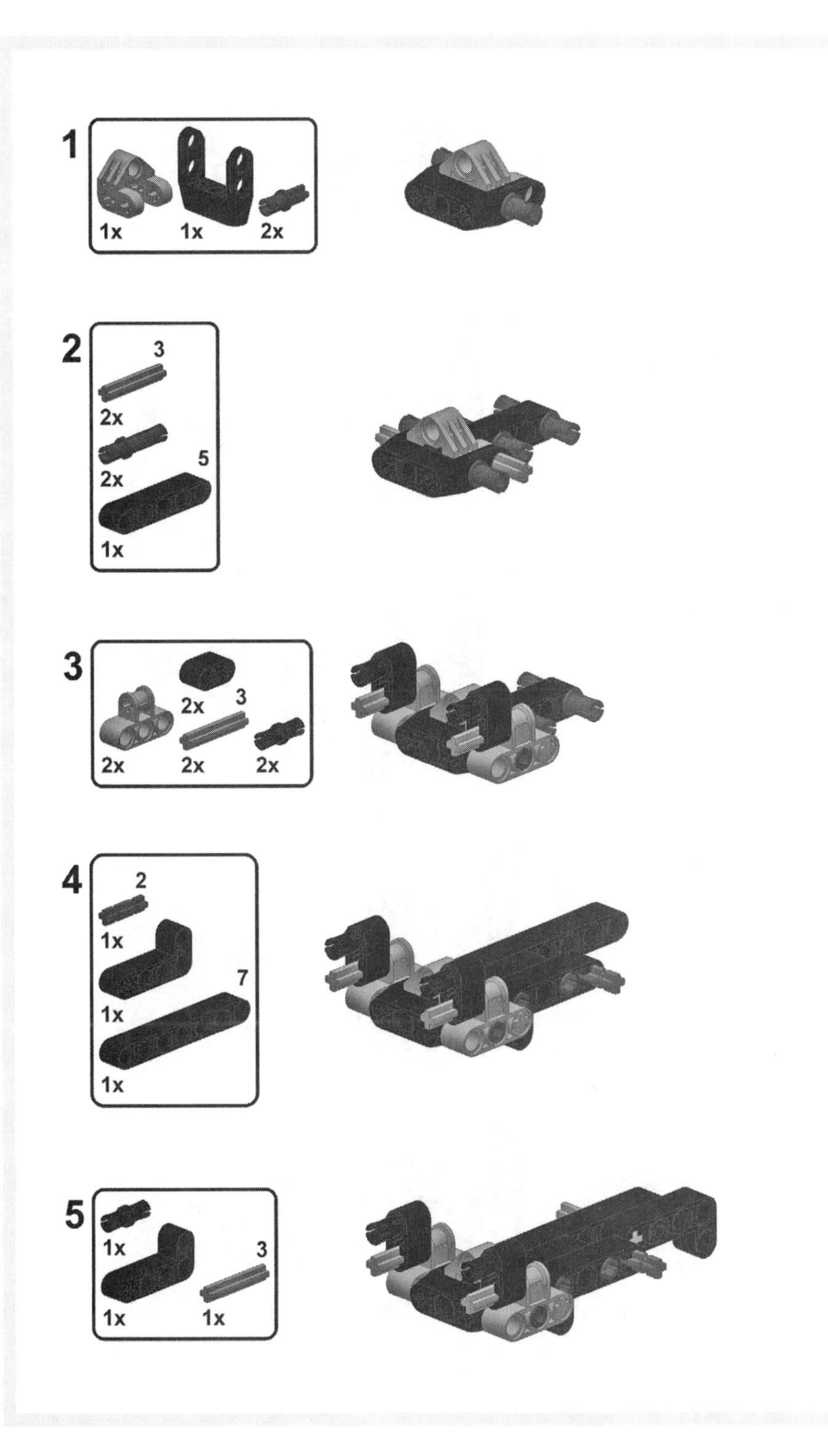
1
1x
1x
2x
2
3
2x
2x
5
1x
3
2x
3
2x
2x
2x
4
2
1x
1x
7
1x
5
1x
3
1x
1x

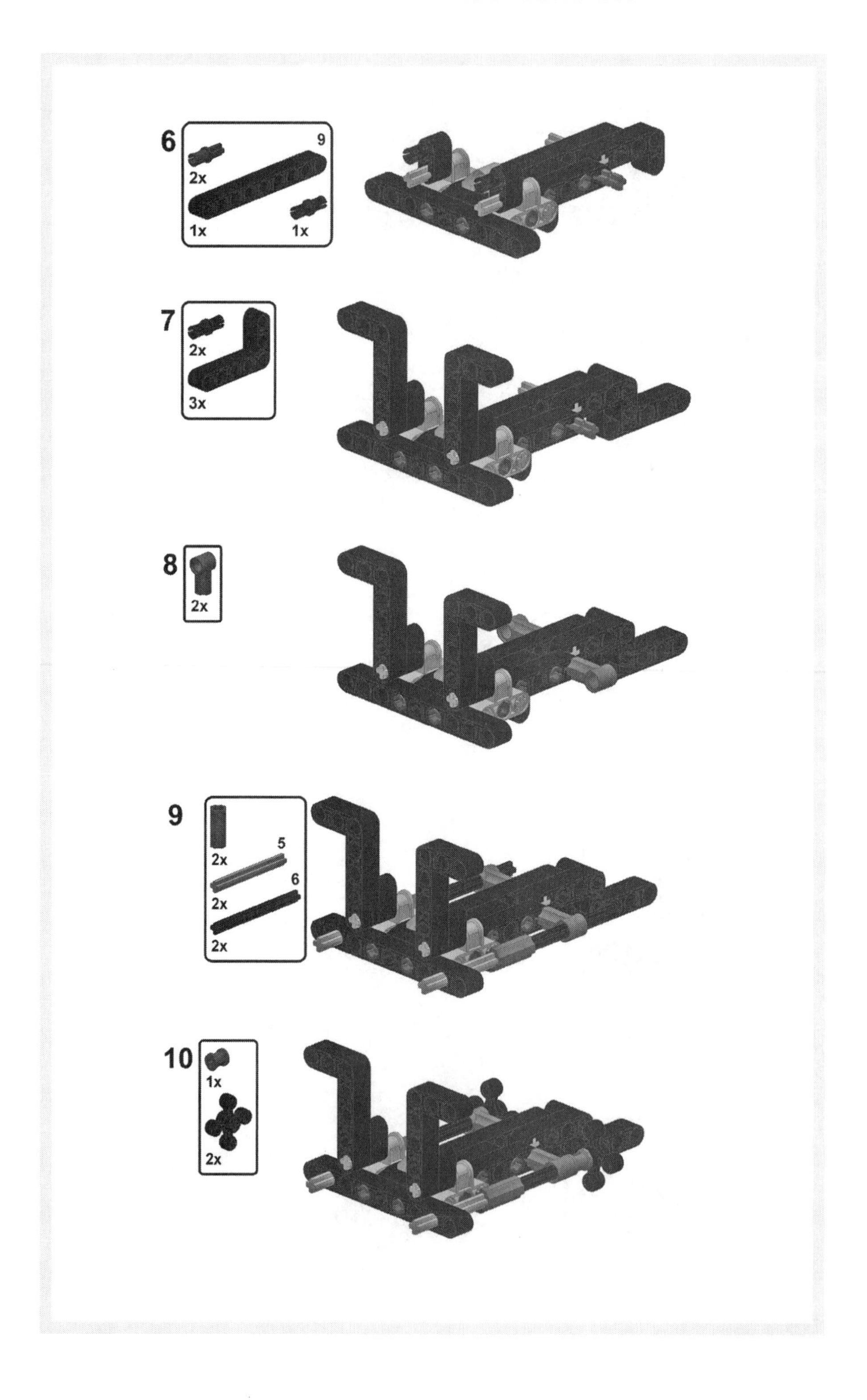
6
9
2x
1x
1x
7
2x
3x
8
2x
9
5
2x
6
2x
2x
10
1x
2x

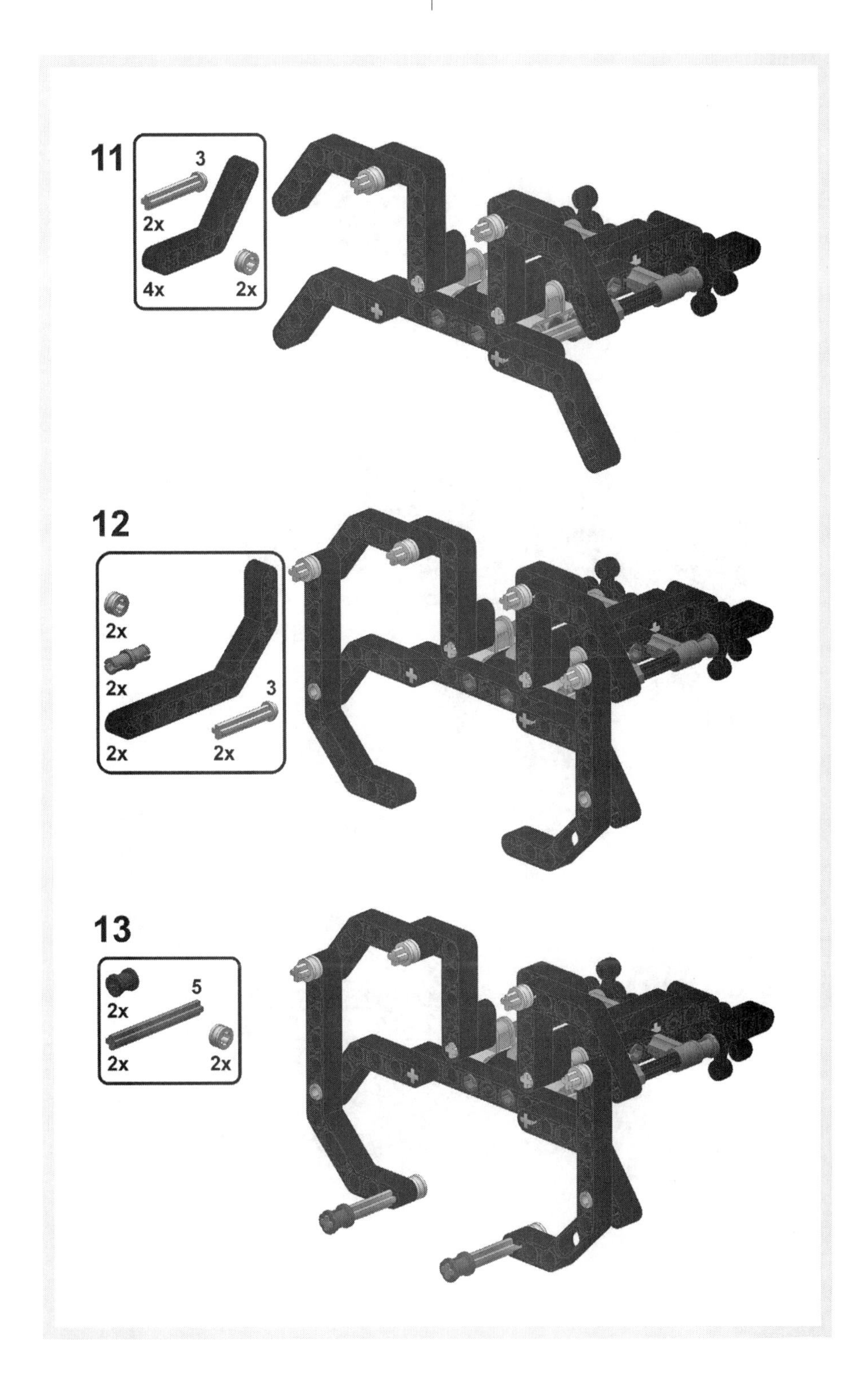
11
3
2x
4x
2x
12
2x
2x
3
2x
2x
13
5
2x
2x
2x

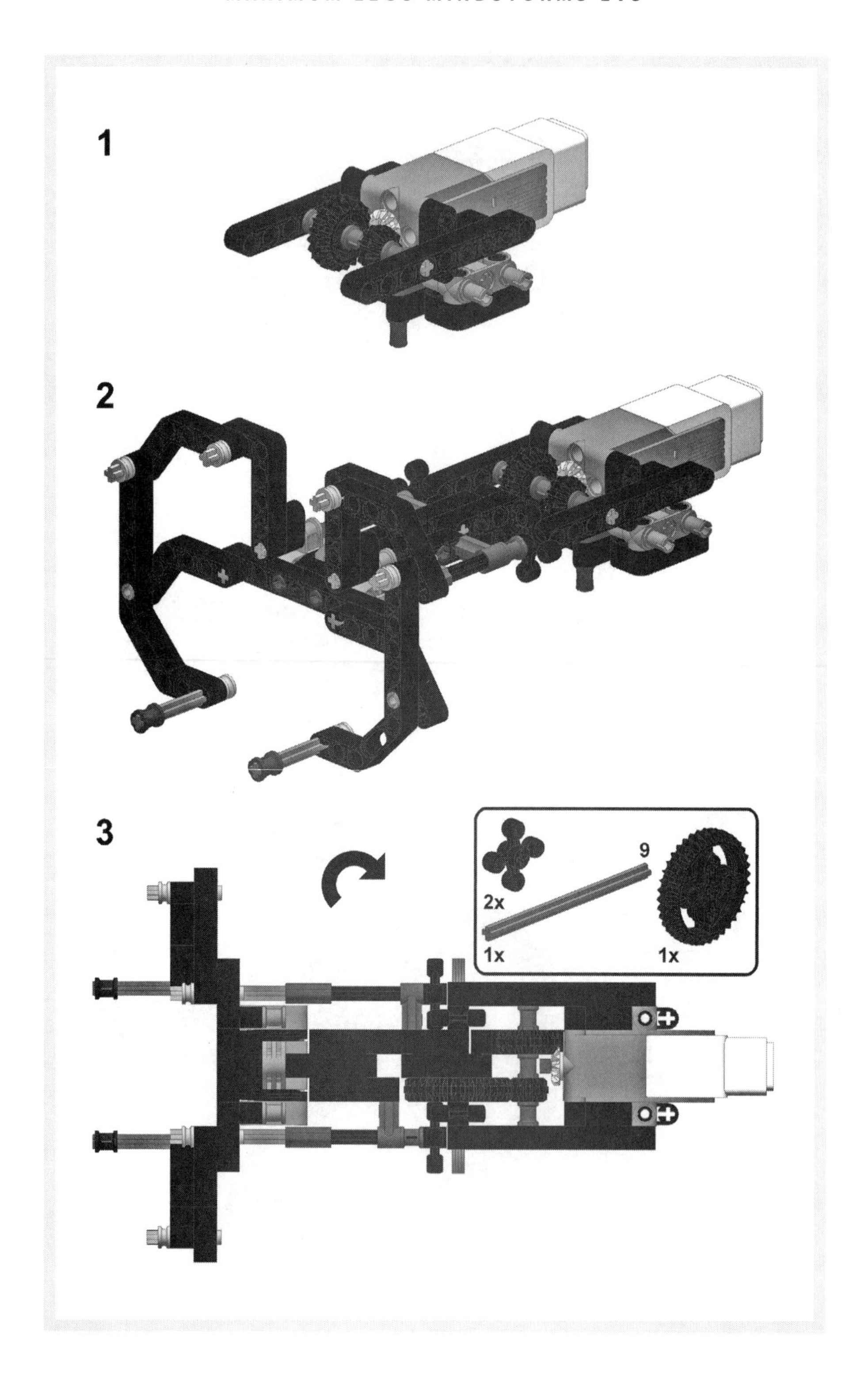
1
2
3
2x
9
1x
1x

15

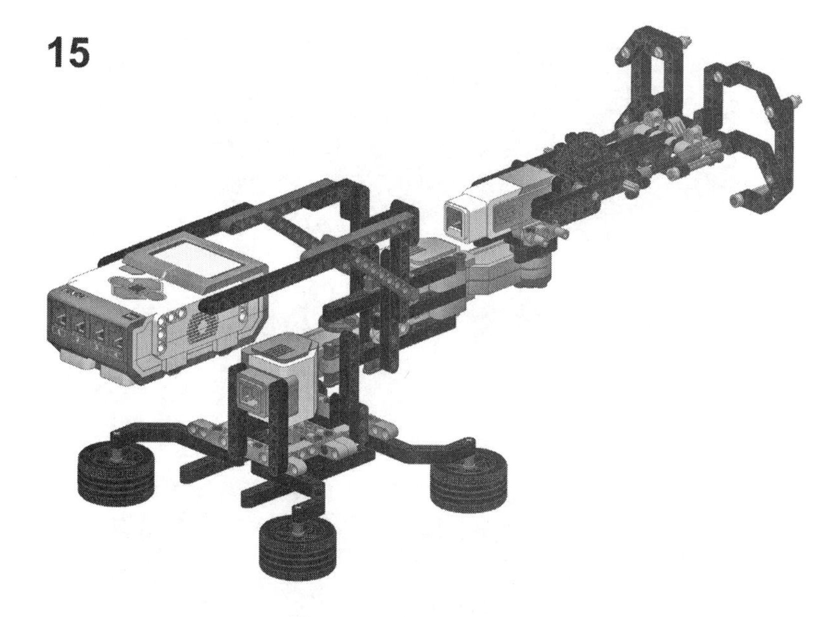

Cables

Cable management for robotic arms is always tricky. Start by plugging a long cable from port A to the base motor. The cable should run forward between the U shaped opening, then around the left to the motor (Figure 19-4). Now plug a short cable from port B to the middle motor (Figure 19-5). Finally, plug a medium cable from port C to the claw motor. This cable should form a loop upwards (Figure 19-6).

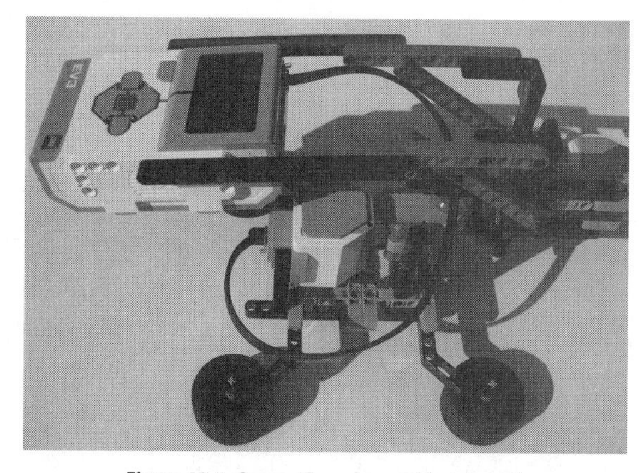

Figure 19-4: Connecting a long cable to port A

Figure 19-5: Connecting a short cable to port B

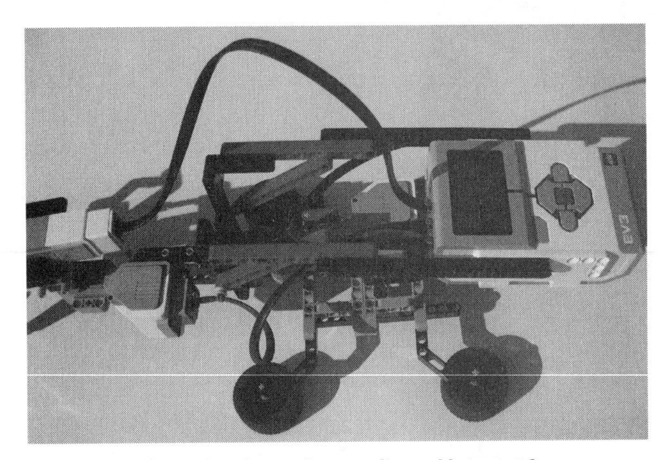

Figure 19-6: Connecting a medium cable to port C

Scarab performs a balancing act, with the EV3 brick and a motor acting as the main counterweight to the arm, and the other motor acting as a counterweight to the claw.

The claw itself is driven by two main axles, one for each side of the claw. Knob wheels transfer the torque of the motor axle to two axles. This also ensures that the fingers on the claw are synchronized.

The most complex part of the arm is the lift mechanism (Figure 19-7). Ideally we would use two motors, one to open and close the claw, and one to lift the wrist. Since we only have one motor, the claw must grab the chess piece and lift the wrist with one shared motor. When the claw closes on a piece (or on itself) the bevel gears can no longer turn. Normally, this would cause the motor to stop.

But notice the axle that allows the wrist to bend upwards (see Figure 19-7). When the claw is closed on an object, the gear train becomes locked up to the wrist joint, while the axle can still rotate against the big gear. This means the axle is the only part that can now move, and it moves upwards until it hits the small stopper piece (see Figure 19-8). With this design, you get two axle rotations for the price of one.

Figure 19-7 The claw clamps down, locking the gear (arrow).

Figure 19-8 The arm lifts.

Programming Scarab

Now we need to give the program several measurements so it can perform the calculations. We must be precise to achieve accuracy. These measurements include:

- The length of the main arm from the shoulder axis to the elbow axis. (15.5 cm)
- The length of the forearm from the elbow axis to the claw. (16.0 cm)

The full code listing is as follows:

```
import lejos.hardware.Sound;
import lejos.hardware.motor.*;
import lejos.hardware.port.*;
import lejos.robotics.RegulatedMotor;

public class SCARA {

  static RegulatedMotor claw;
  static RegulatedMotor fore;
  static RegulatedMotor base;

  // Calibration limiter values:
  static int LIFT_CLAW = 1050;
  static int OPEN_CLAW = 0;
  static int FOREARM_STRAIGHT = -145;
  static int BASEARM_STRAIGHT = 95;

  // length of base arm (shoulder to elbow) cm
  static double ARM_BASE_LENGTH = 15.5;
  // length of fore arm (elbow to claw) cm
  static double ARM_FORE_LENGTH = 16.0;

  public static void main(String[] args) {
    SCARA arm = new SCARA();
    Sound.beepSequenceUp();
    arm.calibrate();

    // Any odd number array size works:
    double [] x = {23.5, 23.5, 27};
    double [] y = {9.4, -9.4, 1.7};

    int i=0;
    for(int r=0;r<10;r++) {
      arm.gotoXY(x[i], y[i]);
      arm.closeClaw();
      i++;
```

```
      if(i>=x.length) i=0;
      arm.gotoXY(x[i], y[i]);
      arm.openClaw();
      i++;
      if(i>=x.length) i=0;
    }
    claw.rotateTo(0);
    fore.rotateTo(0);
    base.rotateTo(0);
    Sound.beepSequence();
  }

  public void calibrate() {
    // Calibrate claw:
    calibrate(MotorPort.C, 20, true);
    claw = new EV3MediumRegulatedMotor(MotorPort.C);
    claw.setSpeed(100);
    claw.resetTachoCount();

    // calibrate base arm:
    calibrate(MotorPort.A, 20, true);
    base = new EV3LargeRegulatedMotor(MotorPort.A);
    base.setSpeed(10);
    base.rotate(BASEARM_STRAIGHT);
    base.resetTachoCount();

    // calibrate forearm:
    calibrate(MotorPort.B, 30, false);
    fore = new EV3LargeRegulatedMotor(MotorPort.B);
    fore.setSpeed(30);
    fore.rotate(FOREARM_STRAIGHT);
    fore.resetTachoCount();
  }

  public void calibrate(Port port, int pwr, boolean
reverse) {

    UnregulatedMotor motor = new
UnregulatedMotor(port);
    motor.setPower(pwr);
    if(reverse)
      motor.backward();
    else
      motor.forward();
    int old = -99999999;
    while(motor.getTachoCount() != old) {
```

```
      old=motor.getTachoCount();
      try {
        Thread.sleep(500);
      } catch (InterruptedException e) {}
    }
    motor.stop();
    motor.close();
    Sound.beep();
  }

  public void openClaw() {
    claw.rotateTo(OPEN_CLAW);
  }

  public void closeClaw() {
    claw.rotateTo(LIFT_CLAW);
  }

  public void gotoXY(double x, double y) {

    double c = Math.sqrt(x * x + y * y);
    double angle1a = Math.asin(y/c);
    double angle1b = Math.acos((ARM_BASE_LENGTH *
ARM_BASE_LENGTH + c * c - ARM_FORE_LENGTH * ARM_
FORE_LENGTH)/(2F *ARM_BASE_LENGTH * c));

    double angle1 = angle1a + angle1b;
    double angle2 = Math.acos((ARM_BASE_LENGTH *
ARM_BASE_LENGTH + ARM_FORE_LENGTH * ARM_FORE_LENGTH
- c * c)/(2F *ARM_BASE_LENGTH * ARM_FORE_LENGTH));

    if(x < 0) {
      angle1 = Math.PI - angle1;
      angle2 = Math.PI - angle2 + Math.PI;
    }

    rotateShoulder(angle1);
    rotateElbow(angle2);
  }

  public void rotateElbow(double toAngle) {
    // Arm starts at tacho 0 which is actually
180 degrees
    // hence subtract 180
    int toCount = (int)Math.
toDegrees(toAngle) - 180;
    toCount *= -1.0; // Elbow motor reversed
```

```
    fore.rotateTo(toCount);
  }

  public void rotateShoulder(double toAngle) {
    int toCount = (int)Math.toDegrees(toAngle);
    toCount *= 1.0; // Shoulder NOT reversed
    base.rotateTo(toCount);
  }

}
```

The code above will pick up, move and put down two chess pieces in a loop. It cycles, so the pieces will continually move onto three squares. It will do it 10 times, although this number could easily be increased or decreased depending on how long you want the demo to last.

In order to get consistent moves by this arm, all the parts need to start in the same position each time. We can ensure they do this through calibration. Calibration can occur by moving an appendage until it hits a limit switch, which indicates to the computer which position it is in. I tried to keep cables to a minimum for this arm, so we will use a slightly different method to calibrate the three parts of the arm.

The calibration method moves the different parts of the arm the full breadth of movement until the appendage contacts a limiter, causing the motor to seize. It can tell is has seized because the tachometer stops changing. It then moves the parts of the arm a certain angle to get then to the straight starting position.

The methods for rotateElbow(angle) and rotateShoulder(angle) allow the program to select an angle to move the arm segment to. This makes it convenient for later, when the program calculates the arm angles required to move to a specific set of coordinates.

The method goTo(x, y) contains calculations for moving the positions of the arm to the appropriate angles. You can find out the details of these equations in Appendix A.

So what is the position of the robot relative to the coordinate space? The origin point (0, 0) is at the axle hole of the base motor (see Figure 19-9). The positive x axis is aligned with the robot arm when it is in the straight position. We measured the length of the base-arm at 15.6 cm and the forearm at 16 cm, meaning when the robot is in the straight position the claw is at $x = 31.5$, $y = 0$. With this in mind, we will position the chess board and chess pieces appropriately when we test it out below.

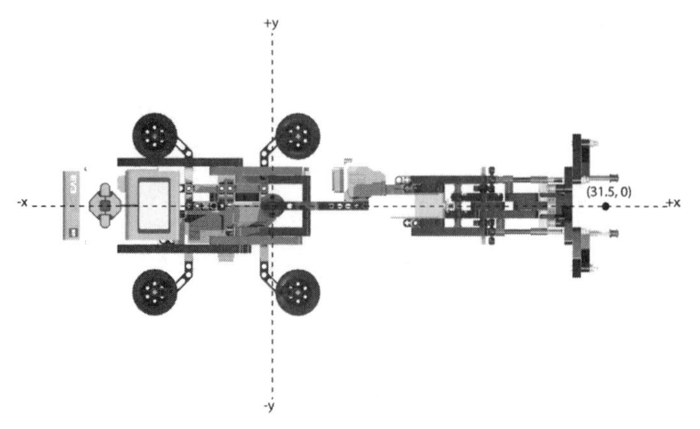

Figure 19-9: Identifying the relationship to coordinate space

Using the SCARA Arm

When you run the code, the first thing the robot will do is perform a calibration routine. As mentioned, it does this by moving each motor until it encounters resistance. It then beeps and moves the motor back to the neutral position.

The demo for this robot is inspired by a SCARA robot in the Tech Museum of Innovation in San Jose, California (see Figure 19-10). A robot in this museum constantly picks up and puts down toy building blocks (it also spells words entered from a keyboard). It does this all day, year round without ever dropping a block. The precision is incredible. In order to give the SCARA arm a good workout, it will try a similar task using chess pieces.

Figure 19-10: A precision robot arm

The demo will pick up and put down pieces from three different coordinates. Before we can test Scarab we must position the arm relative to the chess board—in this case, they are pawns (see Figure 19-11). I put them on a chessboard for effect, but this is not necessary. Only two chess pieces are needed. The robot will pick up at 1, drop at 2, pick up at 3, drop at 1, pick up at 2, etc...

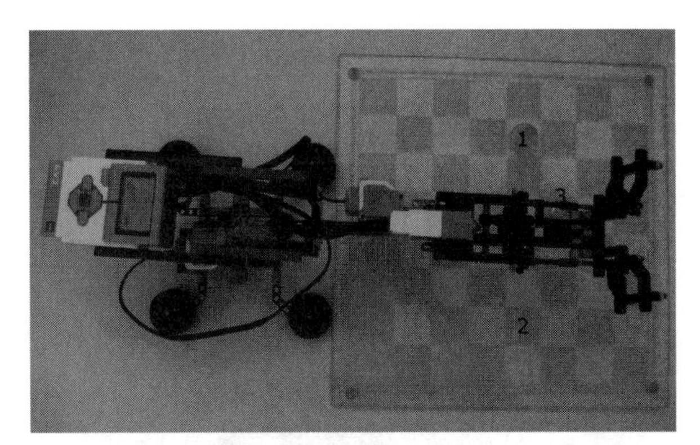

Figure 19-11 Positioning the arm relative to the chess pieces.

You are now ready to try out the arm. To place the pieces at the coordinates indicated in the code, you can try measuring the x and y distances in cm, but there is bound to be some error with the arm. Instead, start the program and place each of the two pieces where the claw opens (this will become more clear when you run the code). Once they are in position, the robot is quite reliable at moving them around.

Each cycle of picking up and putting down a piece will repeat for 10 cycles. You can raise or lower this number. You can also expand the array of x and y coordinates, but make sure to use only an odd number of pairs in the x and y arrays in the main() method. To end the program at anytime press enter and down simultaneously.

> ***Tip!***
>
> If the robot can't reach low enough to pick up the pawns, try raising the chessboard with a thin book or two.

Results

Considering the limitations of building this arm, it actually performs with a high degree of precision. In my tests, once I had the pieces positioned properly it rarely dropped or failed to pick up a piece. However, there were a few issues that brought an unexpected dose of reality to this project.

Lack of Turntable

Ideally the robot would use a turntable piece for each joint, but these are not included in the EV3 kit and therefore I was unable to incorporate it into the design. The turntable offers stable rotation, it doesn't sway, and the cables exert less effect on it (see Figure 19-12).

Figure 19-12: Using the turntable

Counterbalancing

Another effect of the lack of turntable is that the design requires the EV3 brick to counterbalance the weight of the forearm and claw. However, the equilibrium of this balance changes depending on the angle of the elbow and shoulder.

When the elbow is straight, the claw hangs down very low. When the elbow is bent, the claw is higher from the ground. This means the elevation of the claw is inconsistent, making it difficult to pick up and drop pieces with precision.

If you have another turntable piece and want a more accurate control, try working it into the design.

> ### Tip!
> If you are serious about perfecting your chess arm, you can obtain ribbon cables from Mindsensors.com. The ribbon cable is less rigid, and interferes less with the arm movement.

Cables

The cables can actually pull on the arm in some orientations. This can cause the arm to fail to move all the way to its destination. However, if you get the cables oriented just right, as described above, they will have a minimal effect.

The Backlash Effect

As discussed in Chapter 5, the backlash effect can cause imprecision with any LEGO robot. When the direction of a motor is reversed, the gears spin for a moment without causing the motor to rotate. Due to the small space between gear teeth, when the input gear (the one at the motor) rotates, the output gear (the final gear in the gear train) lags behind.

Since our arm is moving back and forth, this backlash effect can cause problems with accuracy. Try holding the red axle part of a motor wiggle it back and forth. Notice there is a lot of play.

We can compensate for this by finding out the number of degrees the motor must rotate when reversing direction before the arm begins rotating. Each time we reverse direction, the code instructs the motor to turn that amount first, then begin the actual calculated rotation.

Try it!

Want to take this project even further? Using an open source Java chess program, hack the code and intercept the calls to the chessboard so that your arm physically moves the pieces. Sourceforge and Google Code are good places to look for open source chess programs

CHAPTER 20

One More Cup of Java

TOPICS IN THIS CHAPTER

- java.lang
- java.util
- Event Model
- Recursion
- Enumerations
- Generics

Earlier in the book we examined fundamental OOP concepts and keywords for programming Java. But in order to really code in Java, you need to rely on some of the core classes in the Java language. These classes have nothing to do with robotics, but rather, they allow you to easily manipulate numbers, text, and perform other important functions. Let's start with the basics.

We'll cover the most important classes and methods in java.lang only. It would be too exhausting to include all of the methods and variables here. However, I consider it worth knowing about java.net (which is covered in chapter 14). Also, java.util contains the collections framework (useful for storing sorted lists), as well as date and time classes.

The java.lang Package

Let's examine the core Java API, which is part of the standard Java classes that come with the Oracle JRE. The first package we will examine is the java.lang package.

I won't bother going over the primitive classes, but it is worth mentioning that all of the eight primitive data types are represented in this package. This includes Boolean, Byte, Character, Double, Float, Integer, Long, and Short. These classes contain helper methods for comparing numbers, converting to and from strings, converting to binary, converting to hex, converting to octal, and other methods.

Now to look at some specific classes in this package.

Math

The Math class is the place to go when you need complex mathematical functions. Some of these functions can be very useful in robotics where it is necessary to keep track of distances, angles, and coordinates.

java.lang.Math

```
public static final double E
```

The double value that is closer than any other to e, the base of the natural logarithms.

```
public static final double PI
```

The double value that is closer than any other to pi, the ratio of the circumference of a circle to its diameter.

```
public static double sin(double a)
```

Returns the trigonometric sine of an angle.

Parameters: a Angle value in radians.

NOTE: *All of the trigonometry functions return values in radians. This means that instead of getting a value of 180 degrees, the value will equal pi, or about 3.1416 (360 degrees equals 2pi in radians). If you prefer working in degrees you can use the Math.toDegrees() method for quick conversions.*

```
public static double cos(double a)
```

Returns the trigonometric cosine of an angle.

Parameters: a Angle value in radians.

```
public static double tan(double a)
```

Returns the trigonometric tangent of an angle.

Parameters: a Angle value in radians.

```
public static double asin(double a)
```

Returns the arc sine of an angle, in the range of -pi/2 through pi/2.

Parameters: a Angle value in radians.

```
public static double acos(double a)
```

Returns the arc cosine of an angle, in the range of 0.0 through pi.

Parameters: a Angle value in radians.

```
public static double atan(double a)
```

Returns the arc tangent of an angle, in the range of -pi/2 through pi/2.

Parameters: a Angle value in radians.

```
public static double toRadians(double angdeg)
```

Converts an angle measured in degrees to the equivalent angle measured in radians.

Parameters: angdeg Angle value in degrees.

```
public static double toDegrees(double angrad)
```

Converts an angle measured in radians to the equivalent angle measured in degrees.

Parameters: angrad Angle value in radians.

```
public static double exp(double a)
```

Returns the exponential number e (i.e., 2.718...) raised to the power of a double value.

Parameters: a Double value.

```
public static double log(double a)
```

Returns the natural logarithm (base e) of a double value.

Parameters: a Double value.

```
public static double sqrt(double a)
```

Returns the correctly rounded positive square root of a double value.

Parameters: a Positive double value.

```
public static double ceil(double a)
```

Returns the smallest (closest to negative infinity) double value that is not less than the argument and is equal to a mathematical integer.

Parameters: a Double value.

```
public static double floor(double a)
```

Returns the largest (closest to positive infinity) double value that is not greater than the argument and is equal to a mathematical integer.

Parameters: a Double value.

```
public static double atan2(double a, double b)
```

The regular atan() method accepts a value calculated by using y/x. The problem is, if either x or y is negative, the result of the fraction will also be negative but will not give a clue to which quadrant the angle is in. The atan2() method converts rectangular coordinates x, y (b, a) to polar (r, theta). This method computes the phase theta by computing an arc tangent of a/b in the range of -pi to pi.

Parameters: a The y value in a coordinate system.

b The x value in a coordinate system.

```
public static double pow(double a, double b)
```

Returns of value of the first argument raised to the power of the second argument.

Parameters: a The base number.
b The power to raise the base number to.

```
public static int round(float a)
public static long round(double a)
```

Returns the closest int to the argument. The result is rounded to an integer by adding 1/2, taking the floor of the result, and casting the result to type int. In other words, the result is equal to the value of the expression:

```
(int)Math.floor(a + 0.5f)
```

Parameters: a Value to round.

```
public static double random()
```

Returns a double value with a positive sign, greater than or equal to 0.0 and less than 1.0. Returned values are chosen pseudo-randomly with (approximately) uniform distribution from that range. When this method is called, it uses a static instance of a pseudorandom-number generator, exactly as if by the expression:

```
new java.util.Random();
```

This method is properly synchronized to allow correct use by more than one thread.

Parameters: a Double value.

```
public static int abs(int a)
public static double abs(double a)
```

Returns the absolute value of an int value. If the argument is not negative, the argument is returned. If the argument is negative, the negation of the argument is returned. Note that if the argument is equal to the value of Integer.MIN_VALUE, the most negative representable int value, the result is that same value, which is negative.

Parameters:	a	An int or double value.

```
public static int max(int a, int b)
public static double max(double a, double b)
```

Returns the greater of two int values. That is, the result is the argument closer to the value of Integer.MAX_VALUE. If the arguments have the same value, the result is that same value.

Parameters:	a	First number.
	b	Second number.

```
public static int min(int a, int b)
public static double min(double a, double b)
```

Returns the smaller of two float values. That is, the result is the value closer to negative infinity. If the arguments have the same value, the result is that same value. If either value is NaN, then the result is NaN. The floating point comparison operators consider negative zero to be strictly smaller than positive zero. If one argument is positive zero and the other is negative zero, the result is negative zero.

Parameters:	a	First number.
	b	Second number.

Object

In Java, all objects are subclasses of the Object class, even if the code does not declare that the class extends Object. If you follow the hierarchy of any class all the way to the top you will find the Object class. There are a eight methods in the Object class, of which six are important to a casual programmer: toString() and equals().

java.lang.Object

```
public boolean equals(Object aObject)
```

The equals method compares two reference variables and tests whether they refer to the same object. To compare two objects, call the method from one of them, and use the other object as an argument:

```
boolean match = firstObject.equals(secondObject);
```

Parameters: aObject Object to compare with.

```
public String toString()
```

Normally this method returns a string representation of an object, but most leJOS classes return an empty string, unless you override this method yourself. Once again, strings are not very important in robotics so this would only waste memory if it was implemented.

Runtime

The Runtime class allows you to check the memory in the EV3. The following two methods show how this works:

```
public static void showFreeMemory() {
  Runtime rt = Runtime.getRuntime();
  int free = (int)rt.freeMemory();
  LCD.drawInt(free, 2, 0);
}

public static void showTotalMemory() {
  Runtime rt = Runtime.getRuntime();
  int total = (int)rt.totalMemory();
  LCD.drawInt(total, 2, 2);
}
```

java.lang.Runtime

```
public static Runtime getRuntime()
```

This method returns an instance of Runtime since the freeMemory() and totalMemory() methods are not static.

```
public long freeMemory()
```

This method returns the amount of free memory in the heap. The heap is the amount of memory the user program has access to.

```
public long totalMemory()
```

This method returns the total memory of the heap.

```
public Process exec(String command)
```

This method executes a command in a separate process.

```
public void exit(int status)
```

This method terminates the currently running Java virtual machine.

String

The String class simply contains an array of char primitives and methods to access those characters. String is given special status in Java in that it is not necessary to use the new keyword to create a String object (although the option to use it is open):

```
String island = "Tahiti";
```

It is also possible to create a string by joining two other strings together as follows:

```
String island = "Pit" + "cairn";
```

java.lang.String

```
public char charAt(int index)
```

Returns the char value at the specified index.

```
public int indexOf(String str)
```

Returns index of the first occurrence of this substring.

```
public boolean equals(Object anObject)
```

Compares two strings to see if they have the same characters.

```
public char[] toCharArray()
```

Returns an array of characters representing the string.

```
public String toString()
```

Returns itself.

```
public int length()
```

Returns the length of the string.

StringBuffer

In Java, a String object is immutable, meaning characters in the string cannot be added or removed once the string is initialized. In order to create a new set of characters a new String object must be created, which uses memory.

The StringBuffer, on the other hand, can be modified after creation, so it is more flexible and has the potential to save memory. Most of the methods in the StringBuffer class have to do with appending characters to a character array. In fact, many of the methods mentioned above are included with StringBuffer.

java.lang.StringBuffer

```
StringBuffer append(boolean aBoolean)
StringBuffer append(char aChar)
StringBuffer append(int aInt)
StringBuffer append(long aLong)
StringBuffer append(float aFloat)
StringBuffer append(double aDouble)
StringBuffer append(String aString)
StringBuffer append(Object aObject)
```

Used to append a data type to the StringBuffer. If an object is used, the string representation of the object is used by calling toString().

Parameters: a Data to append to the String..

```
public char [] toCharArray()
```

Returns an array of characters representing the StringBuffer.

System

The System class allows a programmer to interact with the operating system and retrieve information from it. In this case, the operating system on the EV3 brick is Linux.

The system class also includes instances of in, out, and err streams, which you have probably seen used for outputting lines of text to the screen.

java.lang.System

```
public static long currentTimeMillis()
```

Returns the number of milliseconds since the EV3 was turned on. If you have a WiFi adapter, the milliseconds is updated from a time server on the Internet.

```
public long nanoTime()
```

Returns the most precise available system time, in nanoseconds.

```
public void exit(int status)
```

Terminates the currently running JVM.


```
public void exit(int status)
```

Terminates the currently running JVM.

Threads

Threads allow a program to execute several pieces of code simultaneously. They are very useful for robotics programming because each thread can be used to control a separate behavior. For example, you can use one thread per sensor that needs to be monitored. A single thread could be used to monitor the light in a room, while another could monitor a touch sensor.

You can also use threads to control different parts of your robot. For example, I could create one thread to control a gun turret or robot arm, and another thread to control wheel movement. Threads are very easy to create in Java; you simply extend the abstract Thread class and place the main code for your thread in the run() method. In order to demonstrate this, let's make a program that does two tasks at once - counting to 1000 and playing random music:

```
import lejos.hardware.*;
import lejos.hardware.lcd.*;

class BadMusic extends Thread {

 public static void main(String [] args) {
   new BadMusic().start();
   new Counting().start();
   LCD.drawString("Shatner Rocks", 1, 5);
   Button.waitForAnyPress();
```

```
      System.exit(0);
    }

    public void run() {
      while(true) {
        int freq = (int)(Math.random() * 1000);
        int delay = (int)(Math.random() * 300) + 100;
        Sound.playTone(freq, delay, 40);
      }
    }
  }

  class Counting extends Thread {
    public void run() {
      for(int i=0;i<1000;++i) {
        LCD.drawInt(i, 0, 2);
        try{Thread.sleep(1000);
        } catch(Exception e) {}
      }
    }
  }
```

The example above plays random, disjointed, Shatneresque "music" and is sure to be a favorite for many years to come. As you can see in this code, the program has two threads: BadMusic and Counting. BadMusic plays a random note for a random duration, one after another, in a never ending loop. The Counting class counts from zero to 1000, pausing for a second after each number.

Both of these methods are started in the main() method. Keep in mind that the main() method is its own thread, often called the primordial thread, so when you create a thread it will run concurrently with the main() code. Even though the main() method ends, the program will not terminate because the other two threads are still alive.

java.lang.Thread

`public static Thread currentThread()`

Returns an instance of the current thread, which will of course be the thread the line of code is running in.

`public int getPriority()`

Returns the priority of a Thread object.

`public boolean isAlive()`

Tests if this thread is alive.

`public boolean isDaemon()`

Checks whether this thread is a daemon thread (i.e. not a user created thread).

`public boolean isInterrupted()`

Checks whether this thread has been interrupted.

`public abstract void run()`

This method should be implemented with the main code for the thread.

`public void setDaemon(boolean on)`

Marks this thread as either a daemon thread or a user thread.

`public void setPriority(int priority)`

Sets the priority of the thread. Use the Thread constants to set the priority.

`public static void sleep(long milliseconds)`

Causes the thread to pause for a specific time.

`public void start()`

Begins the thread execution. (Remember not to call run() to start the method!)

`public static void yield()`

Causes the current thread to give way for another thread.

Event Model

leJOS uses the Java event model, which includes listeners and events. The Button and Motor classes use this model, as well as some classes in the navigation and object detection packages (which will be discussed later in this book). This model allows for clean, easy to understand code when the EV3 waits for an event to occur.

In the event model, an object acts as an event source, such as a sensor (imagine a touch sensor). One or more objects can register with the sensor to listen for events. When an event occurs (such as the touch sensor being pressed) all listeners are notified of the event and can respond accordingly (Figure 20-1).

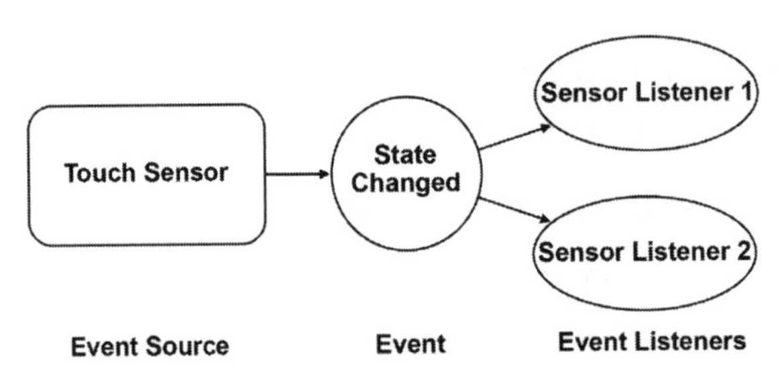

Figure 20-1 Touch Sensor Event Sequence

The following event listener listens for a button press. While it is listening your code can do anything, such as counting to 20 before exiting, as this program does. Try hitting enter to see that it can respond to events while other code executes.

```
import lejos.hardware.*;
import lejos.hardware.lcd.*;
import lejos.utility.Delay;

public class ButtonGalore implements KeyListener {

  public static void main(String[] args) {
    KeyListener listener = new ButtonGalore();
Button.LEDPattern(6);
Button.ENTER.addKeyListener(listener);
    for(int i=0;i<20;++i) {
  LCD.drawInt(i, 0, 2);
  Delay.msDelay(1000);
    }
  }

  public void keyPressed(Key k) {
    Sound.beep();
  }

  public void keyReleased(Key k) {
    Sound.buzz();
  }
}
```

Exceptions

Java uses an error handling scheme to deal with errors (exceptions) when they occur. Exception handling allows error checking code to be separated from the regular program logic. In theory, this makes the code tidy and easy to understand, rather than forming a tangled mess of error checking code and logical code. The common term for this is exception handling.

```
try {
  myThread.start();
} catch(IllegalThreadStateException e) {
  LCD.drawString("Oops.", 0, 1);
}
```

The main virtue of exceptions is that they propagate up the call chain, so it is not necessary to write methods with return values to indicate error codes. For example, exceptions allow you to seamlessly distinguish between a legitimate null return value and a method error.

Additionally, Java programs give stack traces instead of crashing the system, which is useful for locating the source of the error. When you program throws an uncaught exception, you can view the stack trace on the LCD, or output to the console in EV3 Command Center.

Recursion

Recursion is a programming technique that allows a method to call itself. Java (along with C language) allows recursion. Now that leJOS relies on the Oracle JVM, there are no longer limitations on the number of times a method can call itself.

This may seem like a never ending loop, which might be true in some cases. In practice we can check to see if a certain condition is true and in that case exit (return from) our method. The case in which we end our recursion is called a *base case*. Additionally, just as in a loop, we must change some value and incrementally advance closer to our base case.

```
void recursion(int counter) {
  if(counter == 0)
    return;
  else {
    System.out.println("hello" + counter);
```

```
    recursion(—counter);
  System.out.println("unwinding " + counter);
  }
}
```

In the above example, if you call the method with a starting value of 5, it will output hello and a number five times, with the number decrementing each time. Once it is done, however, and the methods start returning, you will see unwinding, with the numbers increasing. As you can see, recursion can be complex to think about.

Enumerations

Languages deal with numbers, but sometimes you just want a constant to represent something that is not necessarily a number. For example, black or red are not numbers, they are colors. Typically in the past, programmers used integers.

```
public static final int BLACK = 0;
public static final int RED = 1;
```

Then if a method required a color, the integer constant was used.

```
public void setColor(RED);
```

However, this was a little strange because you could pass it any number, even invalid numbers. You could also add different colors together, which was nonsensical.

Java introduced enums to cope with this. You can now create a set of enums using the following code.

```
enum Color {BLACK, RED, BLUE, YELLOW, WHITE}
```

Notice no semicolon is required after this statement. Now your new method can accept the Color enum in the method.

```
public class ColorSetter {
  enum Color {BLACK, RED, BLUE, YELLOW, WHITE}
  private Color myColor;
  public void setColor(Color color) {
    myColor = color;
  }
}
```

Feel free to continue using the old method of integers as enumerations if that is your preferred style. The enum type was added to Java after many of our leJOS classes were programmed, which means many of them still use integer constants.

Generics

The Java language has many ways to store collections of data, the most popular being ArrayList. In the old Java language, collections of objects always stored the Object type, and it was up to you to cast the Object into whatever specific object you stored in the collection.

```
String x = (String) list.get(0);
```

This was inconvenient to use a cast every time the collection retrieved an object, and it was also unsafe because code could inadvertently store the wrong object type in the collection, or the code could try to get the wrong type. If it did, it would fail at runtime as opposed to the compiler pointing out the error beforehand.

Java now supports generics, which allows you to specify which types of objects you can store in a collection. To do this, insert the class type within pointed brackets, such as <String>.

```
ArrayList <String> list = new ArrayList
<String> ();
```

It looks a little strange at first if you aren't used to using them, but now you can safely retrieve String objects from the collection.

```
String x = list.get(0);
```

CHAPTER 21

Navigation

TOPICS IN THIS CHAPTER

- Historical Navigation
- From Ships to Robots

Movement is a central concept in robotics. We humans take movement for granted because (after the first year of life) it comes very easily to us, requiring little conscious thought. With robots, it is a lot harder.

There are a surprising number of concepts to master in order to perform moves. We will cover each aspect throughout this book, trying out different techniques that will culminate in a robot that is able to perform reliable navigation.

In 2009, the leJOS developers began to notice a number of significant limitations with our existing navigation classes. We spent a lot of time designing a robust, simple, and reliable navigational architecture. To help us define the roles and responsibilities of basic navigation, we looked to past examples of navigation.

This chapter is a simple, light look at navigation with no code examples or robots to build. Enjoy!

Historical Navigation

The British Navy worked out a surprisingly good object oriented design for navigating the seas by assigning specific roles to officers and personnel. In doing so, they formed a chain of command to accomplish reliable navigation. Understanding these roles will help to understand how navigation occurs.

Navigation has been a vexing problem for most of human history. Untold books and movies have depicted the challenge of navigation, such as *The Bounty* (1984), *Longitude* (2000), and *Master and Commander* (2003).

There are two kinds of navigational players: those in the chain of command and those who merely supply intelligence. Those in the chain of command send action commands to other classes. Intelligence actors merely provide data when asked. We'll start at the bottom of the chain and work our way up.

Vehicle

At the very bottom of the chain of command is the vehicle, or in this example, a ship. The ship, like all vehicles, is capable of propulsion and steering. The ship uses wind and sails for propulsion, and a rudder dragging in the water for steering (see Figure 21-1).

Figure 21-1 A sailing ship

There are a diverse number of vehicles in existence: cars, airplanes, boats, hovercraft, and even futuristic walking vehicles. These vehicles use different methods for propulsion and steering.

Although each vehicle is capable of moving, their methods of control are very different. It seems as though we need an expert to control the vehicle. That brings us to the first player on the ship.

Pilot

A ship of the Navy has a wheelman who steers the ship (see Figure 21-2). The wheelman's specialty is that he knows how to control the movement of a ship.

Figure 21-2 The pilot of the ship

Just like there are lots of different vehicles, there are also lots of different pilots: a race car driver, a jet pilot, or even a futuristic battle-mech pilot. Each one of these players only knows how to control a specific vehicle.

Unfortunately, on his own, the pilot gets lost easily because he does not keep track of the position of his vehicle. We need a slightly more intelligent player to aid the pilot.

Navigator

The navigator is a middle-man who is able to take instruction from the captain and then tell the pilot which moves to make in order to get from one location to another (see Figure 21-3).

Figure 21-3 The navigator

The navigator keeps track of position using a coordinate system. This system is a grid containing lines of latitude and longitude. When the navigator is given a target coordinate, he figures out the moves needed to get to the target, and passes those moves to the pilot.

The navigator can only plot the moves if he currently knows where the ship is located. If he has no idea of the current coordinates of the ship, he'll have no idea how to plot a series of moves in order to get to a destination. To obtain the current coordinates, he needs the expertise of several diverse officers who are adept at pinpointing the current location.

Location Providers

On board the ship are a number of officers who are experts at identifying the location of the ship (see Figure 21-4). It is their job to answer the perplexing question, "Where am I?" At their disposal are a number of navigational instruments to help them obtain a location estimate: a sextant to perform sun sighting, a long rope with knots that drags behind the ship to determine speed, a compass, and of course, the stars. Officers can even use a telescope to identify islands in the distance.

Figure 21-4 Locating current position

So what information do these location providers provide? The pertinent information comes down to coordinate information (latitude and longitude) and heading. But with so many different sources of information (stars, sun sighting, island sighting), each with their own reliability, how do we put this information together into one best-guess at the current location?

Localization Filters

The crew needs one officer to take all the readings and figure out the most accurate reading possible. This officer is called the master's mate. He is the highest ranking officer on this ship, short of the captain. It is up to him to collect data from the officers and use his judgment to synthesize a single reading to present to the captain.

In essence, the master's mate is a localization filter. A localization filter uses a number of strategies to figure out the most accurate reading from the various location providers. In programming terms, a Kalman filter is often used to synthesize several readings into one more reliable reading.

Maps

A cartographer uses observational skills to produce a representation of where the ship has been. Often the cartographer was not on the ship, but his maps were (see Figure 21-5). Maps are important to sailing ships because they allow the captain to avoid dangerous reefs and seek out towns. These map features are known as landmarks.

Once a map is created, the information can be used to plot a complicated route that avoids obstacles and seeks out a specific target. The map is used by the...

Path Planner

The navigator merely gets from one point to another, but he is largely ignorant about what sort of obstacles lay between those points. The captain, on the other hand, is a little more sophisticated. Using map data, the captain plots a series of points that avoid dangerous obstacles. The path planner is the captain (see Figure 21-5).

Figure 21-5 The captain using a map to plot a course

The captain is told by a higher authority where to go, but it is up to the captain to plot the trip so that the crew is not endangered. So who tells the captain what to do?

Mission Planning

At the top of the navigational hierarchy are the admirals. They determine where the ship is going and what the crew will do once they get there. Once they know where they want the ship to go, they give the captain a goal, and it is up to the captain to carry out those orders.

From Ships to Robots

Now that you have some idea of the tasks required to make a complete navigational architecture, let's examine how these might be carried out in code. Some of the goals we had when designing a navigation API are as follows.

1. Compatible with many different types of robots, not just two-wheeled differential steering robots.
2. A general API compatible with EV3 and non-EV3 hardware.
3. Generate information from a wide variety of different sensors such as compass, GPS, tachometers, gyroscope, accelerometer, range finder—the list goes on and on.
4. Ability to synthesize state information (speed, location, heading) from multiple sensor sources.
5. Ability to use different path finding algorithms.

All of these goals ended up in the Navigation API.

Navigation API

The navigation API provides a convenient set of classes and methods to control a robot. These methods allow the robot to control the direction of movement, move to a point and keep track of coordinates. The next several chapters will guide you through using each of these objects one at a time.

The data flow of these objects is the same as the chain of command described above for the British Navy. All commands flow down the chain of command, and never the other way (although data can flow between different objects). The chain of command is shown in Figure 21-6.

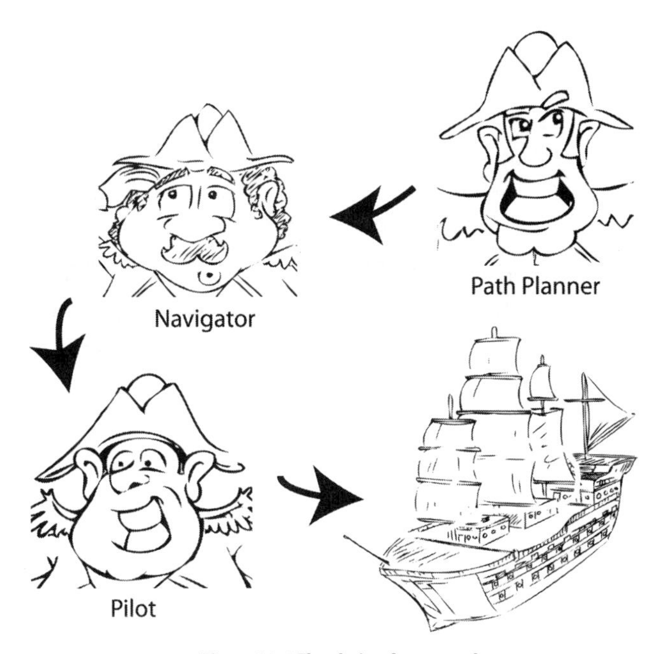

Figure 21-6 The chain of command

As previously mentioned, the data providers do not issue commands. They merely report location data. Data flow moves to the Pilot, Navigator, or Mission Planner.

One mystery remains: who or what is the mission planner? As it turns out, the mission planner is you! You get to set goals for your robot. Maybe the mission is as simple as bringing a plate of cookies from the kitchen to the living room. Perhaps it is more complex, such as controlling a team of three soccer playing robots. You get to use your intelligence and creativity to plan out your own mission. To find out how, continue reading the remaining chapters on navigation.

CHAPTER 22

Moves

TOPICS IN THIS CHAPTER

- Metered Moves
- The Pilot
- Differential vehicles
- Basic Movement Project
- Move Listeners

In this section, we will cover the first and arguably most important step in navigation, which is basic movement. The goal is to create vehicles that can perform precise moves. As briefly noted in the previous chapter, the pilot is responsible for driving, steering, and turning a vehicle.

The actual physical characteristics of the robot are hidden from every class in the navigator package except for the pilot. For example, a robot can roll, walk, jump, fly, swim or float from one location to another, but the external pilot methods to make it move all look the same. The great thing about using pilots is that they allow diverse types of robots to participate in navigation, regardless of their construction.

Metered Moves

As we saw earlier in the book, it is possible to move a robot around using only the motor classes, simply by rotating the motors forward and backward. However, to drive to specific locations, it is necessary to have control over the distances a vehicle moves and the angles that it turns.

One of the primary goals of the navigation classes is to allow precise metered control of vehicle moves. The leJOS developers spent significant time thinking about what moves are and how to represent them with data. Eventually we concluded movement can be represented by four basic types (see Figure 22-1).

1. Straight line travel
2. Arcs (a curved line)
3. On-the-spot rotations
4. Stop

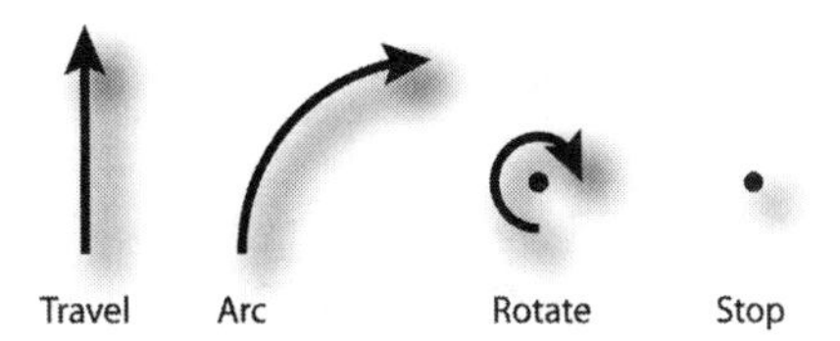

Figure 22-1 Determining the four basic move types

By stringing together a sequence of these moves, a vehicle can travel from one location to another. A robot is always performing one of these four moves. Let's decompose these moves further.

A move is composed of two smaller movements—the amount it travels and the amount it rotates throughout the move. These component moves take place simultaneously to create the four complex moves above. Table 22-1 shows each of these moves.

Move Type	Distance	Angle	Permutations
travel	+ or -	0	2
rotate	0	+ or -	2
arc	+ or -	+ or -	4
stop	0	0	1

Table 22-1 Every permutation of distance and angle

Note that distance and angle can be either positive or negative. This means that there are actually two kinds of travel (forward or backward), two kinds of rotate (clockwise or counter-clockwise), and four kinds of arcs. This produces a total of nine discrete moves.

In leJOS, these moves are represented by the Move class, which is a simple container of move information. You can use the Move class to tell a robot what move to make, or you can retrieve a Move object to indicate what kind of move a robot just made. These are the core methods of the Move class:

- getTravelDistance() – the distance moved, normally in centimeters (we will explain this later in the chapter).
- getTurnAngle() – the angle the robot rotated over the course of the move, in degrees.
- getArcRadius() – the radius of the arc that was travelled.

Now that we have examined all the moves a vehicle is capable of executing, let's find out how to actually perform those moves with a physical robot. We will need to call on our reliable friend, the pilot (see Figure 22-2).

The Pilot

There are many types of robots: two-legged walkers, four-legged walkers, car-steering vehicles, and differential robots (such as Rov3r, which appears later in this chapter) to name a few. In leJOS jargon, a Pilot is a class that controls a specific robot type. A Pilot can only control one type of robot. For example, the DifferentialPilot class can only control a differential drive robot with two drive wheels and a caster wheel for balance.

Figure 22-2 Steering and control courtesy of the pilot

The DifferentialPilot is the workhorse of the Pilot classes. Most people use this type of robot because differential motor control is capable of all the moves and it is simple to build the chassis.

Differential steering has one requirement: the robot must be able to turn within its own footprint. That is, it must be able to change direction without changing x and y coordinates. The robot wheels can be any diameter because these physical parameters are set by the constructor. Once these parameters are set, the pilot class works the same for all differential robots.

The leJOS developers have put in a lot of effort to make the DifferentialPilot as accurate as possible. The internal equations to keep track of movement are very accurate, down to fractions of an inch. One source of inaccuracy is skidding, which occurs when a robot starts, stops, or performs turns. Depending on the surface, skidding can produce large errors. To combat this, the DifferentialPilot attempts to accelerate and decelerate slowly and smoothly. You can even adjust the acceleration and speed using these DifferentialPilot methods:

- setAcceleration() – sets the acceleration of the robot in degrees/second2
- setRotateSpeed() – set the speed at which a robot will rotate, in degrees/second
- setTravelSpeed() – set the speed at which the robot will travel, in units per second

The DifferentialPilot has methods to perform all of the basic moves, which were described earlier in this chapter:

- arc(double radius, double angle)
- backward()
- travel(double distance)
- rotate(double angle)
- forward()

Let's try using the DifferentialPilot.

Basic Movement

In this section we will create a project that can perform some basic moves. We'll use a simple differential drive robot called Rov3r. Rov3r uses differential steering, thus making it 100% compatible with the DifferentialPilot class.

The robot will also be used in later chapters as the workhorse navigating robot for the book, so don't tear it apart quite yet when you are done with this chapter.

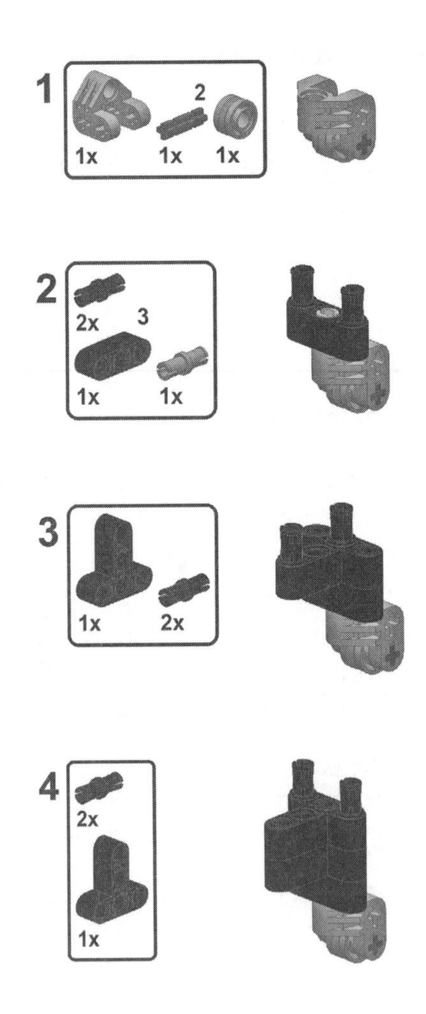

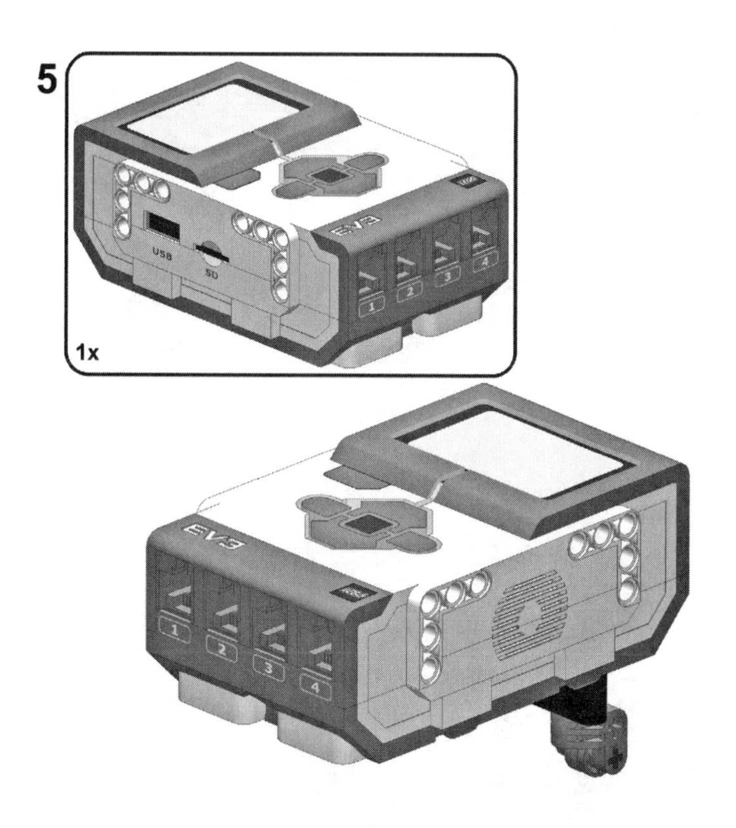
5
1x
USB
SD
EV3
1
2
3
4

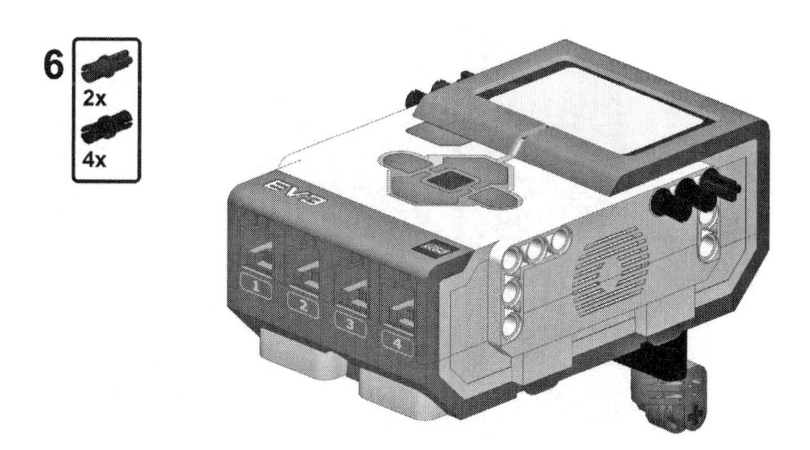
6
2x
4x
EV3
1
2
3
4

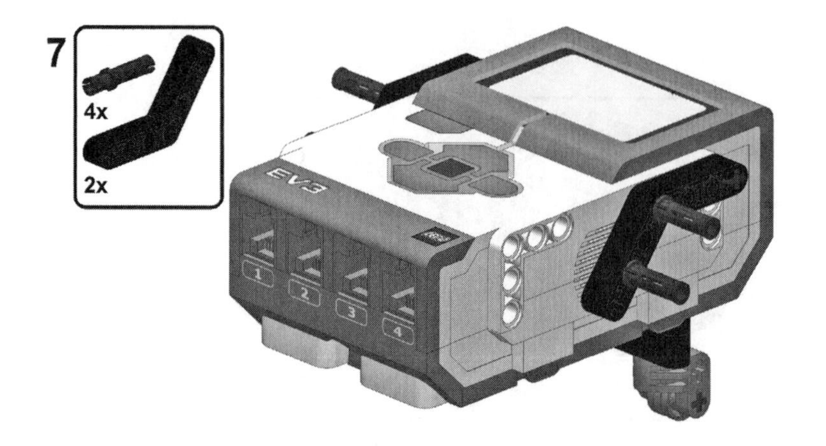
7
4x
2x
EV3
1
2
3
4

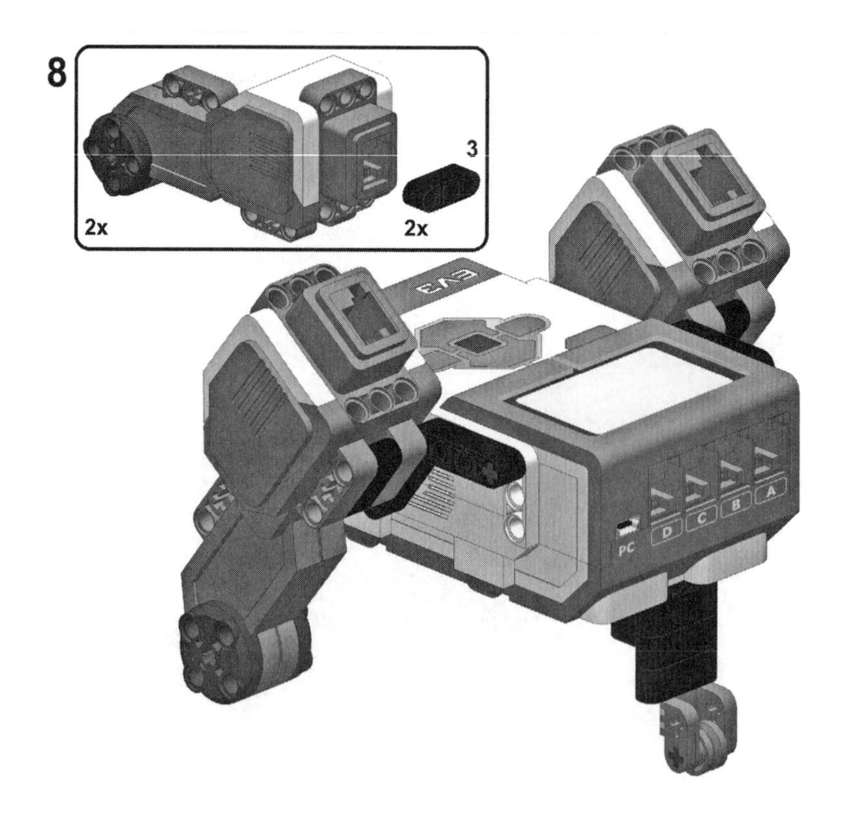
8
3
2x
2x
EV3
PC
D
C
B
A

9

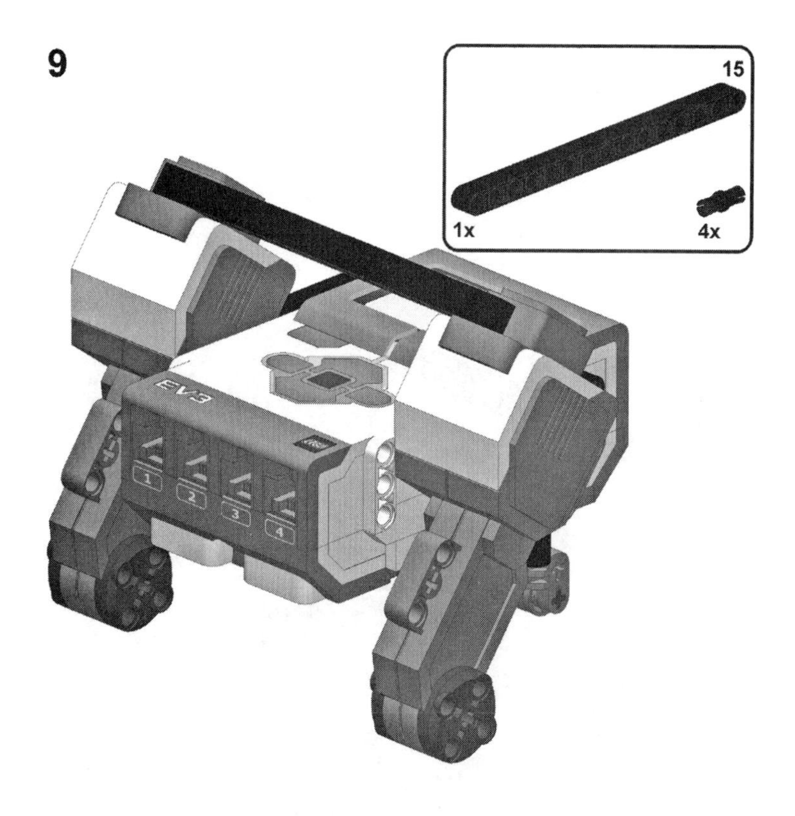

10

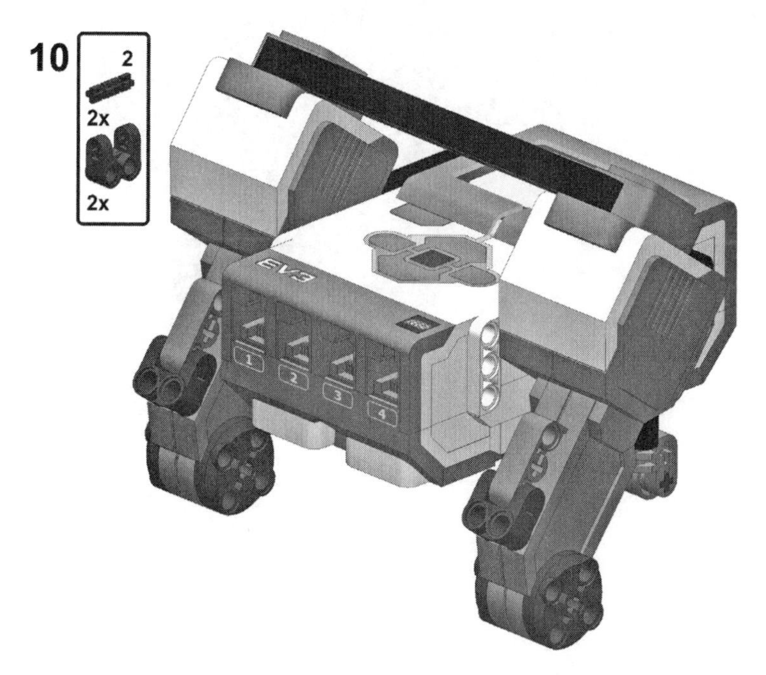

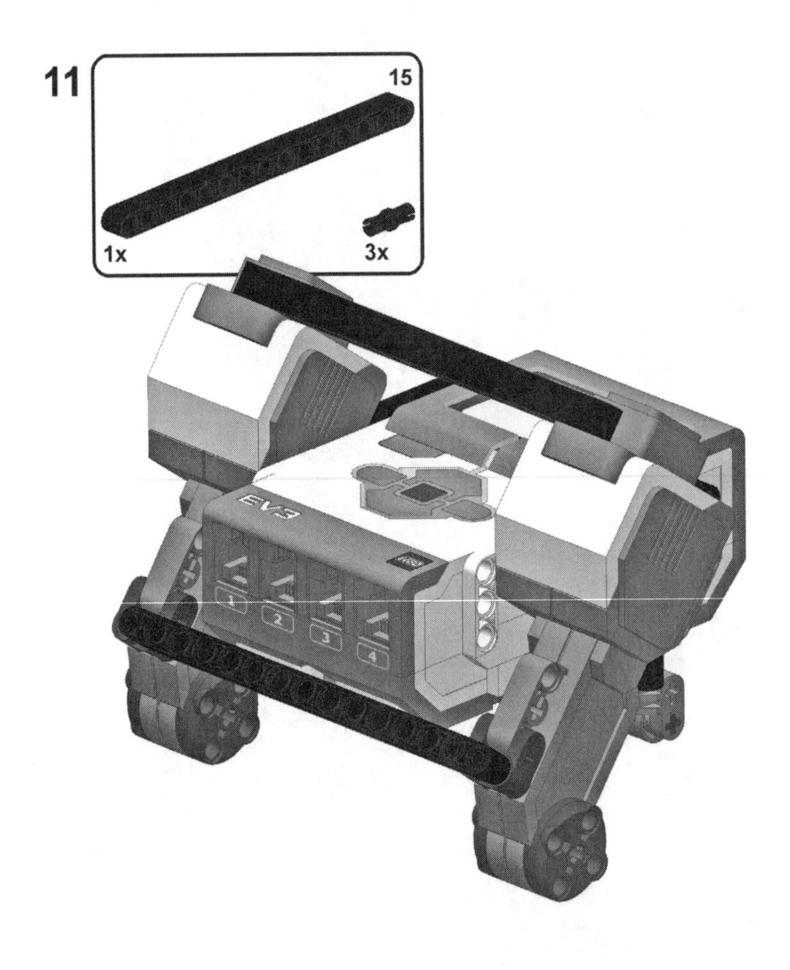
11
15
1x
3x
EV3
1
2
3
4

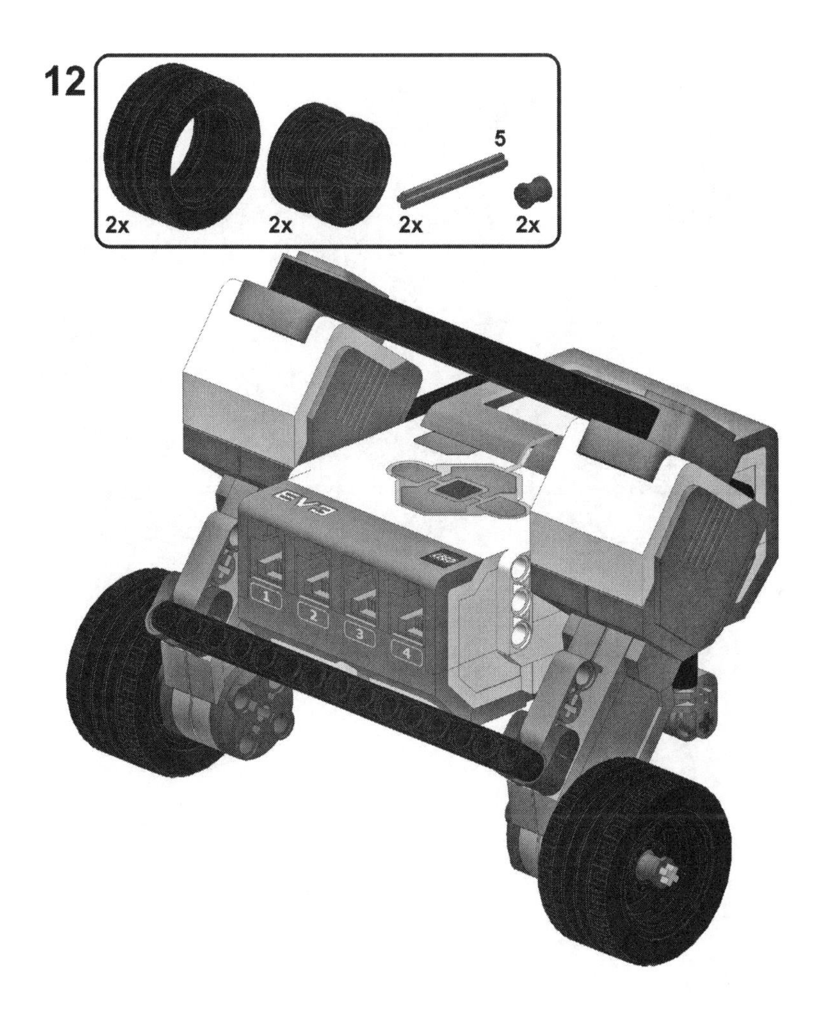
12
5
2x
2x
2x
2x
EV3
1
2
3
4

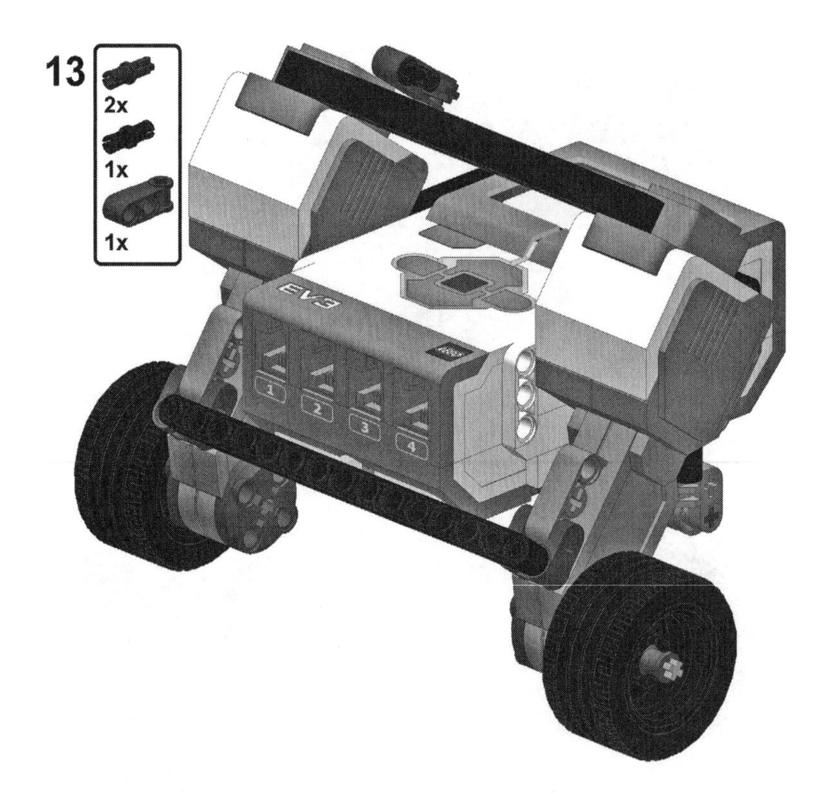
13
2x
1x
1x
EV3
1
2
3
4

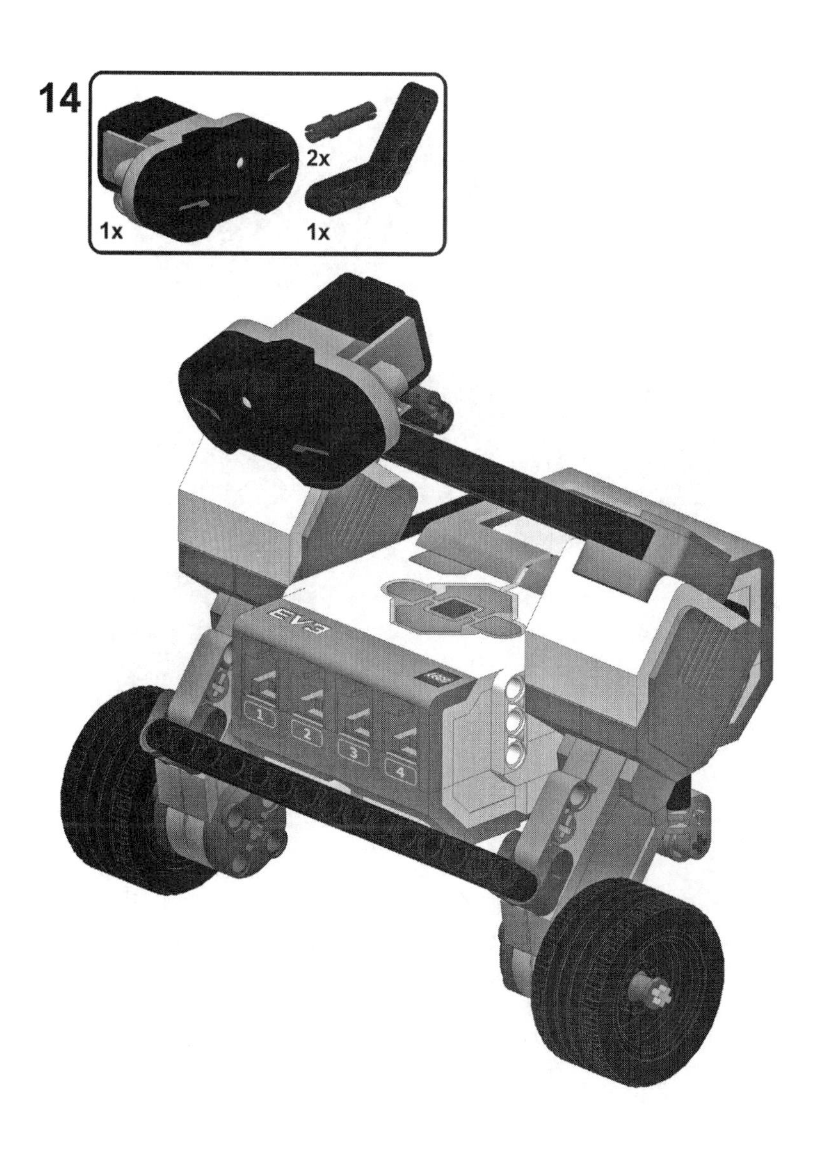
14
2x
1x
1x
1
2
3
4

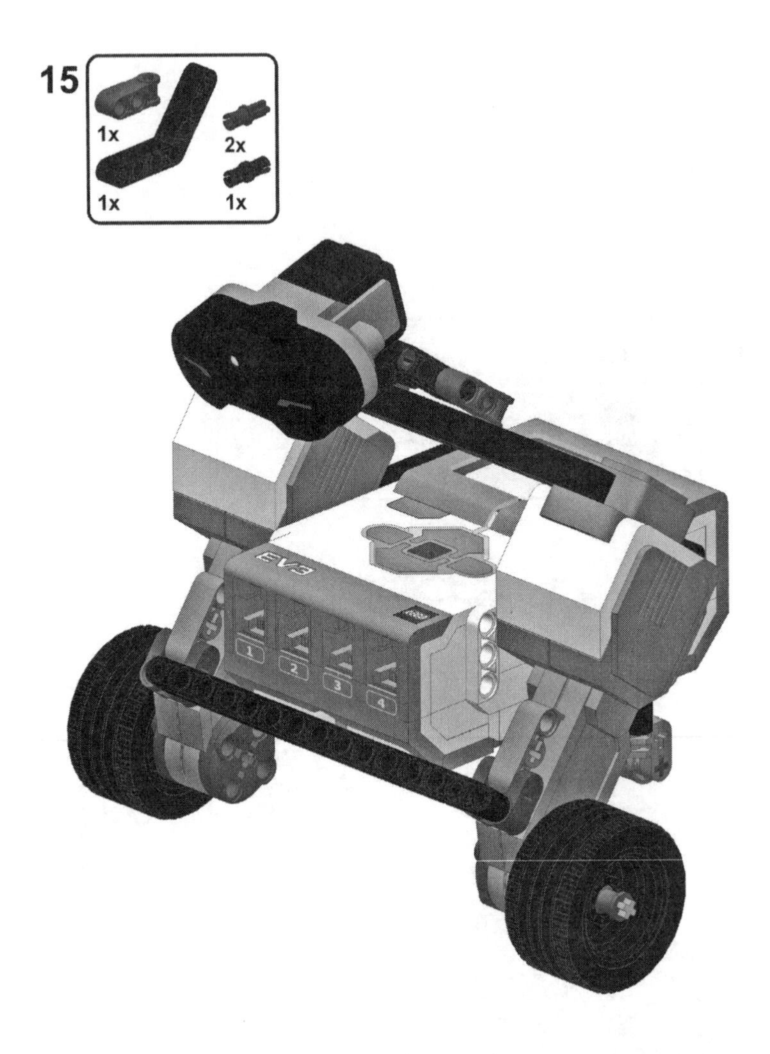

Cables

Connect a short cable from port 4 to the IR sensor. Now plug the right motor into port B, and the left motor into port C using medium cables.

Programming a DifferentialPilot

The DifferentialPilot class requires three basic parameters to work properly:

- Tire Diameter
- Track Width
- Motors

Diameter is the widest measurement from one side of a circle to the other. This is easy to acquire because LEGO prints the diameter right on the tire wall. The tires in the EV3 kit are 4.32 cm and are identical to the tires from the NXT 2.0 kit. These tires look like street racing tires, with a wide surface contacting the floor (see Figure 22-3).

Figure 22-3 EV3 "Street Racing" tires

In order to record accurate rotations when the robot turns, it is important to know the measurement from wheel to wheel, called *track width*. There should be a theoretical point where the wheel touches the ground.

The flatness of the EV3 tires makes it difficult to determine precise track width. Since LEGO tires are symmetrical, the best way is to measure from the center of one tire to the center of the other (see Figure 22-4). Rov3r has a track width of 15.2 cm according to my measurements. However, measuring closer to the inner points of these tires works better due to the nature of these tires. I used 14.5 cm.

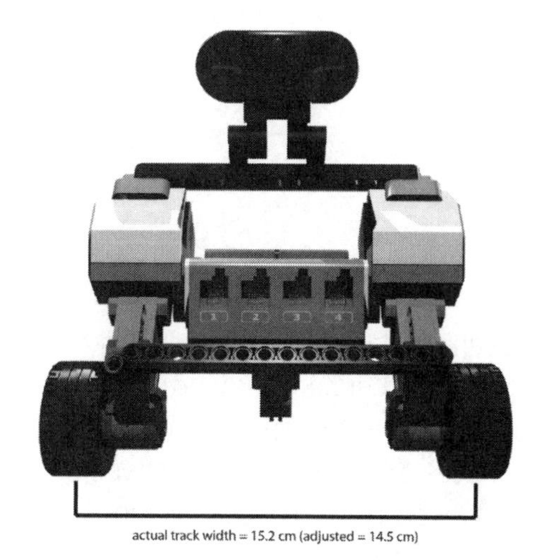

Figure 22-4 Measuring track width of Rov3r

The final parameters for the DifferentialPilot are the motors for the left and right wheels. For the differential pilot, these are the motors plugged into the B and C ports.

Now that we know all the parameters of the Rov3r, let's lay down some code to test the accuracy of distances and rotations. The following program attempts to drive the robot forward 100 cm, and then rotate three times (360 x 3 = 1080 degrees). We could rotate only once, but three times will allow an average value. Before running the program, try measuring out 100 centimeters on the ground to see how accurately it moves. When it is done moving forward, press the center button and monitor how accurately it rotates three times.

```
import lejos.hardware.lcd.*;
import lejos.hardware.motor.*;
import lejos.hardware.Button;
import lejos.robotics.navigation.*;

public class MeasureMoves {

  public static void main(String [] args) {
    double diam = DifferentialPilot.WHEEL_SIZE_EV3;
    double trackwidth = 15.2; // or 14.5

    DifferentialPilot rov3r = new
DifferentialPilot(diam, trackwidth,
Motor.C, Motor.B);
    rov3r.travel(100);
    LCD.drawString("Press ENTER", 0, 3);
    Button.ENTER.waitForPressAndRelease();
    rov3r.rotate(1080);
    Motor.B.close();
    Motor.C.close();
  }
}
```

Results

Hopefully you have been able to refine the track-width enough so your robot performs three full rotations accurately. It is worth trying the robot out on a few different surfaces to notice how the robot behaves. Carpet is much different from smooth floor surfaces. Small differences between hard surfaces like linoleum, tile, glass and hardwood also make a difference. Adjust the track-width value for your particular surface.

You might also notice that the tires are not entirely consistent from one test to the next. The reason is because the EV3 tires are wide and flat, therefore different parts of the tire make contact with the surface over the course of a move. Tires which are rounded and come to somewhat of a "point" are actually better, such as LEGO motorcycle tires or LEGO NXT 1.0 tires.

Just for Fun!

Did you know that the DifferentialPilot allows tires with different diameters? If you have access to larger LEGO tires, substitute one of the tires (see Figure 22-5). Then try this alternate constructor in place of the one used in the code above—but make sure you substitute the correct track-width and wheel diameters:

```
DifferentialPilot robot = new
DifferentialPilot(8.16, 4.32, 17.1,
Motor.B, Motor.C);
```

Normally you would think that having one tire larger than the other would cause it to drive in a circle, even though it is trying to drive straight. However, watch closely and you'll notice the small wheel turns faster than the large wheel. DifferentialPilot takes the different wheel sizes into account and precisely spins the smaller tire faster, producing a straight line. (Yes, there is some serious math hiding in many of the leJOS classes.)

Figure 22-5 Piloting a robot with different tire diameters

NOTE: *Sometimes the motors are reversed, causing your robot to move in directions you did not intend. This is easy to correct:*

Problem	Solution
Goes backward instead of forward	Use a MirrorMotor class to reverse motor directions.
Robot turns right instead of left	Swap the motors in the constructor (or swap the port cables)

Perpetual Rover

Now that we have the correct values for the DifferentialPilot, let's try a more substantial program. The short program below produces a simple perpetually moving robot that will move forward until it encounters an obstacle, at which time it backs up and changes direction.

```
import lejos.hardware.lcd.LCD;
import lejos.hardware.motor.*;
import lejos.hardware.port.*;
import lejos.hardware.sensor.*;
import lejos.hardware.Button;
import lejos.robotics.navigation.*;
import lejos.utility.Delay;

public class PerpetualRov3r {

  public static void main(String [] args) {
    double diam = DifferentialPilot.WHEEL_SIZE_EV3;
    double trackwidth = 15.2;
    DifferentialPilot rov3r = new
DifferentialPilot(diam, trackwidth,
Motor.C, Motor.B);

    EV3IRSensor ir = new EV3IRSensor(SensorPort.S4);
    SensorMode distMode = ir.getMode("Distance");
    int distance = 255;

    rov3r.forward();
    while(!Button.ESCAPE.isDown()) {

      float [] sample = new float[distMode.
sampleSize()];
```

```
distMode.fetchSample(sample, 0);
     distance = (int)sample[0];
     LCD.drawString("DIST: ", 0, 2);
     LCD.drawInt(distance, 3, 5, 2);

     if(distance < 70 && distance !=0) {
       rov3r.travel(-20);
       rov3r.rotate(45);
       rov3r.forward();
     }
     Delay.msDelay(100);
   }
   ir.close();
  }
}
```

Make sure to substitute the appropriate tire size and track width for your robot and environment.

Move Listeners

One final topic before we move on is the MoveListener interface. The MoveListener is primarily used for communication between the navigation classes, but there are several practical scenarios where you might want to use a MoveListener.

For example, if you have a robot that has an infrared sensor pointed forward and mounted on a motor, you might want it to rotate slightly left or right if the vehicle begins steering around a corner so the sensor is pointed where the robot is traveling. A MoveListener could listen for arc movements and rotate the sensor appropriately.

So how does it work? First, create a class that implements the MoveListener interface. There are two methods your class must implement:

```
moveStarted(Move event, MoveProvider mp)
moveStopped(Move event, MoveProvider mp)
```

Once your class has these implemented, you can add an instance of this class to the pilot, which is a MoveProvider. All pilots are MoveProviders; they make moves and are therefore capable of reporting the moves they just made. The MoveProvider method to add a MoveListener is as follows:

```
addMoveListener(MoveListener listener)
```

Let's alter the example above to include a MoveListener that will output the results of every move to the LCD display:

```
import lejos.hardware.lcd.LCD;
import lejos.hardware.motor.*;
import lejos.hardware.port.*;
import lejos.hardware.sensor.*;
import lejos.hardware.Button;
import lejos.robotics.navigation.*;
import lejos.utility.Delay;

public class PerpetualListener implements
MoveListener {

  public static void main(String [] args) {

    double diam = DifferentialPilot.WHEEL_SIZE_EV3;
    double trackwidth = 15.2;

    PerpetualListener listener = new
PerpetualListener();
    DifferentialPilot rov3r = new
DifferentialPilot(diam, trackwidth,
Motor.C, Motor.B);

rov3r.addMoveListener(listener);

    EV3IRSensor ir = new EV3IRSensor(SensorPort.S4);
    SensorMode distMode = ir.getMode("Distance");
    int distance = 255;

    rov3r.forward();
    while(!Button.ESCAPE.isDown()) {

      float [] sample = new float[distMode.
sampleSize()];

distMode.fetchSample(sample, 0);
      distance = (int)sample[0];

      LCD.drawString("DIST: ", 0, 2);
      LCD.drawInt(distance, 3, 5, 2);

      if(distance < 70 && distance != 0) {
        rov3r.travel(-20);
        rov3r.rotate(45);
        rov3r.forward();
      }
```

```
      Delay.msDelay(100);
    }
    ir.close();
  }

  public void moveStarted(Move move,
MoveProvider mp) {}

  public void moveStopped(Move move,
MoveProvider mp) {
    LCD.drawString("Moved " + (int)move.
getDistanceTraveled() + " cm ", 0, 4);
  }
}
```

There are lots of other scenarios in which you might find the MoveListener interface useful. If you are outputting information to a GUI, you might want to display all the moves the robot made in real-time. You could also have multiple robots reporting moves to a single MoveListener GUI (one path drawn in red and one in blue). In this scenario, the MoveListener code would need to differentiate the robots from each other by checking the MovementProvider variable and drawing the appropriate color.

Just for Fun!

The simplest type of Pilot implements only the MoveController interface (see the API docs to explore this interface). This interface allows movement forward and backward only, with no means to rotate. Why did we incorporate such a simple interface? Do such vehicles exist in real life? There are certainly slot cars that only move forward or backward. In the game *Portal 2*, the robot character Wheatley travels all over an enormous research complex on a guided rail system. Of course, the most popular vehicle of this type is a locomotive. All locomotives have no steering mechanism, and thus only forward or reverse movement is possible.

Let's build a useful MoveController! Imagine a baby plant with high hopes of becoming a large plant someday. This plant requires sun. To help the plant achieve its dreams, program a MoveController that uses a light sensor to find the brightest spot under a window; that is, it tracks the beam of sunlight throughout the day as the sunlight moves across the floor. The robot can move the plant forward and backward to keep it in the brightest spot under the window.

CHAPTER 23

Coordinates

TOPICS IN THIS CHAPTER

- Cartesian Coordinates
- Navigator API
- The Rambler

Human language can describe a location in words: "I am in the living room. You are at the corner of 5th Avenue and 3rd Street. She is in Texas." However, these descriptions mean nothing to your LEGO robot because it has no understanding of words. Instead, your robot can understand numbers. In this chapter, we will use a Cartesian *coordinate system* to describe locations using numbers.

Cartesian Coordinates

A two-dimensional coordinate system keeps track of two numbers, x and y. Numbers grow larger and smaller along the x and y axes. Both of these axes start at zero and include positive and negative numbers (Figure 23-1). The x and y axes divide the system into four quadrants. Any *point* in a two dimensional area can be plotted on this grid using values of x and y.

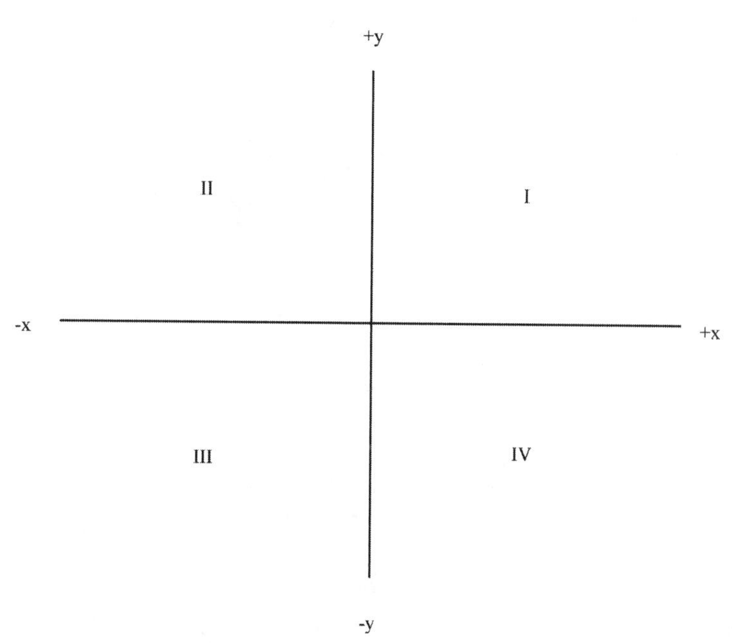

Figure 23-1: A Cartesian coordinate system.

Furthermore, in a Cartesian coordinate system, rotations to the left (counter-clockwise) are designated as positive rotations. Perhaps on some other planet in the universe, there is a population that decided to use clockwise rotation as positive. Not this planet, however, so we are stuck with it.

If you rotate +90 degrees, it means you rotate counter-clockwise one-quarter turn. Likewise, a rotation of -90 degrees is equivalent to rotating clockwise one-quarter turn.

The Navigator API

Now that we know a little bit about coordinates, how do we make a robot drive to a specific location? Earlier in this book we learned about pilots, and how pilots allow a robot to perform moves and drive specific distances. Using a pilot, there is another class called Navigator that tells the pilot how to drive to a specific location (see Figure 23-2).

Figure 23-2 The navigator thinking up some moves

The Navigator accepts a MoveController (any kind of pilot object) in its constructor. It calculates a series of moves to drive from one location to another. The Navigator knows nothing of how the robot works and it does not care how the pilot figures out how to move around. Instead, it just asks the pilot to execute the moves.

The following code (which assumes a pilot exists) instantiates a Navigator and drives to the target coordinate x=100, y=200.

```
Navigator nav = new Navigator(pilot);
nav.goTo(100, 200);
```

It is pretty simple so far. The target coordinates are sometimes known as waypoints. In fact, if you coded a program to generate a series of waypoints, you could feed the coordinates to the navigator using Waypoint objects:

```
Waypoint wp = new Waypoint(100, 200);
nav.goTo(wp);
```

Sometimes you need to plot a predetermined path of waypoints for a robot to follow. In this case, you can feed the waypoints to a navigation queue using the addWaypoint() method.

```
for(int x=0, x<1000, x+=100)
  nav.addWaypoint(new Waypoint(x, 20));
```

Rambling Around

To demonstrate coordinate navigation, let's take the trusty Rov3r robot from Chapter 22 and put it through a randomly generated obstacle course. The following code will make the robot venture from its starting point to a random location and back again. Each time it will display the coordinates so you can see where it is headed. It will wait for you to press the square enter button between moves.

> ***Just for Fun!***
>
> How many ways can you store x and y? You might have noticed leJOS has a lot of containers that deal with x and y coordinates. We have Point, Pose, Coordinate, Waypoint, and Node. Why so many? Each container serves a specific task, and normally has a set of helper methods that deal with that task. For example, the Node container has methods for adding and removing other nodes from the set (more on nodes later).

```
import lejos.robotics.navigation.*;
import lejos.utility.Delay;
import lejos.hardware.Button;
import lejos.hardware.Sound;
import lejos.hardware.lcd.LCD;
import lejos.hardware.motor.*;

public class Rambler {

  public static final int AREA_WIDTH = 200;
  public static final int AREA_LENGTH = 200;
  public static boolean exit = false;

  public static void main(String[] args) throws
Exception {
```

```
    DifferentialPilot p = new DifferentialPi
lot(DifferentialPilot.WHEEL_SIZE_EV3, 15.2,
Motor.C, Motor.B);
    Navigator nav = new Navigator(p);

    // Repeatedly drive to random points:
    while(!exit) {

      LCD.drawString("Target: ", 0, 1);
      double x_targ = Math.random() * AREA_WIDTH;
      double y_targ = Math.random() * AREA_LENGTH;
      LCD.drawString("X: " + (int)x_targ +
" ", 0, 3);
      LCD.drawString("Y: " + (int)y_targ +
" ", 0, 4);
      LCD.drawString("ENTER to drive", 0, 6);
      LCD.drawString("or press EXIT", 0, 7);
      Sound.beepSequenceUp();
      Button.LEDPattern(4);
      int button = Button.waitForAnyPress();
      Button.LEDPattern(0);

      if(button == Button.ID_ENTER) {
        nav.goTo(new Waypoint(x_targ, y_targ));
        nav.addWaypoint(new Waypoint(0, 0, 0));
        while(nav.isMoving())
          Delay.msDelay(50);

      } else exit = true;
    }
  }
}
```

Testing and Results

You need an area of about four meters square for this course (two by two meters). If you have more or less than that, you can change the two area constants at the start of the program (these values are in centimeters). Try placing a dime at the starting location so you can see how much accuracy it loses each time it returns. It is important to have Rov3r precisely lined up along the imaginary x-axis at the start (see Figure 23-3). You could even place a dime at the randomly generated target each time to see how close it gets.

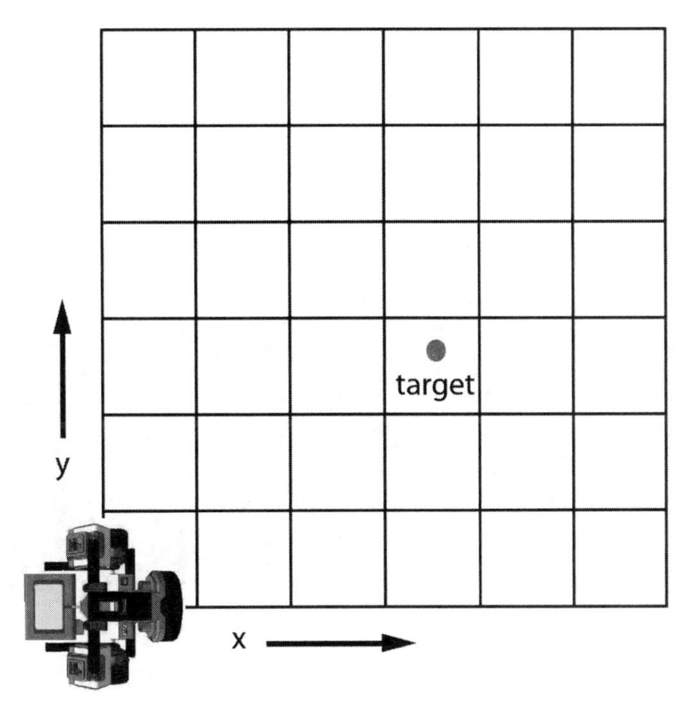

Figure 23-3: Lining up the robot along the x-axis

Rov3r does a good job of measuring distances but the robot has a weakness when it rotates. You will probably notice that after the first waypoint the accuracy is off slightly, which causes it to miss the second waypoint even further, and so on.

The errors are cumulative, so the longer your robot travels, the further off course it becomes until the location coordinates are essentially meaningless. This is a chronic problem with tachometers. You can minimize drift problems by running the robot on a smooth, hardwood floor, but you can also seek to improve rotation accuracy.

CHAPTER 24

Arcs

TOPICS IN THIS CHAPTER

- Arcs
- Navigation with Steering

The most popular robot chassis is the differential steering robot. A version of this chassis, Rov3r, was introduced earlier in this book. The reason for its popularity is because these robots are easy to build and capable of performing a diverse set of moves.

However, the most popular type of vehicle on earth, the automobile, uses an analog steering mechanism. This method of travel is ideal for humans because steering cars are stable at high speeds and they are easy to operate.

It would be a big omission if the navigation package left out the most popular vehicle on the planet. However, steering poses a special challenge because a steering vehicle is unable to rotate within its footprint. Instead, it can only change direction by driving an arc.

Arcs

When driving a car, you rotate the steering wheel to make a turn. This causes the vehicle to drive an arc. If you turn the wheel slightly, it causes a shallow arc and if you turn the wheel sharply, it causes a tight arc (see Figure 24-1).

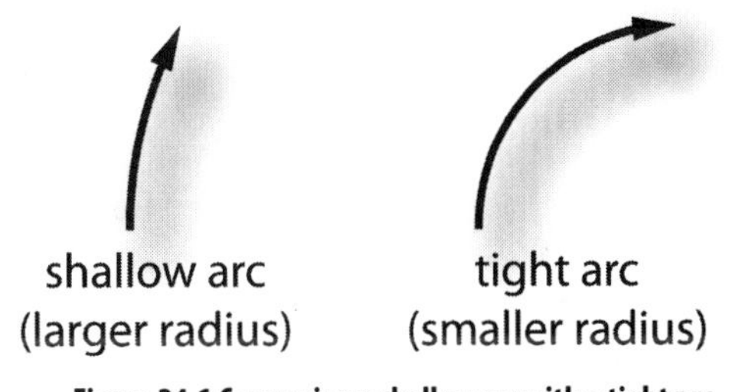

Figure 24-1 Comparing a shallow arc with a tight arc

What is an arc and how can we tell a robot how to drive a specific arc? Several components define an arc, notably the radius and angle (see Figure 24-2). The arc angle (degrees) can also be expressed as a length in units, such as centimeters. An arc with a larger radius will produce a shallow arc, and an arc with a smaller radius will produce a tight arc (revisit Figure 24-1 above).

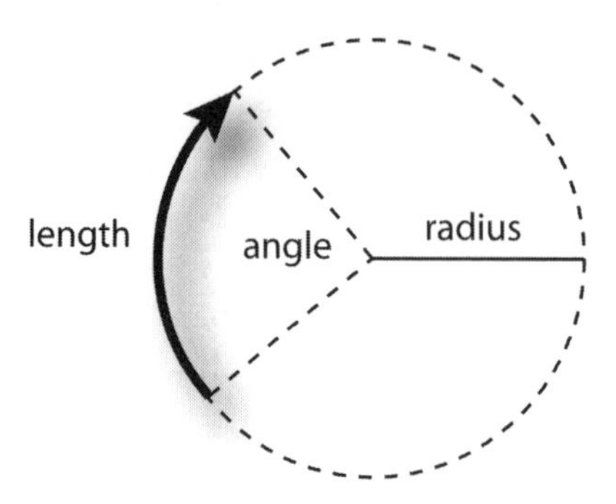

Figure 24-2 Reviewing the components of an arc

In chapter 22, we examined four types of moves: straight travel, arc, rotate, and stop (see Figure 22-1). We also noted that within the arc move, there are actually four sub-types. These types are determined by the direction of movement around the arc, and whether the center of the circle lies to the left or right of the robot when it is performing an arc (see Figure 24-3).

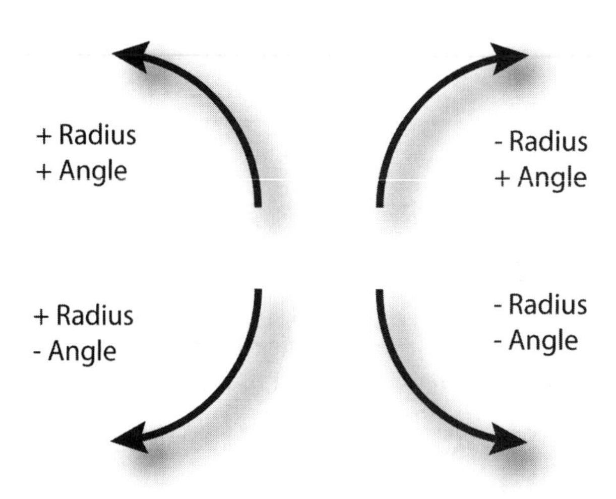

Figure 24-3 The four kinds of arc moves

NOTE: *This is sure to confuse you even more. Recall earlier that in a Cartesian coordinate system, positive angles are counter–clockwise. Therefore, technically the right side of Figure 24-3 is incorrect. When a robot travels forward and right along an arc, the angle should be negative. However, we use the opposite approach with leJOS because it is confusing to give commands to a pilot class using a true Cartesian system. In leJOS, a positive angle moves the robot forward while a negative angle moves it backward.*

For example, if you ask a robot to travel an arc of 110 degrees and a radius of -25, it will drive forwards and to the right. If we ask it to arc 110 degrees with a radius of 25, it will arc to the left. If we make the radius larger, such as 50, the arc will not be as tight. Now that we are able to describe an arc with numbers, let's see how we can navigate with arcs.

Analog Steering

Any class that implements the ArcMoveController interface is capable of driving an arc. The principle method in this interface is the arc() method, which requires the two parameters described above: radius and angle.

Several pilot classes implement the ArcMoveController interface, such as DifferentialPilot and SteeringPilot. The method of movement is irrelevant to the API. Typically a wheeled robot is performing movement, but to the API, it does not matter if it is a walking robot or even a hovercraft. As long as the vehicle can fulfill the basic movements, it is acceptable to the API.

Just for Fun!

All of the four moves (travel, arc, rotate and stop) can be represented in mathematical terms by an arc. To travel a straight line, the radius size is infinity. To perform an arc, the radius is greater than zero and less than infinity. And to rotate on the spot, a radius of zero is used (reducing the arc to a point).

Steering robots are a little more finicky than differential robots because the steering requires calibration. The next chapter contains instructions to build and program a steering robot, but for this chapter, we just want to explore navigation with arcs, so the trusty Rov3r is used instead.

We can force the Rov3r to emulate an analog steering vehicle. We merely need to increase the minimum turning radius of the DifferentialPilot from zero to a positive number using the setMinRadius() method.

```
myPilot.setMinRadius(34);
```

Now if you pass the DifferentialPilot object to a Navigator, it will treat the robot as though it has analog steering capable of turning with a radius of 34 centimeters.

Steering and Navigation

The previous chapter demonstrated differential steering navigation. To move from one coordinate to another, a differential robot simply rotates in place until it points at the destination coordinate, then it drives the appropriate distance (see Figure 24-4a).

Steering robots do not have the luxury of turning in place, so the navigator must calculate a path using arcs and straight travel to move to the destination. The algorithms are contained in ArcAlgorithms, a special class just for performing these calculations. Moves are plotted by driving an arc until it points at the target, followed by a straight-line travel to the target (see Figure 24-4b).

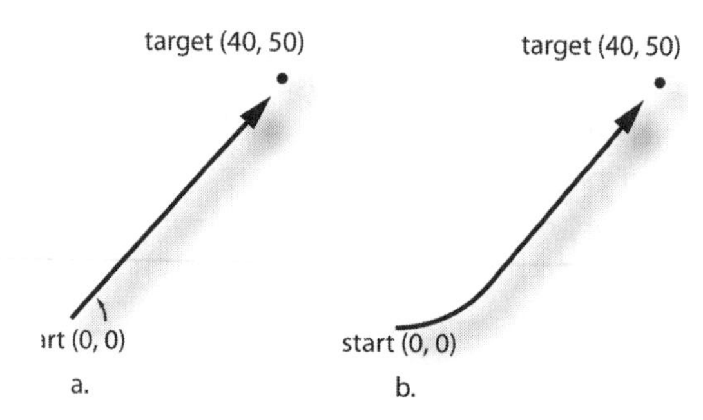

Figure 24-4 Comparing differential and analog steering

In the previous chapter we constructed a pilot and then gave it to the Navigator class as a parameter in the constructor. The same constructor applies to steering vehicles. The following few lines of code create a Navigator object and cause it to move to the point in Figure 24-4.

```
Navigator nav = new Navigator(arcPilot);
Nav.goTo(40, 50);
```

Heading

The pose of a robot represents a position of a robot at any given time (in the past, at present, or in the future). This pose includes coordinates and the heading of the robot (the direction it is pointing). There are two classes in the navigation package that contain heading data: the Pose class and the Waypoint class (more on these later).

With steering vehicles, such as automobiles, the final pose of the vehicle is critical. When you park a car in a driveway, you cannot leave the vehicle crossways or diagonally, otherwise your neighbors will worry about you. It is especially important to line up the vehicle before attempting to enter a garage, otherwise the vehicle could hit the entryway. In other words, when giving the robot a destination, the final pose of the vehicle is important.

Using the Navigator class, you can specify the final pose of the vehicle as follows.

```
nav.goTo(40, 50, 90);
```

The first two numbers in this method specify the Cartesian coordinates, while the number 90 indicates the final Cartesian angle.

You can also indicate the final pose using a Waypoint object, as follows.

```
Waypoint target = new Waypoint(40, 50, 90);
nav.goTo(target);
```

Putting it all together

Let's create a full program to navigate to several points in coordinate space. This will help us to see some of the nuances with steering navigation.

```
import lejos.hardware.Button;
import lejos.hardware.lcd.LCD;
import lejos.hardware.motor.*;
import lejos.robotics.navigation.*;

public class Steering {

  public static void main(String[] args) {
    // Make sure to use correct tire size and
track-width!
    ArcMoveController pilot = new
DifferentialPilot(4.32, 15.2, Motor.C, Motor.B);
    pilot.setMinRadius(34);

    Navigator nav = new Navigator(pilot);
    nav.goTo(40, 50, 90);
    nav.goTo(0, 0, 0);

    LCD.drawString("Press ENTER", 0, 3);
    Button.LEDPattern(4);
```

```
        Button.ENTER.waitForPressAndRelease();
        Button.LEDPattern(0);

        nav.goTo(40, 50, 180);

        Button.LEDPattern(4);
        Button.ENTER.waitForPressAndRelease();
        Button.LEDPattern(0);

        Motor.B.close();
        Motor.C.close();
    }
}
```

Using the Steering Code

When you run this code, it will attempt to get to the target of 40, 50 and a heading of 90 degrees (see Figure 24-5a). The path it travels is predictable and expected. When you press the square enter button it will drive back to the original starting point.

The second time you press the button, it will drive to the same coordinate but it will arrive with a heading of 180 degrees (Figure 24-5b). Surprised at the convoluted path it took to get there? This path was generated by the ArcAlgorithms class. It calculates all the different path permutations and chooses the shortest path possible to get to the destination. As it turns out, the solution it found involves a lot of reverse arc traveling.

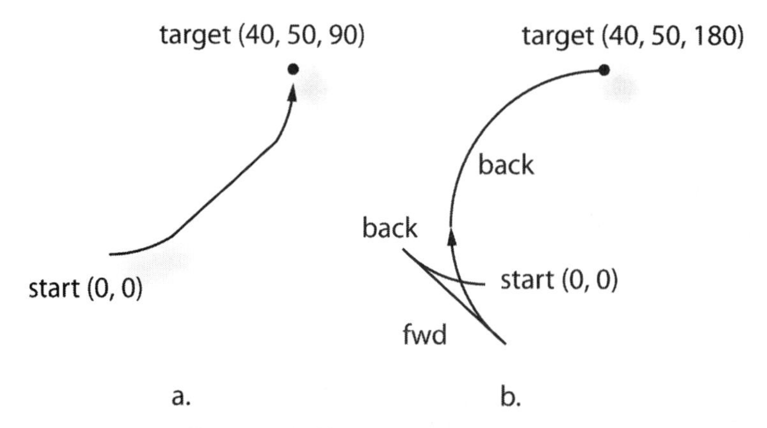

Figure 24-5 Final heading affects the path driven

One other wrinkle with the navigation algorithms is that the robot will always drive forward in between two arc movements. It will never travel backwards for the straight portion of the path. This leads to some seemingly odd behavior. In the code above, try substituting the first destination with this code.

```
//nav.goTo(40, 50, 90);
nav.goTo(20, 0, 0);
```

As expected, the first move just drives forward 20 centimeters. However, to get back to the start, you would expect it to back-up 20 centimeters. This cannot happen, however. Remember, it can only drive forward. Instead, the robot executes a complicated move to get there.

The reason it will never drive backwards for the straight travel (which can be the longest part of the move) is because we felt vehicles should drive forward for long stretches at a time, especially given that most sensors will point forward with actual robots.

You are now familiar with arc navigation. The next chapter contains a project using a sophisticated steering robot to perform navigation.

CHAPTER 25

Ackerman Steering

TOPICS IN THIS CHAPTER

- Differential Drive
- Ackerman Steering
- Steering Robot
- Auto Calibration
- SteeringPilot
- Navigation with Steering

In the previous chapter, we used a differential steering robot to perform arc movements, which simulated the behavior of a steering vehicle. This chapter contains instructions to build and control a true steering vehicle. Like the previous example, this vehicle is capable of full coordinate navigation.

Differential Drive

Steering vehicles seem simple on the surface. After all, it looks like it just needs two wheels to drive the vehicle and two to steer. In fact, there are a few complexities if you want to do it right.

The first complication has to do with the rear drive wheels. When two wheels exist on the same axle, things work fine as long as the vehicle drives straight. However, if you try to steer, the inner wheel will drive a smaller circle than the outer wheel (see Figure 25-1).

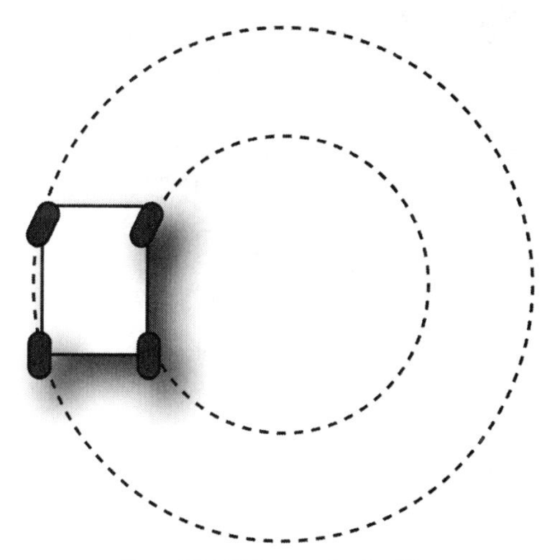

Figure 25-1 Different wheel radii

Why does this pose a problem? The reason is because the wheels are both on the same axle. As you can see in Figure 25-1, the inner wheel travels a shorter distance than the outer wheel. The axle can only turn at one speed, therefore the axle will either break (unlikely) or the wheels will skid on the surface, wearing out the rubber tires and causing wider turns.

Normally a differential is used to allow both drive wheels to turn at different rates. However, since the EV3 kit does not include a differential gear train, we cannot use this solution. Instead, we will remove this problem by using one rear wheel instead of two. The vehicle in this chapter uses a single rear wheel for simplicity (see Figure 25-2).

Figure 25-2 Driving with a single rear wheel

Ackerman Steering

The second problem is similar to the first, except it has to do with the steering mechanism. When the vehicle is making a turn, normally the inner wheel drives a smaller circle than the outer wheel (see Figure 25-1 again). You might not think this is a problem, because the steering wheels are not connected by a single axle. However, if both tires are angled in the same direction, they will both try to steer the same sized circle.

As you can see in Figure 25-1, the inner steering wheel travels a smaller circle than the outer wheel, so it should be turned at a sharper angle to compensate.

If the vehicle does not compensate for this, lateral skidding will occur and the vehicle will not turn as sharply or accurately.

The solution is to use Ackerman steering. This mechanism allows the axle of each wheel to point directly at the center of the turning circle, producing tighter turns with less slippage (see Figure 25-3).

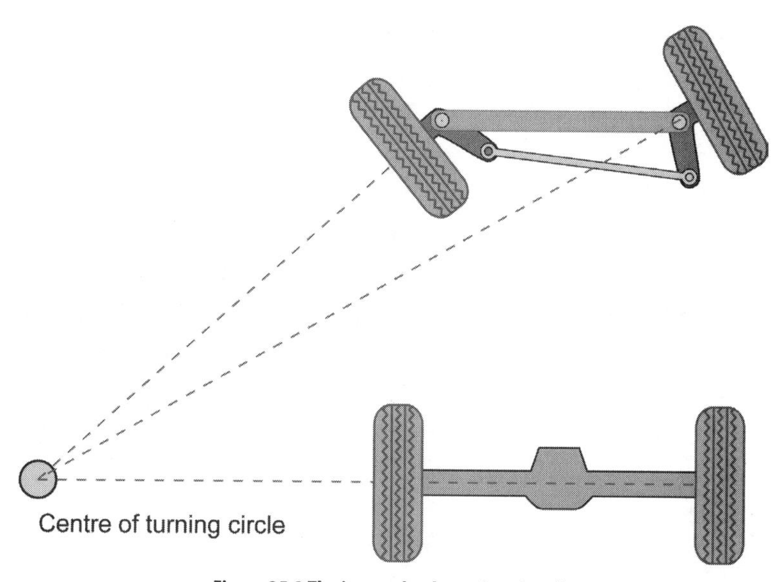

Figure 25-3 The inner wheel steering sharply

Steering Robot

Let's try building an analog steering vehicle. This vehicle is called Ackerbot in honor of the inventor of the steering mechanism.

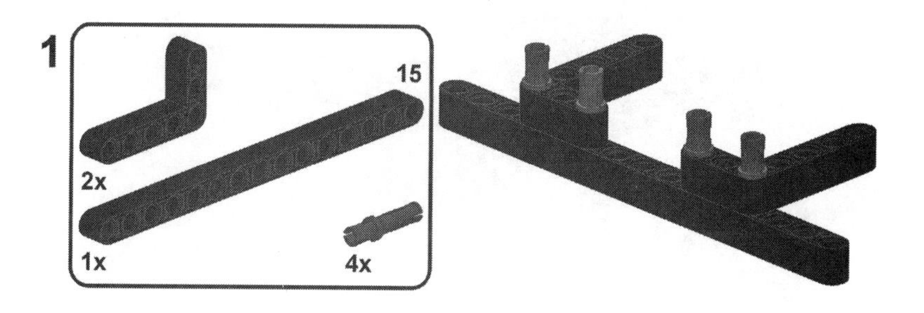

2

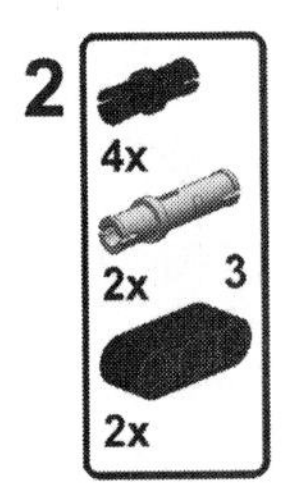

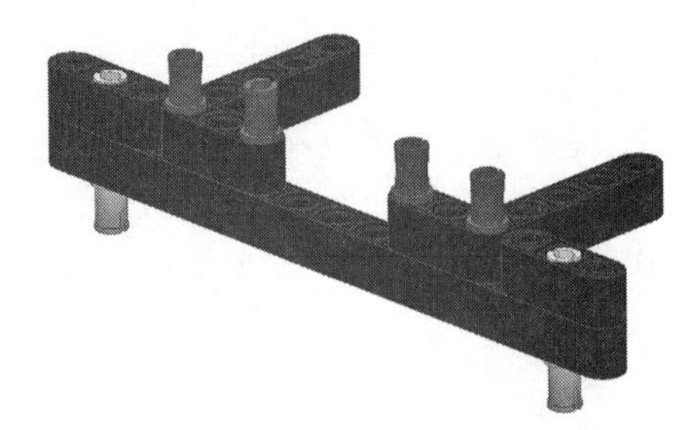

3

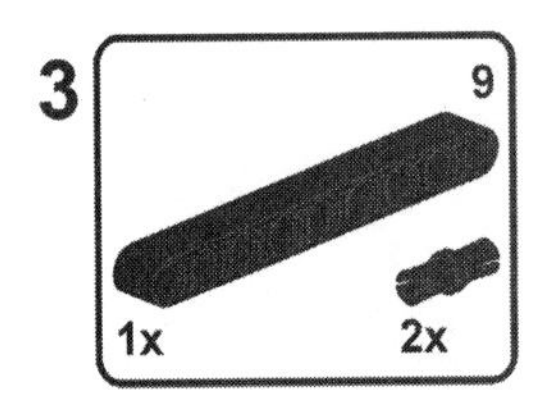

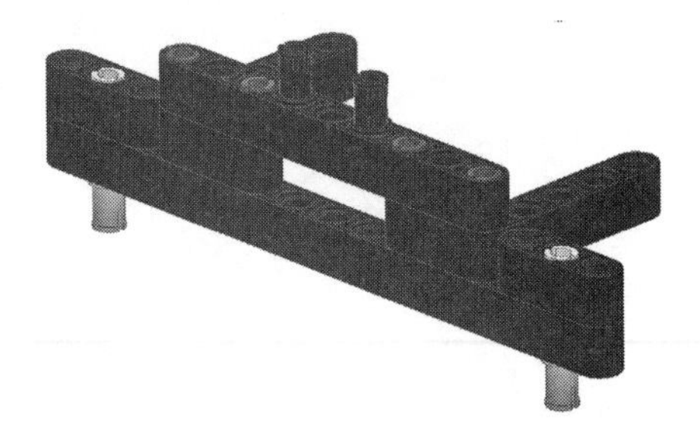

4

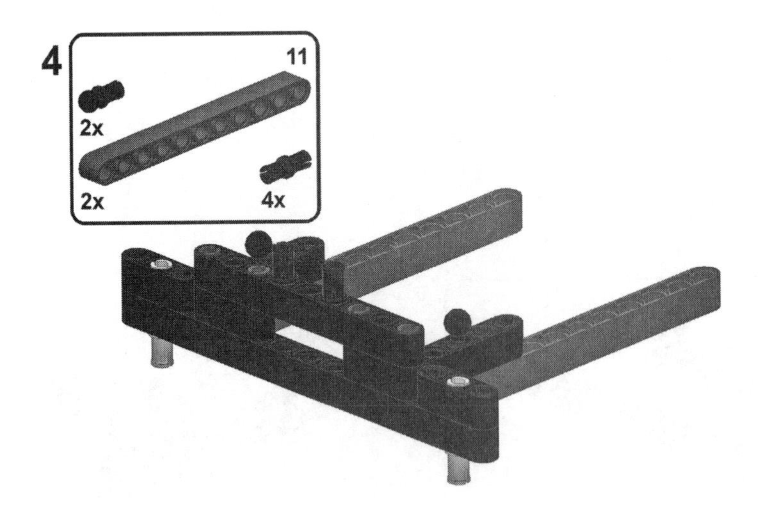

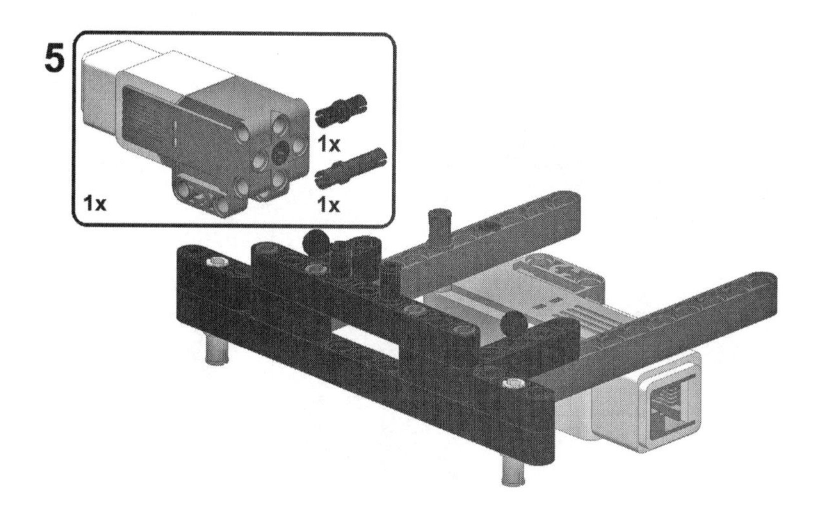
5
1x
1x
1x

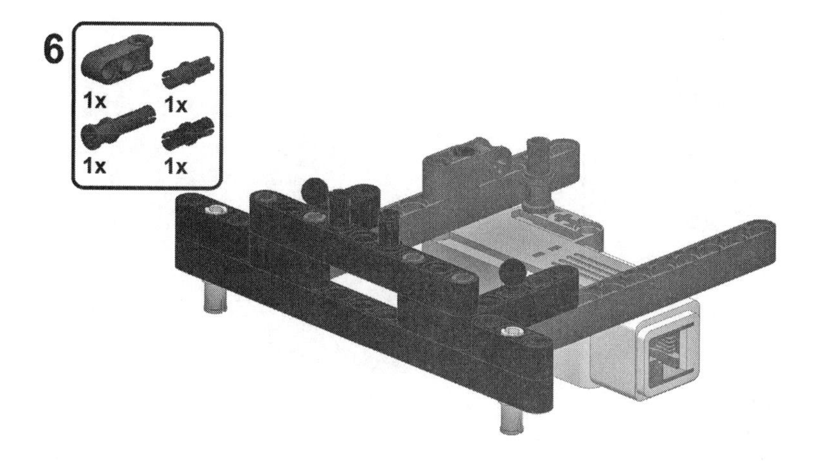
6
1x
1x
1x
1x

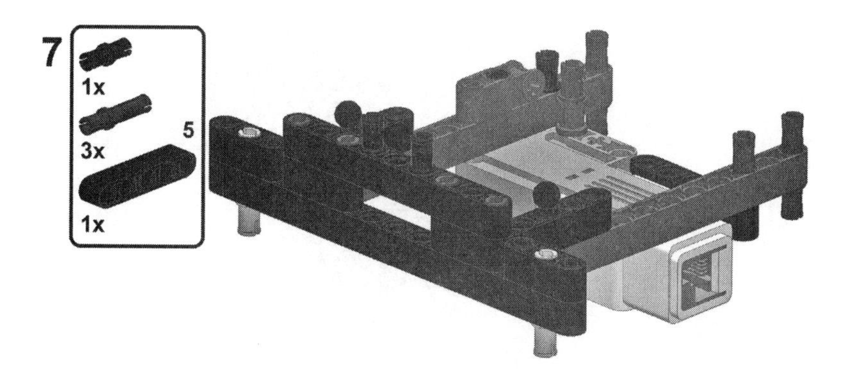
7
1x
3x
5
1x

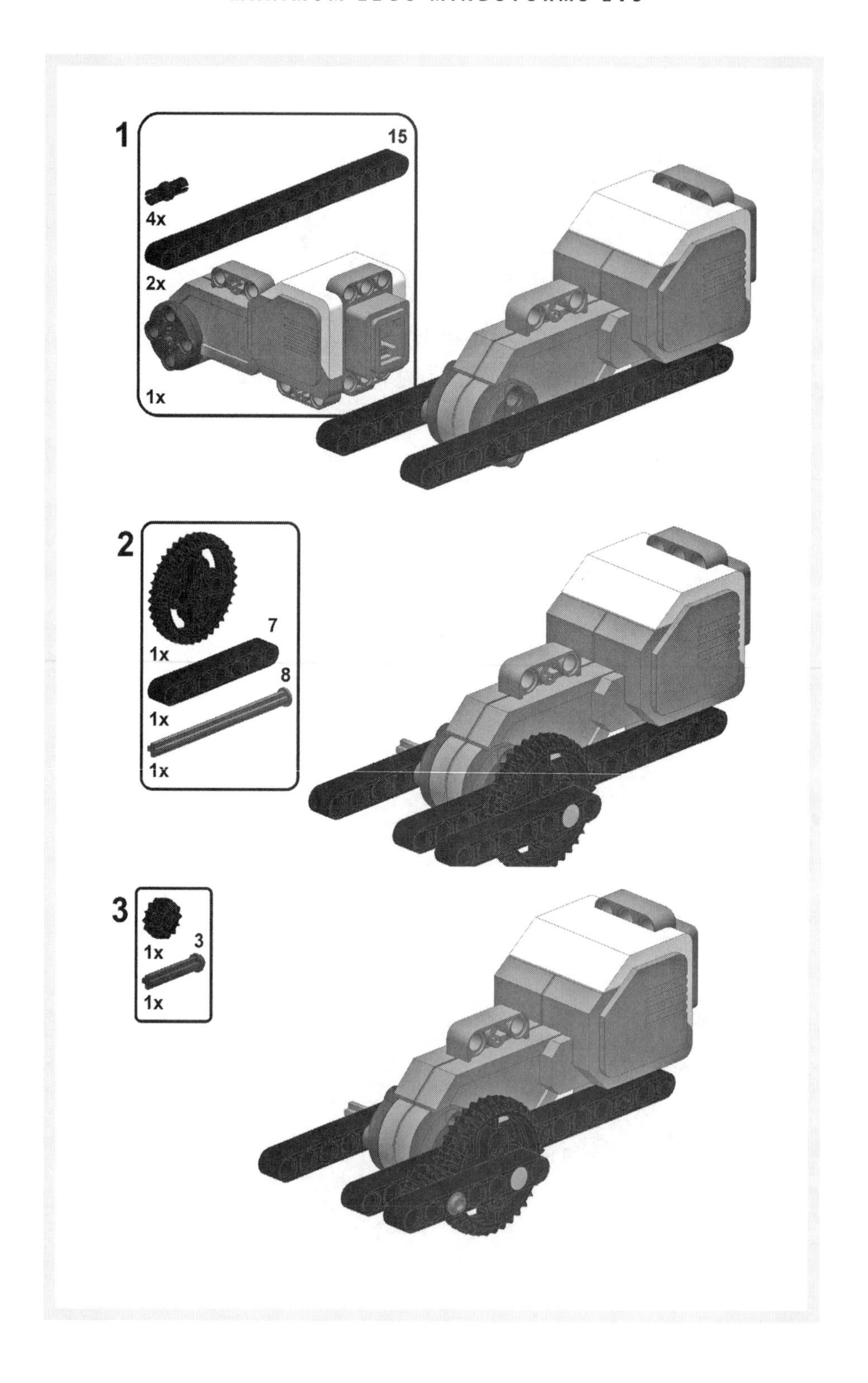
1
15
4x
2x
1x
2
7
1x
8
1x
1x
3
1x
3
1x

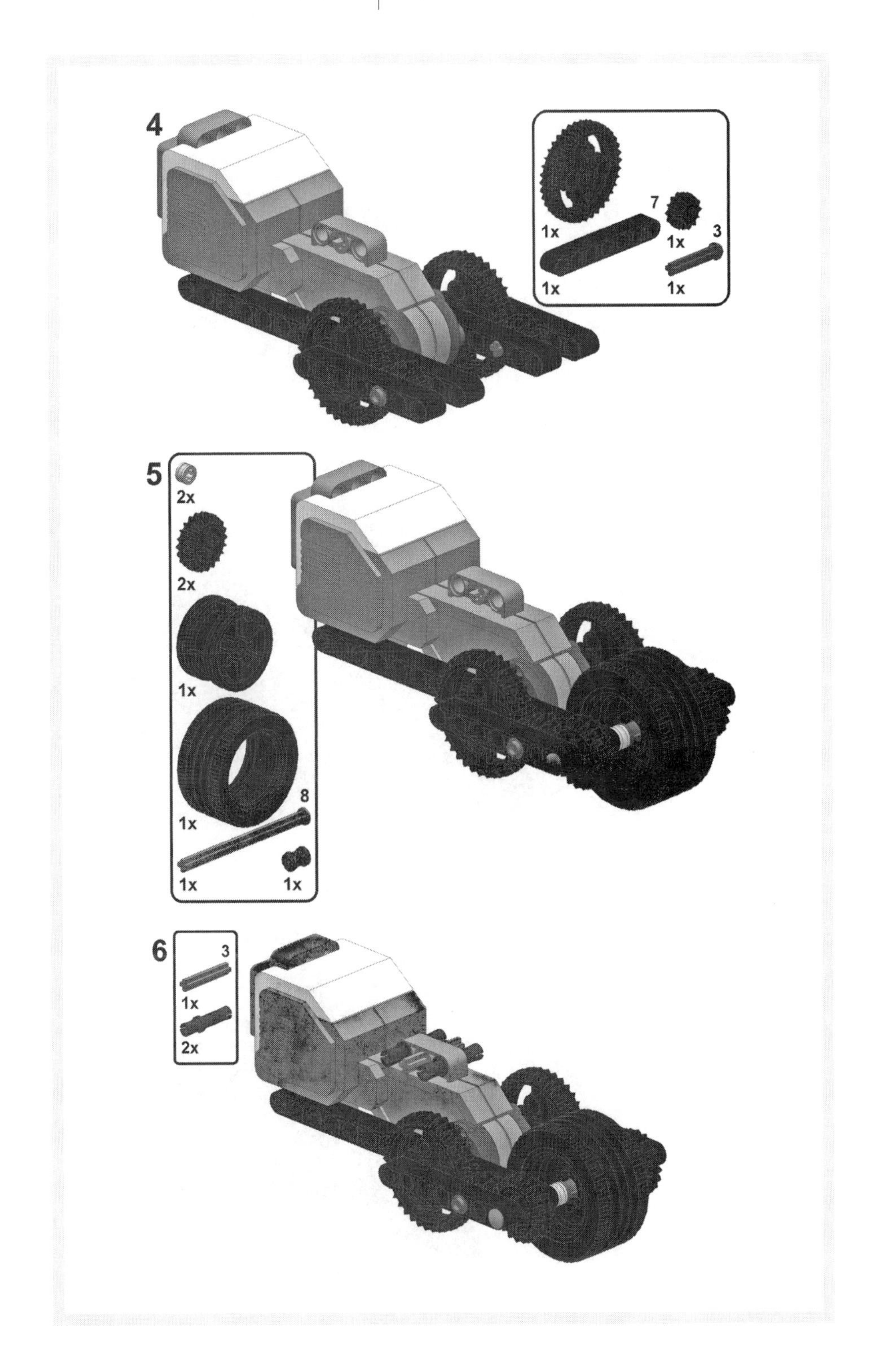
4
1x
7
1x
3
1x
1x
5
2x
2x
1x
1x
8
1x
1x
6
3
1x
2x

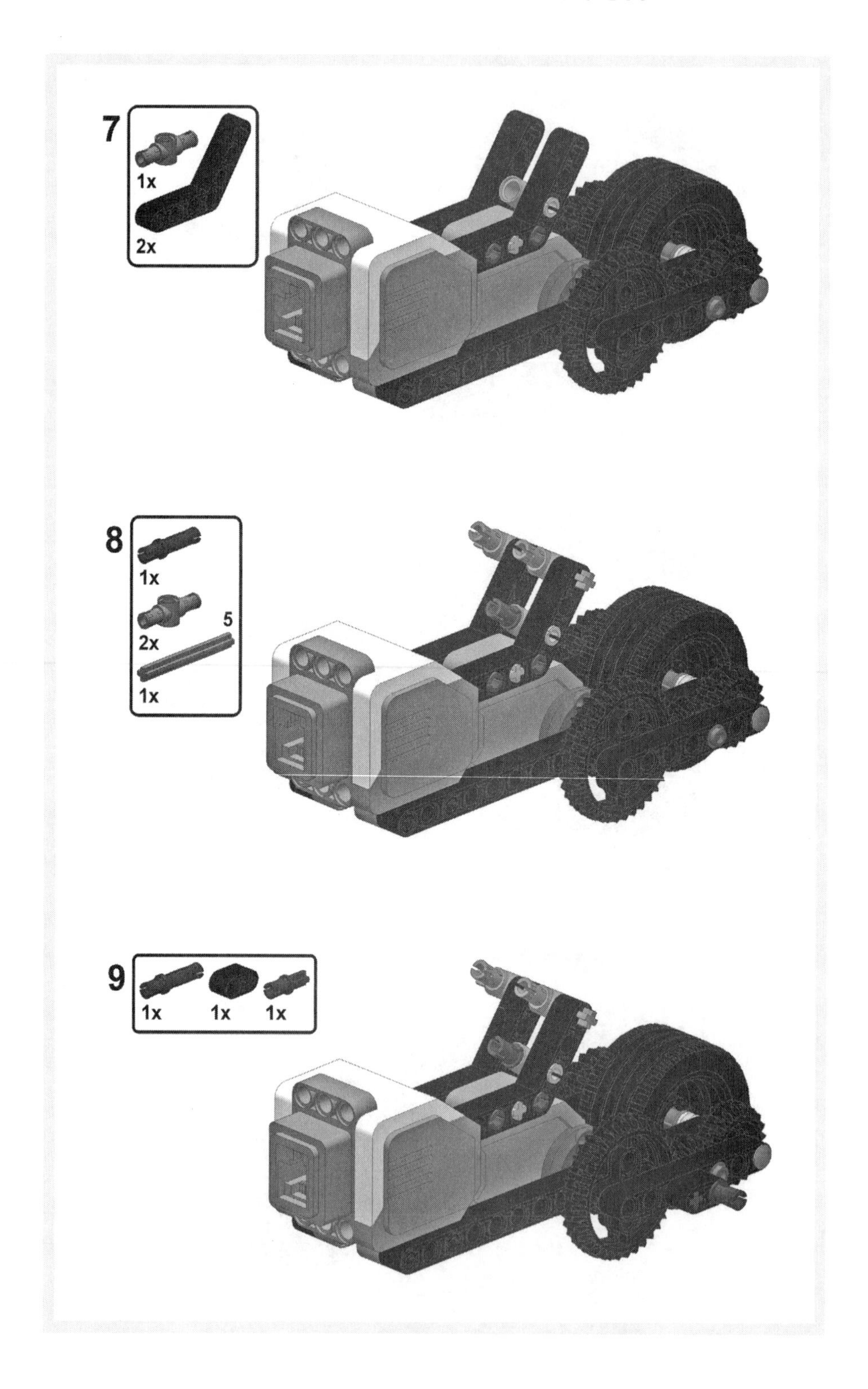
7
1x
2x
8
1x
2x
5
1x
9
1x
1x
1x

8

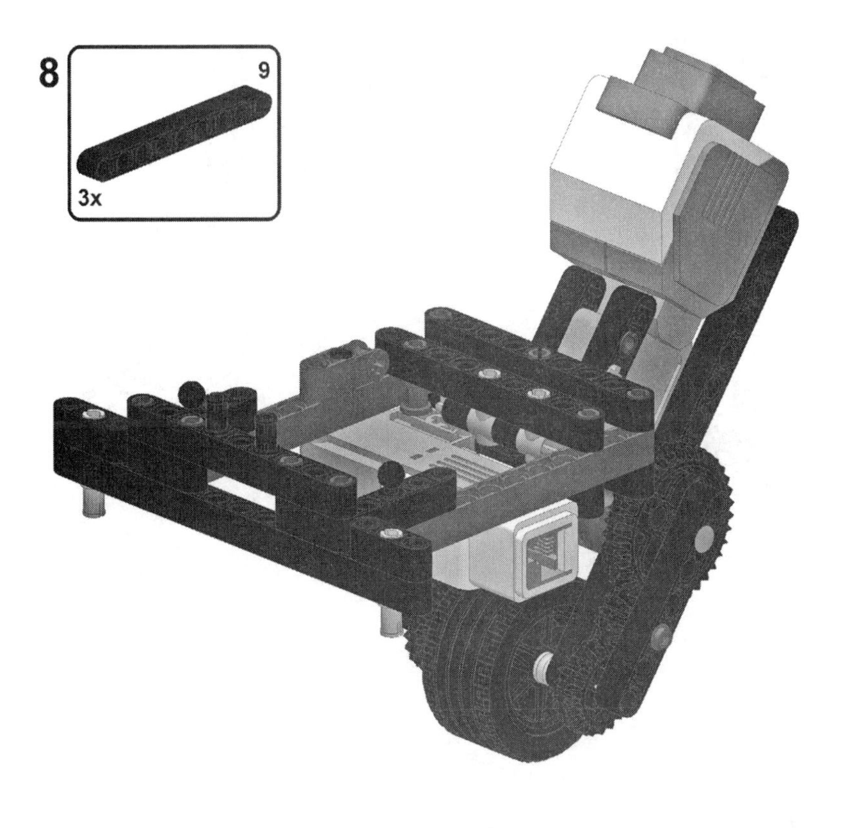

9

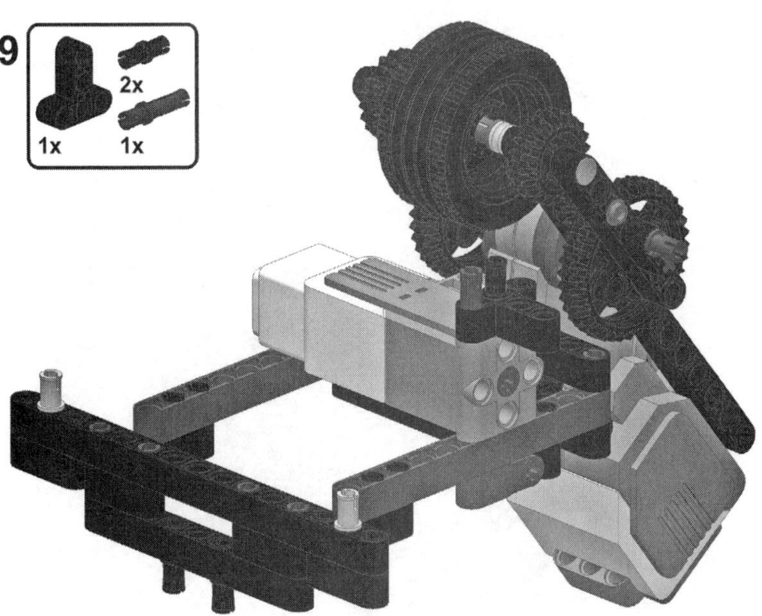

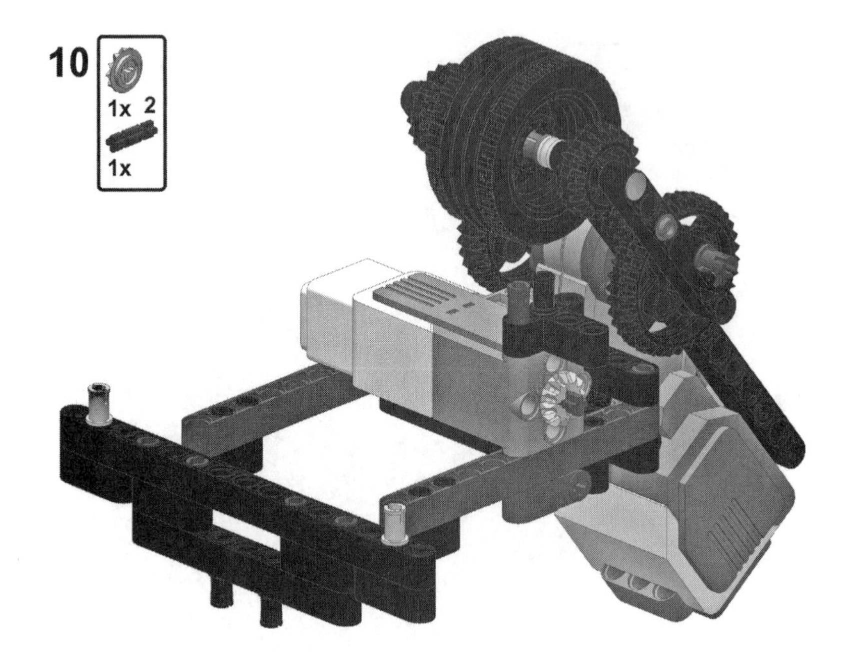
10
1x 2
1x

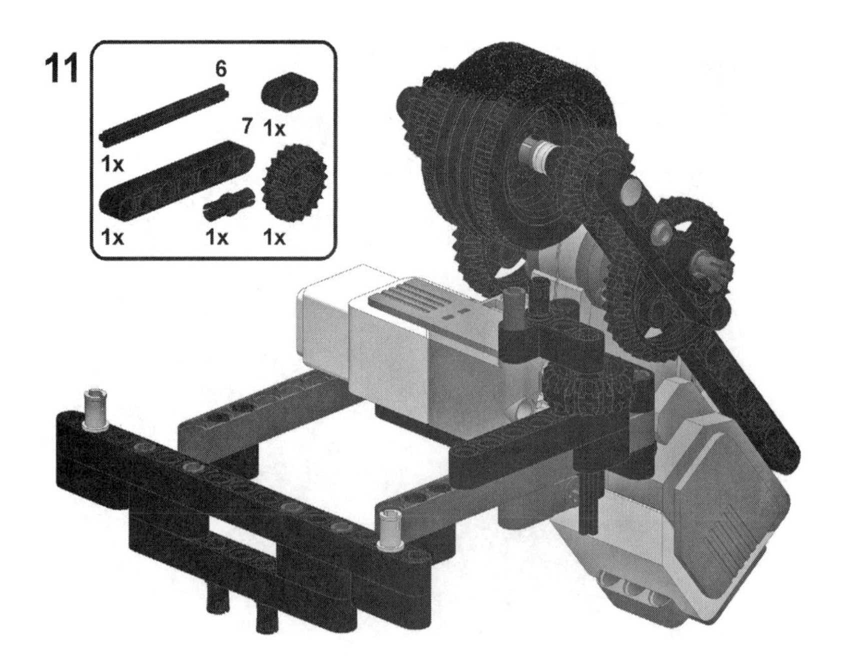
11
6
1x
7 1x
1x
1x 1x

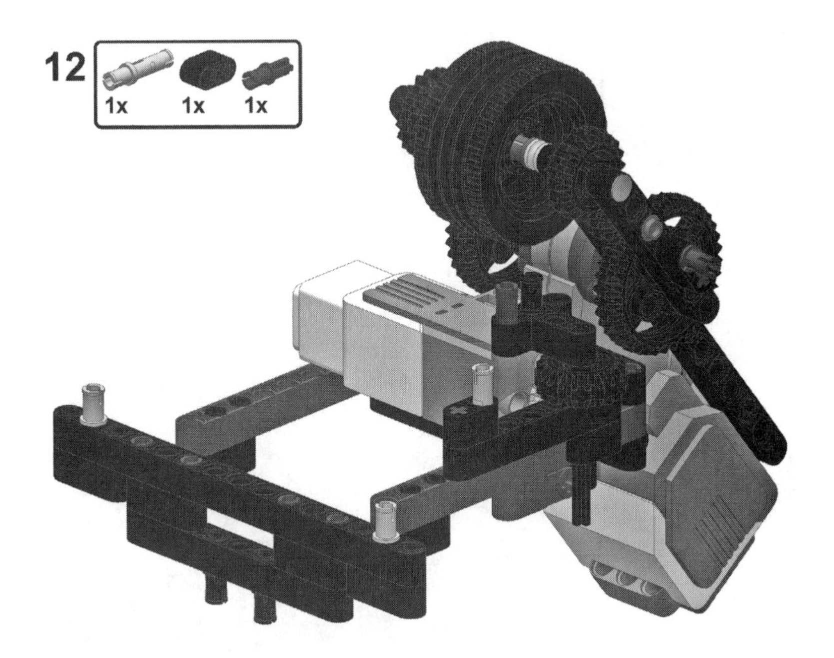
12
1x
1x
1x

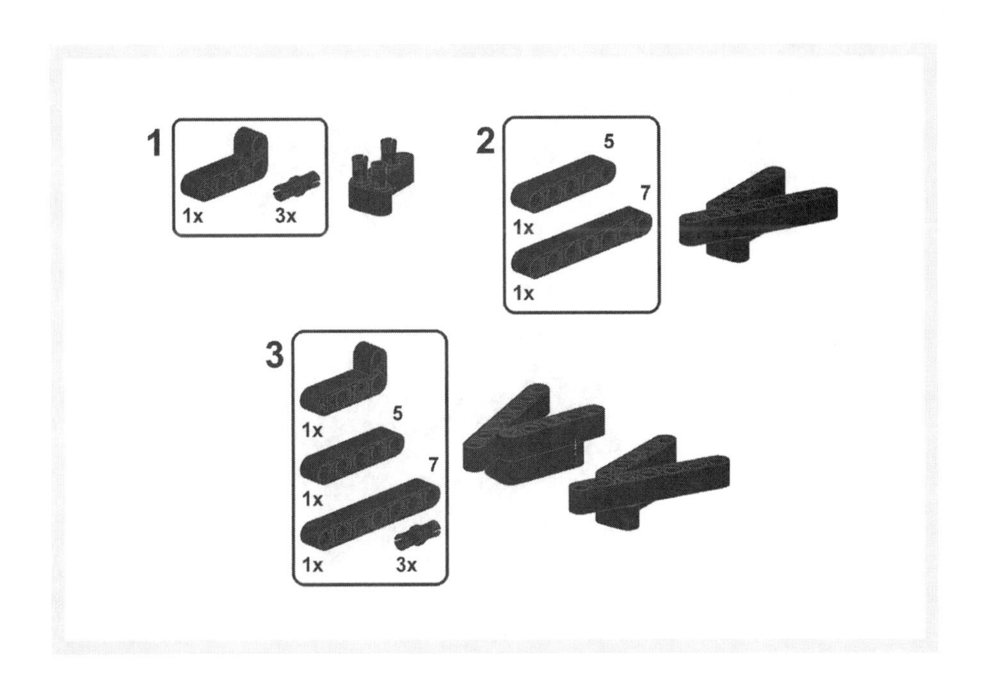
1
1x
3x
2
5
7
1x
1x
3
1x
5
1x
7
1x
3x

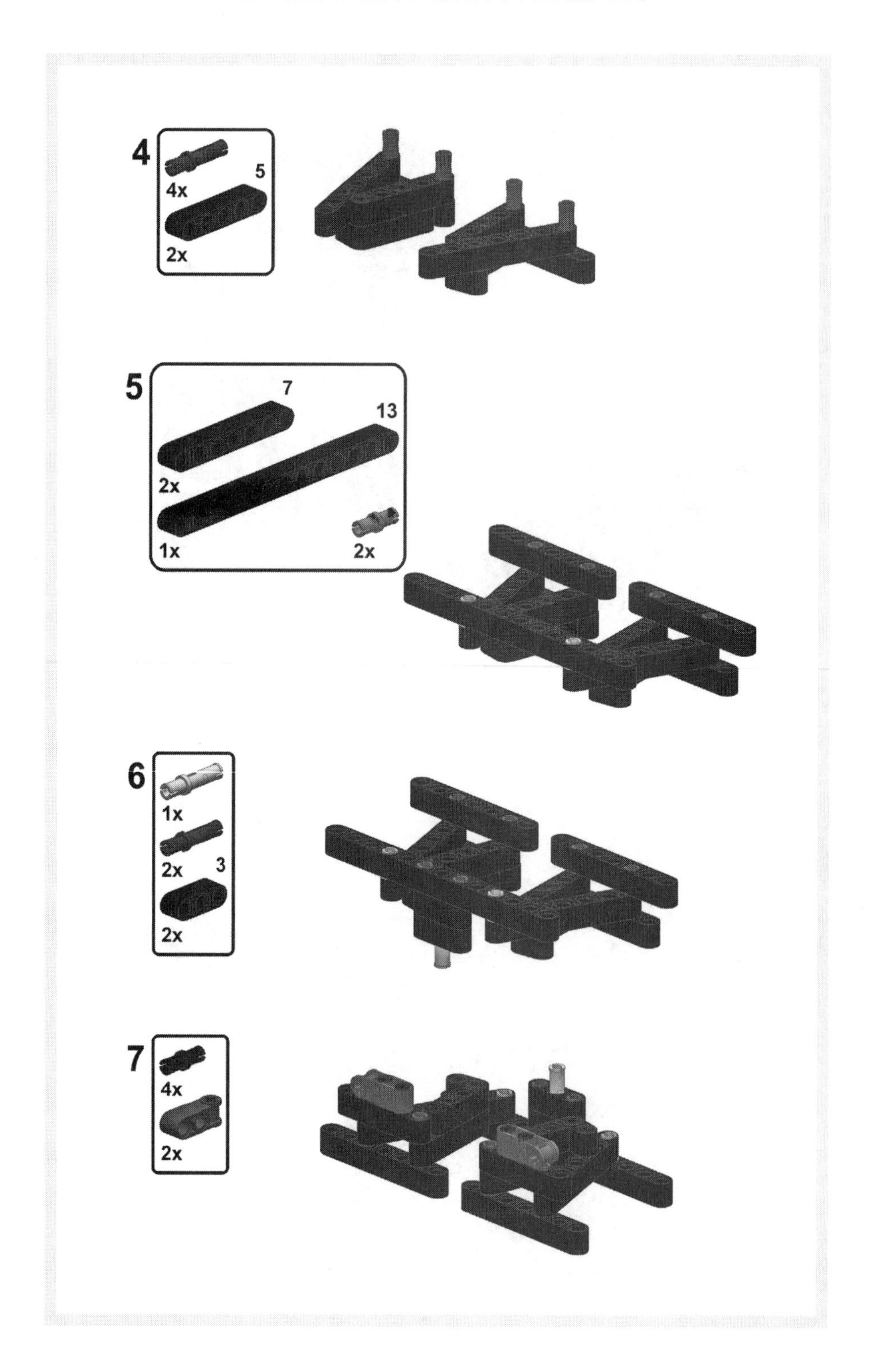
4
4x
5
2x
5
7
13
2x
1x
2x
6
1x
2x
3
2x
7
4x
2x

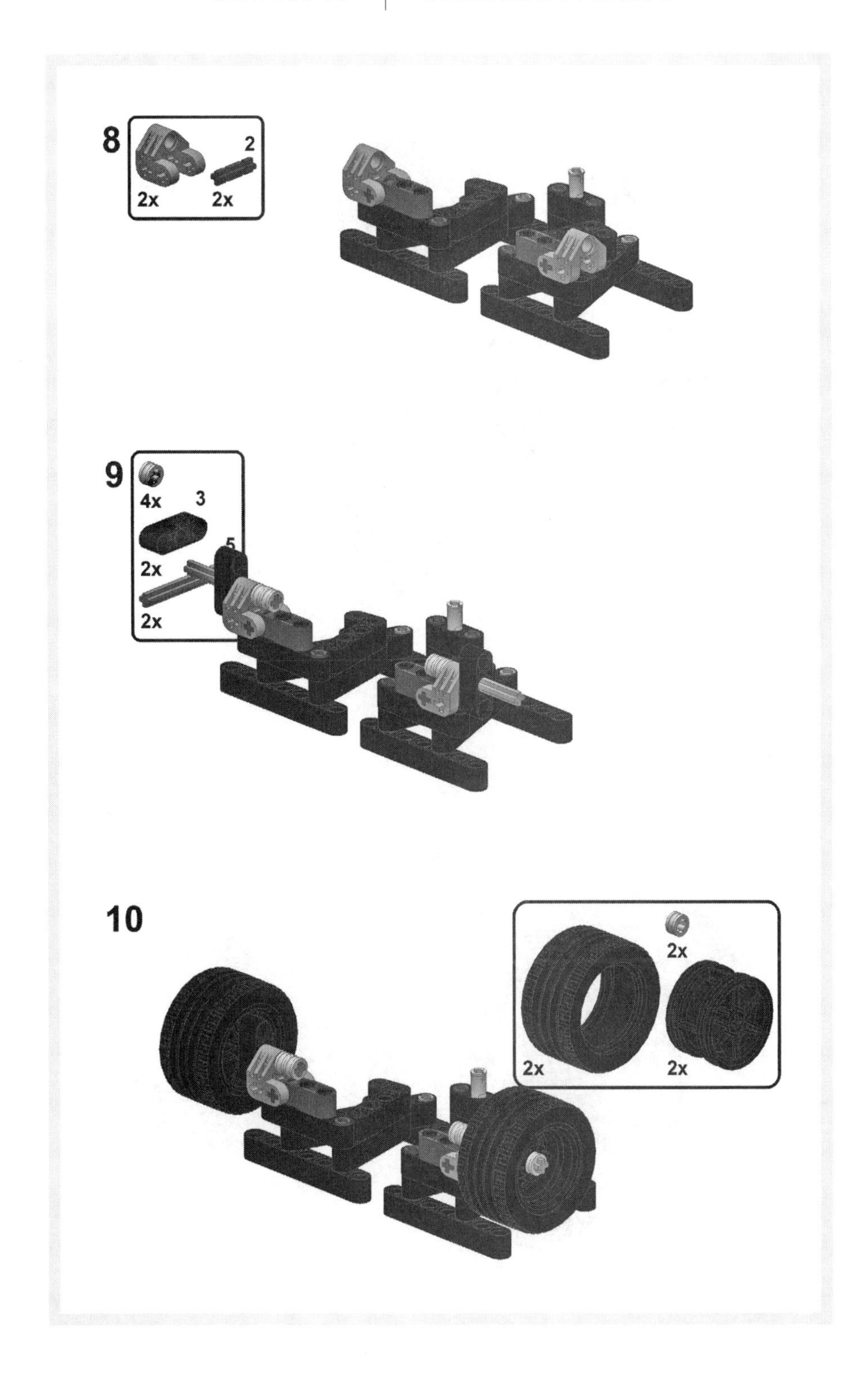
8
2
2x
2x
9
4x
3
2x
5
2x
10
2x
2x
2x

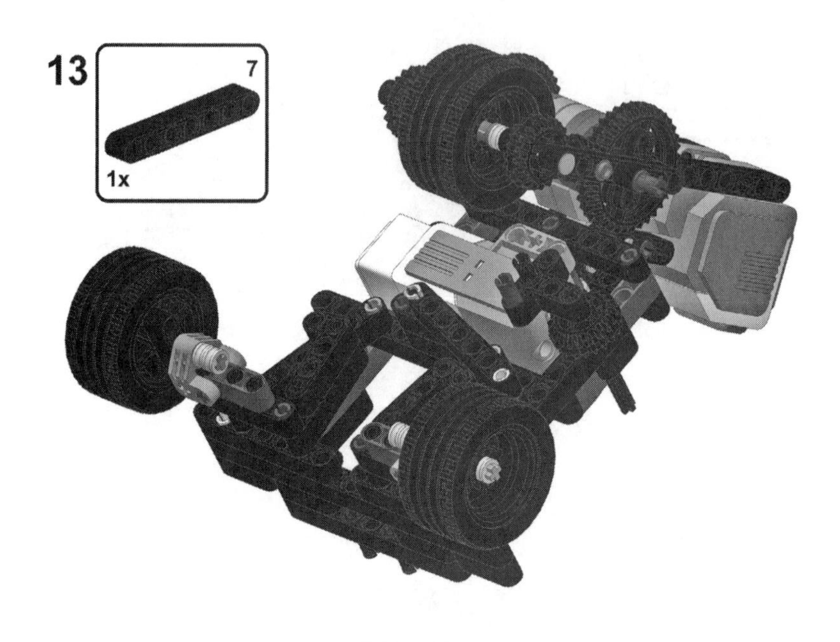
13
7
1x

14
1x
1x

15
1x
1x
1x

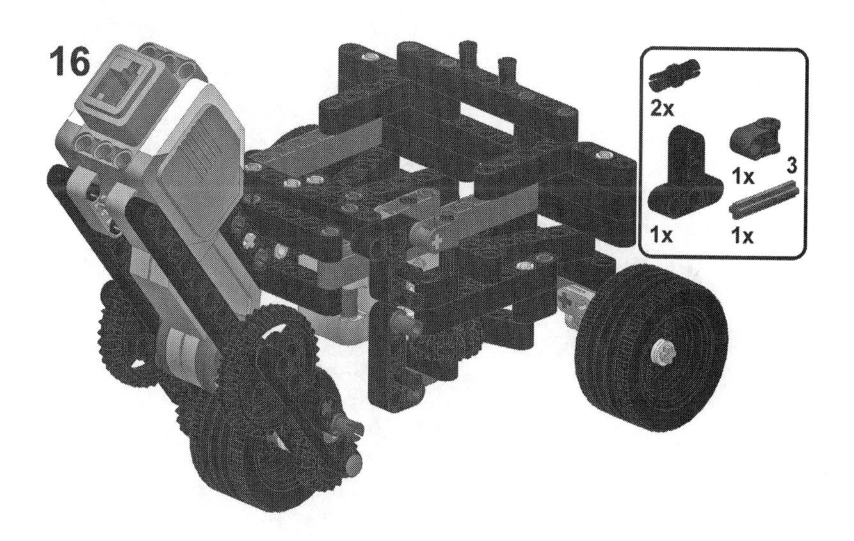
16
2x
1x
3
1x
1x

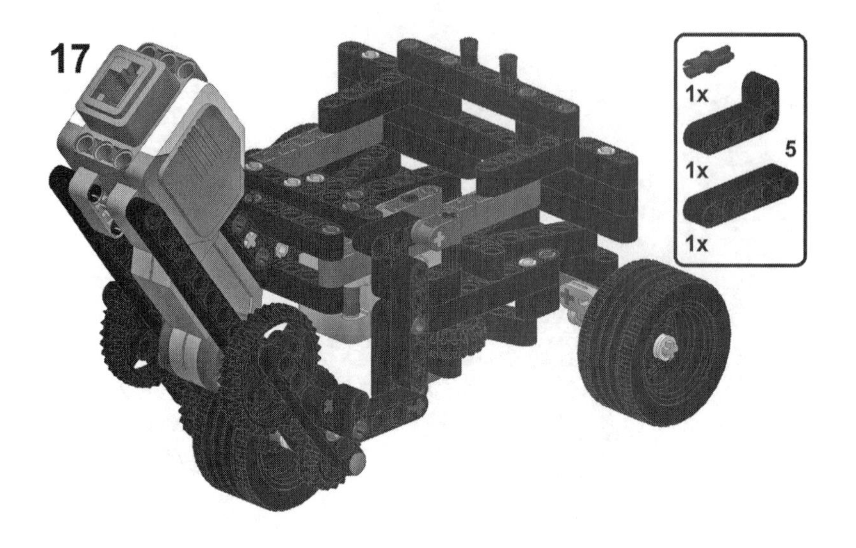
17
1x
1x
5
1x
1x

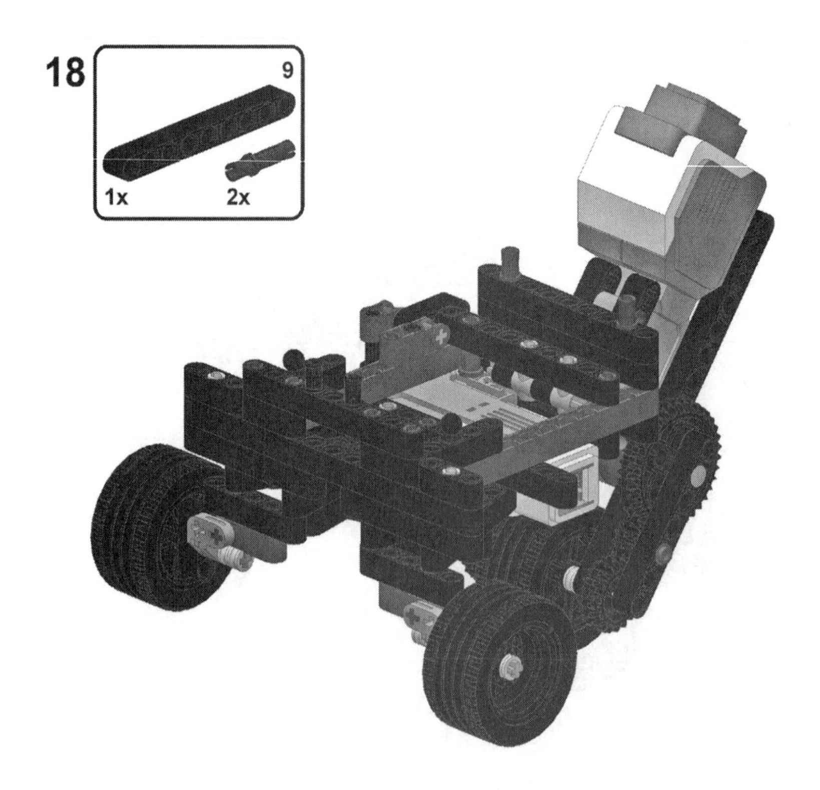
18
9
1x
2x

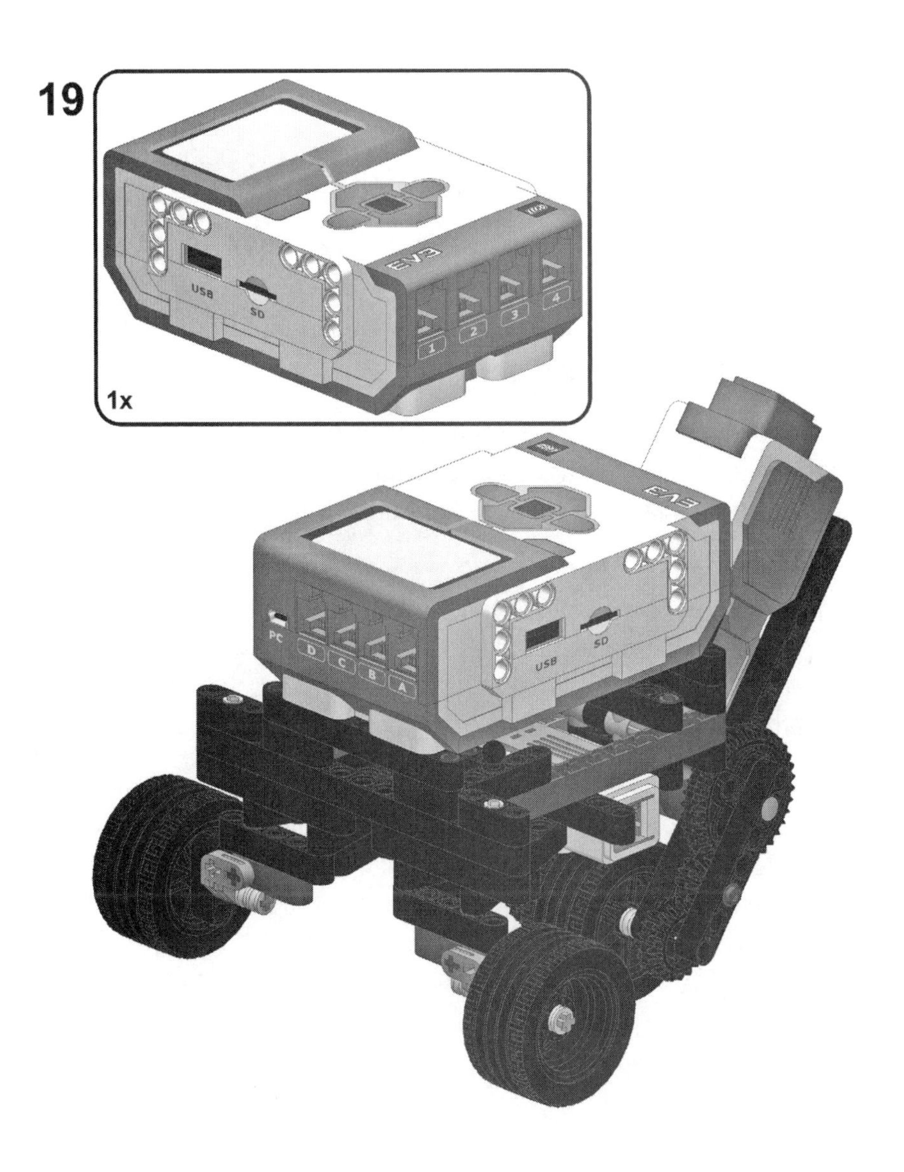
19
1x
USB
SD
EV3
1
2
3
4
PC
D
C
B
A

Cables

Connect the large motor to port B with a medium cable. With another medium cable, connect the medium motor to port C.

Auto-Calibration

With steering vehicles, it is important for the wheels to point straight ahead when the vehicle attempts to drive forward. If they do not, the robot will be off course after only a few moves.

You could manually align the steering using your eye, but it is faster and more accurate to have the EV3 perform alignment using a specialized calibration routine involving the tachometers.

A typical calibration routine rotates the wheels all the way in one direction until it hits a limiter switch. In our case, we will rotate the wheels until the motor encounters resistance from the steering frame. It then records the tachometer reading.

Now it knows how much it needs to turn the motors to turn all the way in one direction. The calibration routine then does the same for the other direction. It also needs to know the center, so it averages the two tachometer readings to find out the value in the immediate center.

This calibration routine is part of the SteeringPilot class, in the method calibrateSteering(). However, since our robot uses Ackerman steering with an asymmetrical steering arm, this method does not quite work as described. Instead, we will have to write our own custom calibration method, using a similar methodology. It basically rotates all the way to the right, then rotates a predetermined number of degrees back to center and then resets the tachometer on the motor to 0, so that center is zeroed. You can view the calibration method in the code sample below.

SteeringPilot

Ackerbot is technically capable of steering a whole range of arcs, with the smallest arc radius of about 28 cm (the minimum radius). However, the SteeringPilot class is only capable of doing either a left arc or a right arc—no middle ground. It cannot perform a shallow arc, only minimum radius arc (the tightest arc).

The reason the SteeringPilot is incapable of a range of arcs is because it is too difficult to calculate the entire range of steering radius given the steering angle. Therefore the code for the steering robot performs only three steering angles: left, straight, right.

Adjustments

Because we are using standard LEGO kits with almost identical parts, your robot should drive with about as much accuracy as mine. However, the surface the robot travels on (carpet vs. hardwood) makes a big difference. If you want to try honing the steering to make your robot even more accurate, you can test out some patterns to see how it performs.

First, let's test the turn radius. The following code calibrates the steering, then drives the robot in several 360 degree circles:

```
import lejos.hardware.motor.*;
import lejos.hardware.port.*;
import lejos.robotics.*;
import lejos.robotics.navigation.*;

public class Ackerbot {

  public static double GEAR_RATIO = 36.0/20.0;
  public static RegulatedMotor medium = null;
  public static RegulatedMotor large = new EV3Large
RegulatedMotor(MotorPort.B);

  public static int CENTER = -74;
  public static int RIGHT = -CENTER;
  public static int LEFT = -51;
  // carpet = 28, hardwood = 26:
  public static double MINTURN_RADIUS = 28;

  public static void main(String[] args) throws
Exception {

Ackerbot.calibrate(MotorPort.C);
    medium = new EV3MediumRegulatedMotor
(MotorPort.C);
```

```
    Ackerbot.recenter(medium);
    medium.setAcceleration(200);

    // Multiply wheel size by gear ratio
    double size = SteeringPilot.WHEEL_SIZE_EV3 *
GEAR_RATIO;

    SteeringPilot p = new SteeringPilot(size, large,
      medium, MINTURN_RADIUS, LEFT, RIGHT);
    large.setAcceleration(1000);

    p.travel(100);
    p.arc(MINTURN_RADIUS, 720);

    medium.rotateTo(0);
    large.close();
    medium.close();
  }

  public static void calibrate(Port port) {
    UnregulatedMotor m = new UnregulatedMotor(port);
    m.setPower(50);
    m.forward();
    int old = -999999;
    while(m.getTachoCount() != old) {
      old = m.getTachoCount();
      try {
        Thread.sleep(500);
      } catch (InterruptedException e) {
      }
    }
    m.flt(); // TODO: Unneeded?
        m.close();
  }

  public static void recenter(RegulatedMotor
steering) {
    steering.setSpeed(100);
    steering.rotateTo(CENTER);
    steering.flt(); // TODO: Unneeded?
    steering.resetTachoCount();
  }
}
```

As you can see in the code, the starting minimum turning radius of this robot is 28. The gear multiplication of the robot complicates things. Notice that when the motor rotates 360 degrees, the actual wheel rotates 648 degrees? This is because the 36 tooth gear is connected to the 20 tooth gear (the small idler gear between them doesn't do anything). The code compensates for this by multiplying the wheel size by the gear ratio.

Ok, let's find out how well the robot works and make some adjustments to some code variables. Set it down, start the program and watch Ackerbot calibrate the steering. Once complete it will drive two circles. Press the square enter button after the first circle.

Ideally, you want Ackerbot to drive a full circle and end precisely where it started. If your robot falls short, try increasing the MIN_RADIUS value until it does a complete lap. Once you have the MIN_RADIUS value properly adjusted, try measuring the radius of the circle it is driving (radius is diameter divided by two). Does the measured radius in fact end up the same as MIN_RADIUS?

Table 21-1 offers some suggested values for MINTURN_RADIUS on different surfaces. Note these values can be tweaked to match up with the surfaces you are using your robot with.

	Hardwood	**Short Carpet**
MINTURN_RADIUS	26 cm	28 cm

Table 21-1: Programming values for different surfaces.

Using the Ackerbot

Now that the SteeringPilot constructor parameters are honed, let's try using it with a Navigator. Comment out the code in the main() method that makes it drive 100 cm and then drive two circles. Replace this code with the following code to see how well it navigates.

```
Navigator nav = new Navigator(p);
nav.goTo(40, 50, 90);
nav.goTo(0, 0, 0);
```

Just for Fun!

There is one other analog steering vehicle in this book in Chapter 14. It was built primarily as a remote control vehicle, but try converting it to an autonomous robot by adapting the code in this chapter.

CHAPTER 26

Localization

TOPICS IN THIS CHAPTER

- Localization
- Odometry
- OdometryPoseProvider
- Odometry Error

Localization is the ability to determine where you are. Since primitive times, inventors have come up with a number of localization methods (see Figure 26-1). As mentioned earlier, the most serious early efforts were to estimate the correct position of a ship at sea with very few visual landmarks—perhaps the occasional island. This book will examine several methods of localization. First, let's find out more about how people determine where they are.

Figure 26-1 Locating position

Localization

Earlier, this book introduced the concept of a coordinate system, which is capable of indicating the x and y position of a robot. Using coordinates, we were able to tell the robot to move to a specific position. But how do we find the current position of the robot?

Localization can be determined using two main strategies: odometry and landmark navigation. Imagine yourself in a very large and empty gymnasium. There are no visual marks on the floor, except for two points marked A and B at opposite ends of the room (see Figure 26-2).

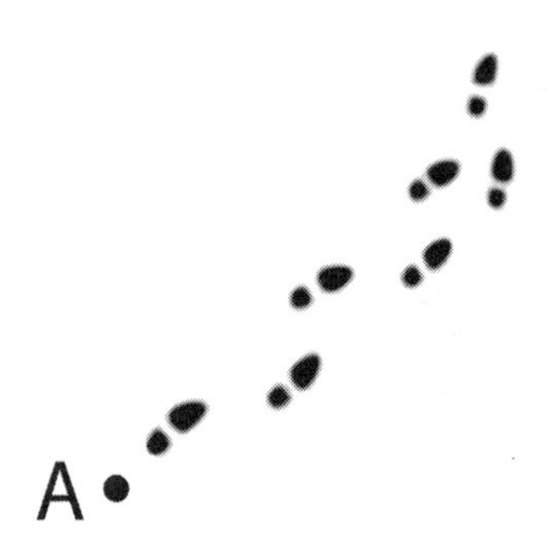

Figure 26-2 Traversing an environment with limited data

Now imagine that you are placed on point A, blindfolded, then told to make your way to point B. To navigate, all you would have at your disposal is *proprioception*, the ability to feel the orientation of your legs, which would allow you to estimate the length of each step. This is a biological form of odometry. By counting the number of steps you make, you would be able to roughly indicate when you were near point B.

As you might guess, the estimate when you arrived at point B is likely to be somewhat off the actual mark. In fact, try it right now in your living room. Place a tiny piece of paper at the far end of the room, place your hand over your eyes, and try to walk until you are right on the mark (make sure not to feel for the paper with your toes, and try not to use audio landmarks such as a noisy air vent or open window). Over a distance of about 20 feet you'll probably be off about a foot.

The second way to navigate is by using landmarks. Imagine the same gymnasium, except that it has a grid pattern of poles spaced five feet apart (see Figure 26-3). If you note beforehand that point B is four poles north and two poles east, you can then close your eyes and feel your way along the poles until you get to the destination.

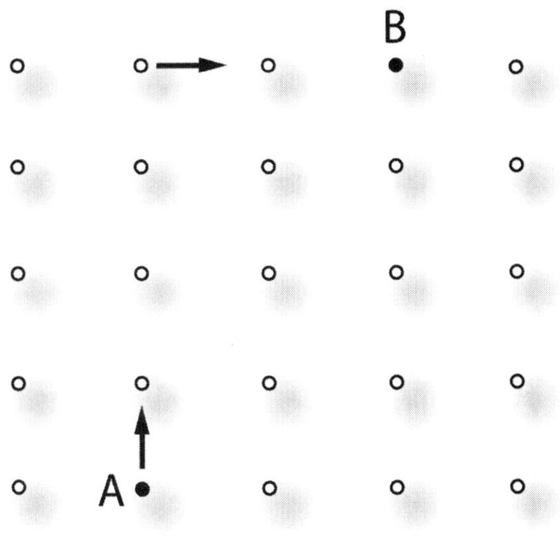

Figure 26-3 Traversing an environment with fixed landmarks

So which navigation system is more reliable? As you might guess, the second system, landmark navigation is by far the more reliable method. If you use blind odometry to get from point A to point B, and then keep repeating this back and forth without ever peeking, you will be hopelessly lost after a few rounds. However, you could navigate around the poles for hours and still maintain your location, as long as you accurately note how many poles you have traversed and in which direction.

Landmark navigation works consistently for long periods of time. In the gymnasium example above, instead of poles you could have performed landmark navigation merely by keeping your eyes open and walking to the B on the floor. In fact, you can navigate using many different kinds of landmarks, including audio (a radio), visual (a light house), smell (a rose bush), temperature (a heat vent) and radio waves (GPS satellites).

You use landmark navigation every time you walk to the fridge. Humans use fixed landmark navigation by looking at local landmarks, such as your neighborhood supermarket, a tree, street signs, or any number of notable objects. GPS uses this method by noting a position relative to four or more satellites.

Landmark navigation doesn't worry about where you have been, just where you are at a particular moment. Odometry, on the other hand, estimates where you are based on your previous position. Let's examine odometry.

Odometry

The most basic localization method is odometry. Using odometry, a robot keeps track of every movement it makes starting from a point of origin (see Figure 26-4). Normally odometry relies on tachometers to keep track of wheel or leg movement—a form of proprioception.

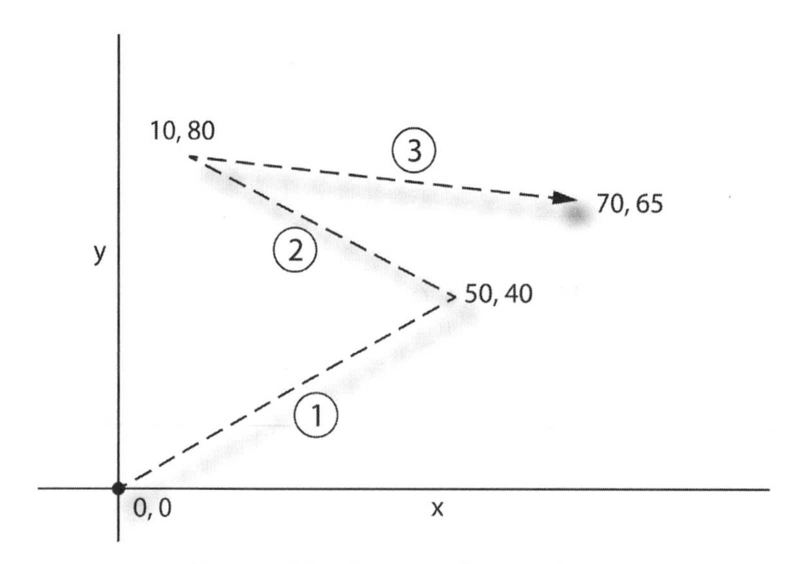

Figure 26-4 Using odometry to estimate coordinates

It is important to stress that odometry is an estimate of the current position. It cannot tell with certainty an exact location.

Humans have used a similar process called *dead reckoning* (also sometimes called orienteering) for a long time. In fact, this method of navigation is used to some degree by all animals, including humans. The art of dead reckoning has been refined by sailors, pilots, geologists, forest rangers, and hikers. The only information needed for dead reckoning is direction and distance.

Direction is usually obtained using a magnetic compass. However, direction can also be determined by recording the rotation of wheels using a tachometer. For example, when one wheel rotates forward and one rotates backward, the robot rotates. By measuring the amount each wheel rotated we can approximate the angle it rotated. Likewise, distance can be approximated by keeping an accurate record of wheel rotations while travelling.

Just for Fun!

Do you know the difference between direction and heading? They are in fact two different things. Direction is the direction your face is pointing. If you are looking at the North Star, your direction is north. However, if you sidestep to the east, your heading is east, even though you are still facing north (see Figure 26-5). Heading does not occur when you are standing still, only when you are moving. GPS is only capable of providing heading. If you try to get direction from a GPS while standing still, the direction value is more or less meaningless.

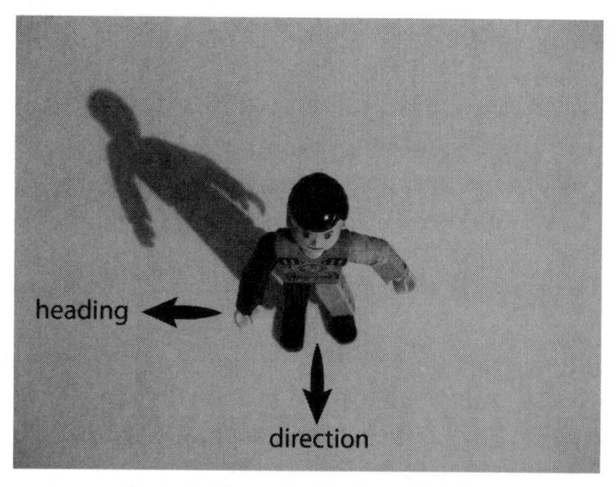

Figure 26-5 Comparing direction and heading

There are some advantages and disadvantages to using odometry with a robot. With smooth wheels, a flat floor, and a good tachometer, a robot will outperform a human in estimating the distance it has traveled. The accuracy of calculating and storing coordinate points is also better in robots than in humans, although a pad of paper often helps humans to keep track of previous travel.

There is a downside to robots, however. Most robots are unable to self-correct their course by analyzing a situation. The robot in this chapter has no ability to visually recognize a target or landmark.

To update coordinates, the robot must perform trigonometry calculations. Luckily for us, leJOS comes with navigation classes to perform the math.

OdometryPoseProvider

Odometry calculations are performed in the OdometryPoseProvider class. This class updates coordinate data by monitoring the motors and using geometry to perform the calculations.

The OdometryPoseProvider class requires a MoveProvider in the constructor. Internally, the MoveProvider updates the OdometryPoseProvider every time a move is made. By updating it with all the latest move data, the class can keep track of the estimated location. All of the Pilot classes implement the MoveProvider interface. That is, when they make a move, they can report the move data to the OdometryPoseProvider via a Move object.

A MoveProvider does not necessarily need to be a pilot. Imagine if a programmer named Gerald created a class called GeraldPoseProvider. Every time he starts or stops a move, he types the information into the computer. This may sound strange to have a MoveProvider that relies on human input, but it would be a true MoveProvider which could be used in the OdometryPoseProvider constructor to estimate Gerald's pose, as long as he accurately provided updates on his movements.

Let's examine an example that moves a robot around while keeping track of the pose using an OdometryPoseProvider. Once again, we'll use the workhorse of this book, the ever reliable Rov3r from Chapter 22.

```
import lejos.hardware.Button;
import lejos.hardware.lcd.LCD;
import lejos.hardware.motor.Motor;
import lejos.robotics.localization.*;
import lejos.robotics.navigation.*;

public class Odometry {
  public static void main(String[] args) {
    double diam = DifferentialPilot.WHEEL_SIZE_EV3;
    double width = 15.2;

    DifferentialPilot robot = new
DifferentialPilot(diam, width, Motor.C, Motor.B);
    OdometryPoseProvider pp = new
OdometryPoseProvider(robot);
    LCD.drawString("Start: " + pp.getPose(), 0, 1);

    robot.rotate(90);
    robot.travel(100);
```

```
    robot.arc(30, 90);
    robot.travel(50);

    LCD.drawString("End: " + pp.getPose(), 0, 2);
    LCD.drawString("PRESS ESC", 0, 4);

  Button.ESCAPE.waitForPressAndRelease();
   }
  }
```

After you run this program, it should perform a series of moves. When complete, it will output the following approximate final pose.

- x = -81 cm
- y = 130 cm
- heading = 180 degrees

Notice this example does not use the Navigator class? It only uses DifferentialPilot, yet it can report the coordinates after each move.

In earlier chapters, we instantiated the Navigator class. What you probably did not realize is that it automatically created an OdometryPoseProvider internally. You can access this by using the Navigator.getPoseProvider() method, which in turn allows you to check the current pose. There is also an alternate Navigator constructor which accepts a PoseProvider as a parameter, allowing you to specify the type of PoseProvider to use.

Odometry Error

The leJOS developers have tried to eliminate as many sources for error as possible from the navigator package, but some error will always remain. It is a good idea to be aware of the source of errors so you can do your best to eliminate them.

- Inaccurate measurements of the track width (distance between the two wheels) can result in inaccurate turns. This can be especially hard to measure when the contact area of the floor and the tire is wide (see Figure 22-4).

- Surface texture of the floor can impact accuracy. On a small scale, the ground can have unevenness, such as the spacing between tiles, the spaces between

Try it!

Localization like this could be useful for a remote control car that uses odometry to keep track of its coordinates. You could try navigating this car blindly through your house using a map, even if it is in another room, by having it report back the coordinates. You could also program the robot to find its way back to the starting position.

hardwood floor slats, or lumpiness of carpets. This can affect rotation and distance characteristics significantly. Try measuring movement on carpet and then compare with the same movement on hardwood. You will see a noticeable difference.

- Wheel slippage can occur when a robot is starting and stopping. This is partly a function of surface texture and tire grip. It also has to do with how fast a robot accelerates. You can minimize this effect by lowering the acceleration speed of the pilot using the setAcceleration() method.
- Backlash is a problem that occurs due to the tiny spaces between gears. Even the internal gearing of EV3 motors contain backlash. You can feel the backlash by lightly applying torque to the motor axle with your fingers. The backlash is a dead-spot where the axle turns freely without any resistance.
- Differences between tires can have a surprising effect on accuracy. This is most obvious when you build two robots and attempt to drive them through the same path. Variation caused by differences in the tires can add up to small variations in the accuracy of the robot. Trying different combinations of tires can help minimize this effect.
- Weight distribution of the robot can also have an effect. When the robot rotates in one spot, the weight distribution might be more towards the front or back of the robot. If too much weight is placed on the castor wheel, you can get rotational wobble.
- Robot construction issues can affect accuracy. If one or more wheels on the robot are not level and driving straight ahead, it can alter the robot course. Saggy wheel supports or wobbling wheels can do the same.

These errors will always plague dead reckoning, which makes landmark navigation more appealing.

CHAPTER 27

Walking Robots

TOPICS IN THIS CHAPTER

- Balance Theory
- Bipedal Locomotion
- Building a Walker
- Walking the Walk

Balance is something we humans take for granted because we are so good at it. Most animals walk on four or more legs, giving them chair-like stability when at rest. A chair maintains static balance; it doesn't have to actively reposition itself to remain upright. We know that a chair with only two legs tips over, yet humans are able to stand with two legs. This is because we maintain *dynamic balance* by constantly shifting our weight to avoid falling. This chapter explores static balance concepts as they apply to walking.

Theory of Static Balance

Static balance applies to objects at rest. Even when a robot moves, however, static balance concepts can still be used to determine if the robot will remain stable. For example, a vehicle with four wheels does not tip over while it is stationary or while moving. The same principle that gives it stability at rest also gives it stability while in motion.

Chapter 6 emphasized that it is desirable to build a stable robot; robots that constantly tip over are not usually considered successes. This section will tell you why a robot is unstable and how to make changes to make it stable.

Every part of a rigid body is composed of particles. Gravity exerts a downward force on each particle. The sum total of the force on each particle produces a single direction of force, called the center of gravity, or center of mass (see Figure 27-1). (The center of gravity is another term for center of mass when operating in a uniform gravity field. Since all our experiments are taking place on earth, we can use the term center of gravity.)

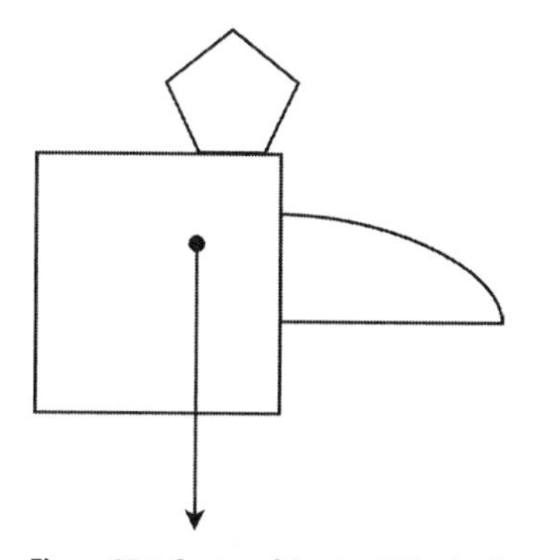

Figure 27-1 Center of Gravity (side view).

To maintain stability, you must keep the center of gravity within the center of a polygon defined by the wheels or legs of your robot. Each part of a robot that touches the ground is a point, so a robot with four wheels defines a rectangle with four points. To maintain stability, the center of gravity must reside within these points (see Figure 27-2A). If the center of gravity is outside these points, the robot will tip over (see Figure 27-2B).

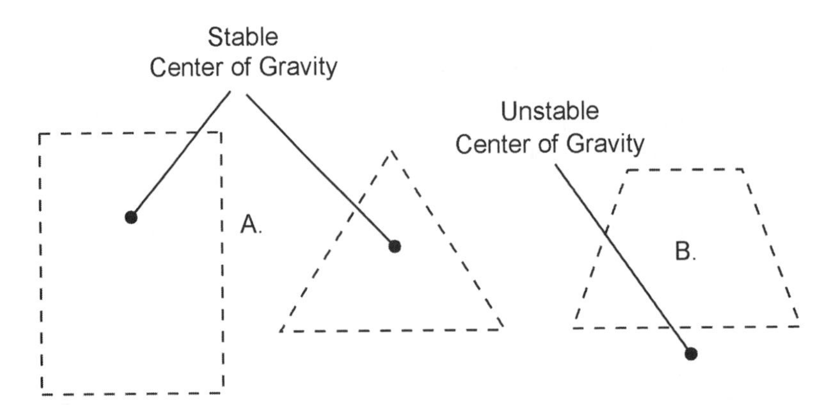

Figure 27-2 Stable and unstable center of gravity (top view).

I've sometimes built robots that were stable when stationary but unstable while moving. The three wheel robot in Figure 27-3A is an example. The center of gravity was too close to the boundary and it was a top-heavy robot. When the robot accelerated or decelerated, it introduced momentum, which shifted the center of gravity outside the polygon formed by the wheels (Figure 27-3B).

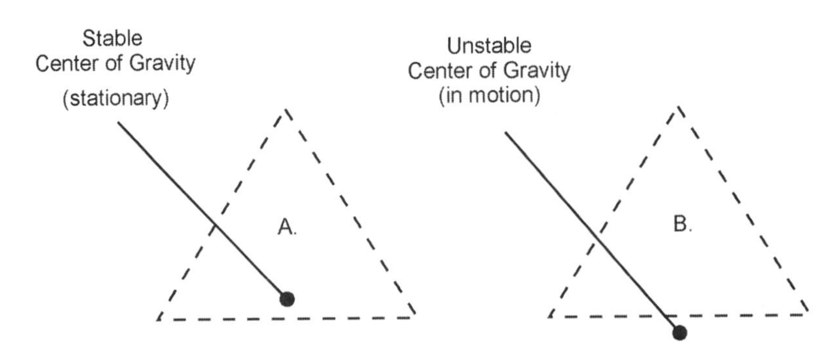

Figure 27-3 Stable while at rest, unstable in motion.

Momentum is a force, just like gravity. It is calculated by multiplying mass and velocity. An object at rest has zero velocity, therefore it has no momentum.

Gravity is a force, determined by acceleration (9.8 ms^2 towards the center of earth) and mass. Momentum is a force determined by the velocity of your robot and the direction of travel. So when these forces are combined, it effectively moves the center of gravity.

The full equation to determine whether your robot will tip over is very complicated and beyond the scope of this book. It would have to take into account the height of the robot, the mass at the top of the robot versus mass at the bottom, and the sizes and angles of its various parts. Instead, we can break this down into some simple rules for stability:

- Top heavy robots are less stable than bottom heavy robots
- Taller robots are generally less stable than short robots
- Robots with high acceleration are more likely to tip over than robots with low acceleration
- Robots with mass centered over their base are more stable

Now that you know something about center of gravity, let's explore how this will affect bipedal robotics.

Bipedal Locomotion

Birds, man, and some lizards are bipedal. Flight is the primary form of locomotion for birds (barring ostriches), and bipedal locomotion is their secondary form, leaving man and some lizards as the only animals with primary bipedal locomotion. In other words, bipedal locomotion is an unpopular form of travel in nature, probably because it isn't optimal. Despite this, robot enthusiasts are drawn to creating bipedal robots; it allows us to mold creations that are similar to ourselves.

In mammals, the most popular base for a leg is the paw, which contains four cushy pads and retractable claws, making them robust for gripping different surfaces. The least popular base is the foot, which offers bipedal creatures additional balance.

Most LEGO walking robots use two motors, because of the low number of motor ports on MINDSTORMS bricks. Each motor allows a joint to move back and forth along one plane, otherwise called the *axis of rotation*.

Now look at one of your own legs. Your leg has three active joints: the hip, knee, and ankle joints. The hip has a ball joint and can move within about a 90 degree range in any direction, forming a cone of movement (see Figure 27-4). The purpose of this movement is to allow your leg to stop you from falling over should you start to lean in the wrong direction. If you tilt in any of the 360 degrees while standing on one leg, the other leg can move to the opposite point to reestablish the center of gravity. Your hip joint allows more movement in the forward direction than the backward direction, which makes it easier for you to travel forward than backward.

Try it!

Stand up and keep your legs stiff. Concentrate on your hip joint and try to keep your leg moving only along one plane, like a robot. To help you do this, keep your arms by your side. Furthermore, lock your heel so you are not using your foot for balance. Also, and this is the hardest one to do, don't tilt your pelvis by rotating your hip socket sideways; in other words, keep your pelvis horizontal with the floor. Notice how you have to scrape your foot along the ground because you can't lift it using your hip sockets? The best you can do is shuffle by leaning back and forth with your upper torso and tilting your torso side to side. Likely your legs will have to break one plane of rotation to maintain balance. It's not so easy to walk on two joints, is it? This is the problem presented by having only three motors in the EV3 kit.

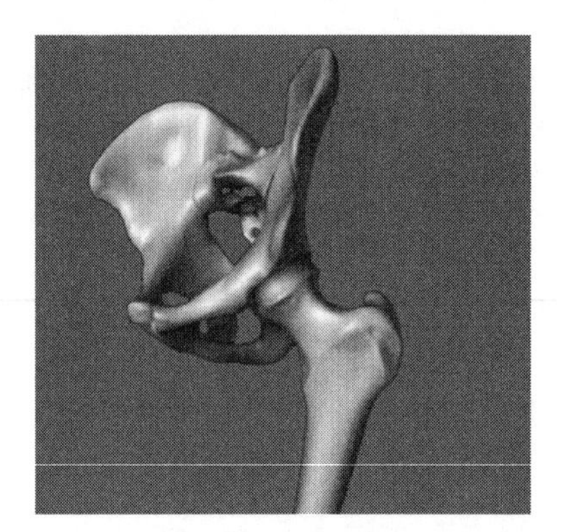

Figure 27-4 Hip joint.

The knee is different from the hip joint because it can only rotate along a single plane. The knee allows you to walk up stairs or other platforms. When you walk, your knee bends in coordination with the hip, allowing you to fold up your leg so it doesn't scrape the ground, then reposition your foot at a different level.

Your ankle joint is somewhat like your hip joint in that it can rotate along more than one plane, though it is more limited. This allows you to place your foot parallel to the surface you are traveling on. The ankle joint is more important for standing in one place than for moving, though it can position the foot to absorb shock when you are moving. (Without feet you would have a very jarring run.)

A shuffler is the most common walker to build with a limited number of motors. Ev3rstorm is a great example of this type of robot. It uses sophisticated engineering, allowing the ankles to move along two axes, just like your own. The downside is that it doesn't use interesting or sophisticated programming.

Several companies have been able to produce impressive bipedal robots. For example, a French company named Aldebaran introduced a bipedal robot with impressive movement capabilities (see Figure 27-5). This robot uses over 20 servo motors to achieve lifelike movement. Obviously a robot like this is beyond the capability of EV3 kit, with it's 3 motors. Even with 20 motors, Nao is still turns somewhat clumsily, shimmying around to turn in a circle.

Try it!

Try walking up some stairs using only your hip joints, as described in the last 'Try it'. Obviously you can't. Now try walking up the stairs by allowing your knee to raise your leg. Study this motion, thinking of ways you can create it in robotics.

Figure 27-5 Aldebaran's Nao robot (copyright 2014 Aldebaran)

Try it!

Try standing on one foot with your other leg raised about six inches off the ground in front of you. While doing this, watch the ankle of the foot that is still on the ground (take off your shoe and sock so you can see your ankle at work). Notice that the ankle is making very small movements? Your ankle is constantly moving your center of gravity from one point to another to keep you from falling. If you move your arms or if there is some wind, your ankle adjusts the center of gravity to compensate.

This chapter will explore walking robots using the three motors in the EV3 kit. Unfortunately, bipedal robots are difficult to build using just three motors. The robot in this chapter uses the limited number of EV3 motors, but it relies on the principles of balance discussed above in order to present itself from tipping over.

A Walking Robot

There is an old saying, "You need to crawl before you can walk." The robot in this chapter capitalizes on that philosophy. Physically, the robot's movement is achieved by something resembling a crawling movement. It lifts itself off the ground, plants the walking foot, and then inches itself forward. By repeating this motion it gains distance, slowly.

One of the challenges of making a walking robot is that it must produce linear motion, such as up-down, and forward-backward movements. However, motors provide circular motion, so the robot needs to convert circular motion into linear motion. This robot can move up and down, forward and backward, and it also has a third motion of rotating in one spot.

Each of these movements required a lot of thought in order to put into practice. It might look simple now, but this is the end result after trying several failed attempts for each movement and coming up with the simplest.

Another challenge was adding each of these three motions physically close together on one robot. These three modules must reside on the same robot in much the same space, making it difficult to get them to work together without the parts coming into contact with each other.

Building the Robot

In this section we will build the walking robot, known as the Shambler.

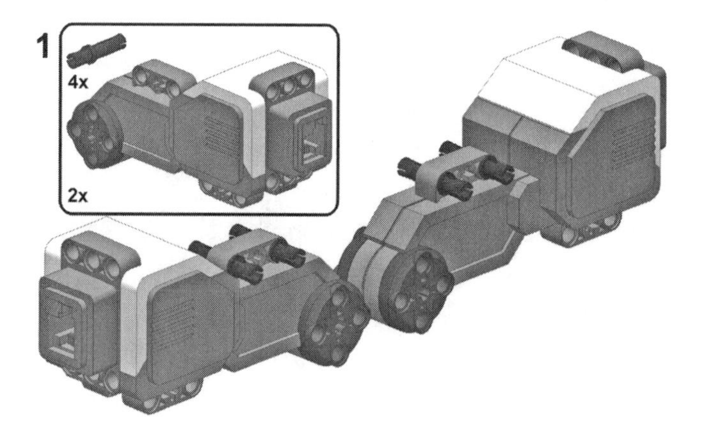

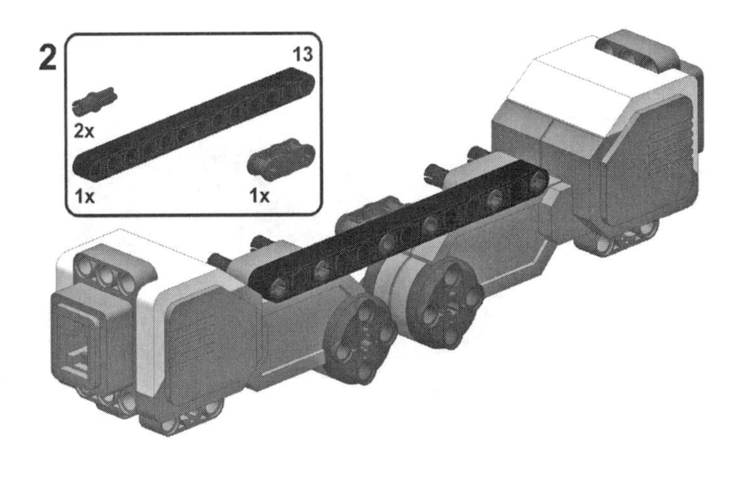
2
13
2x
1x
1x

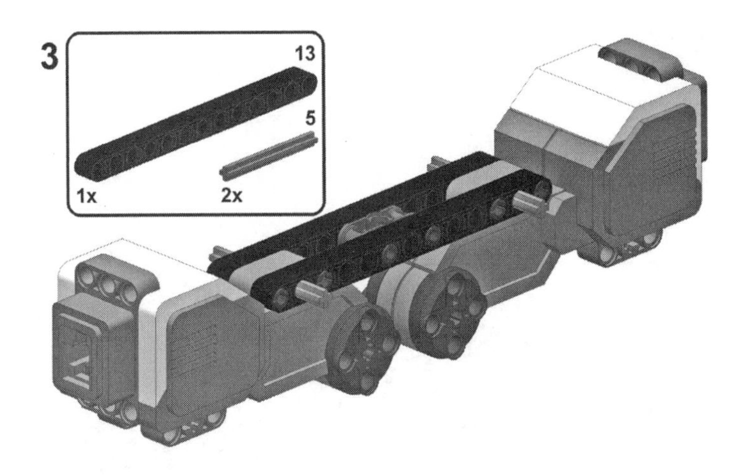
3
13
5
1x
2x

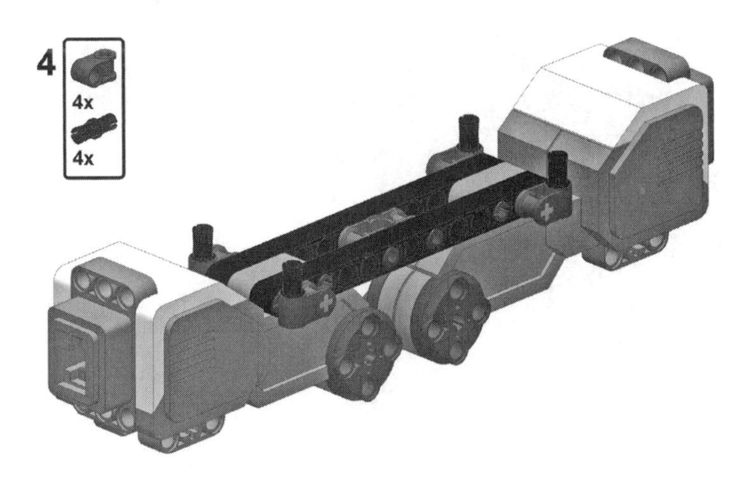
4
4x
4x

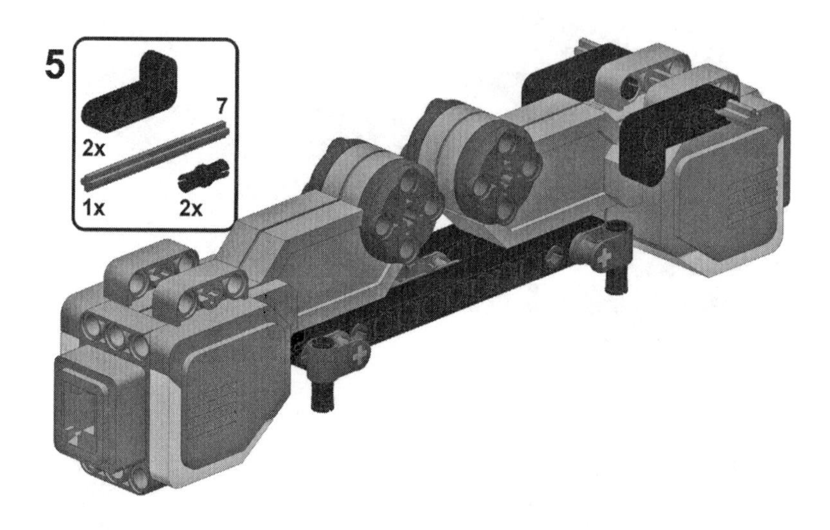
5
7
2x
1x
2x

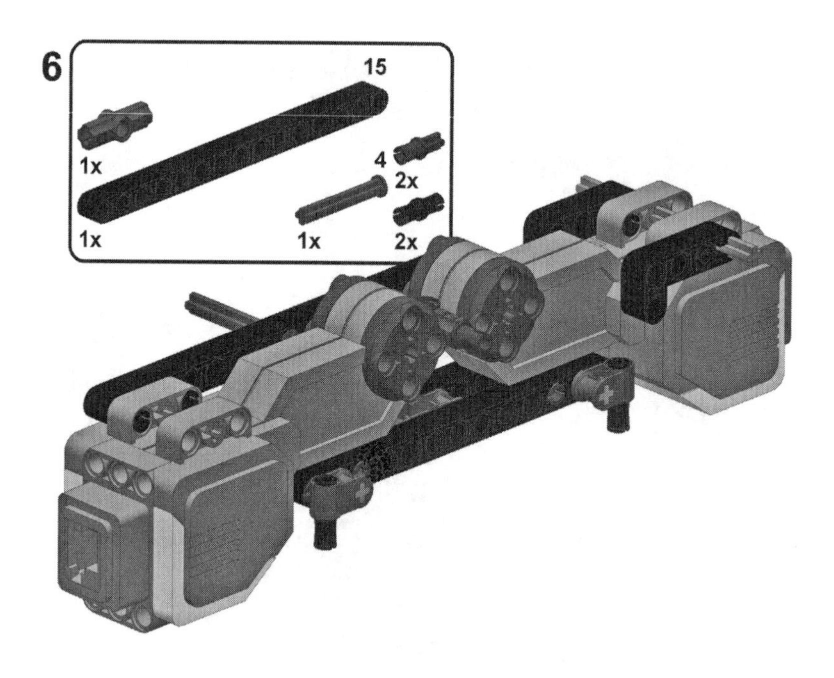
6
15
1x
4
2x
1x
1x
2x

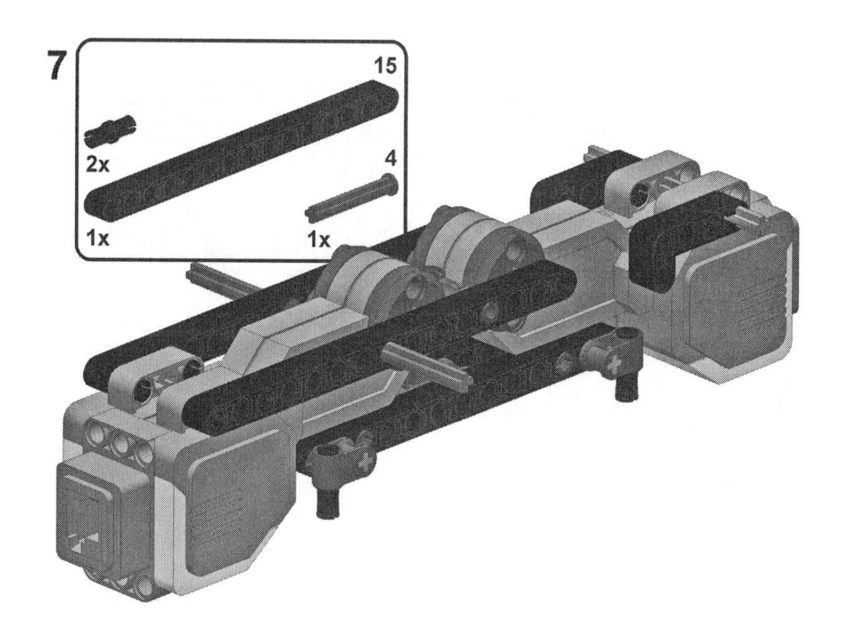
7
15
2x
4
1x
1x

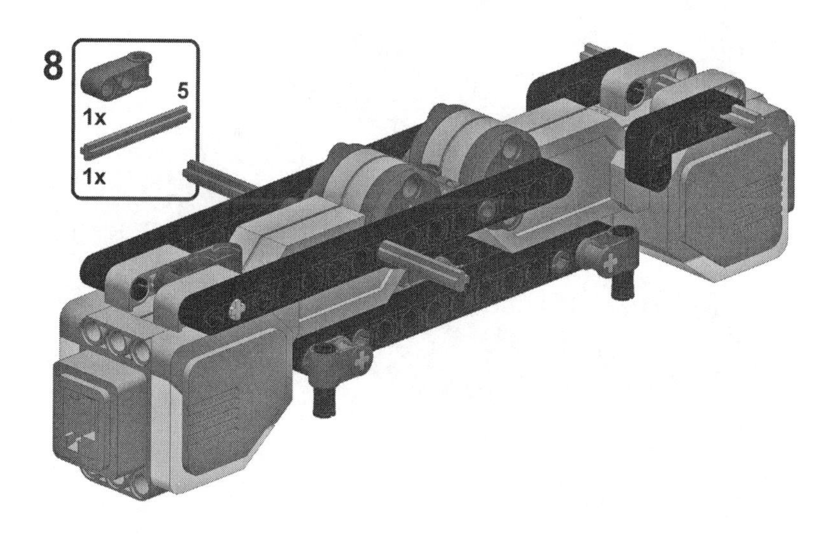
8
5
1x
1x

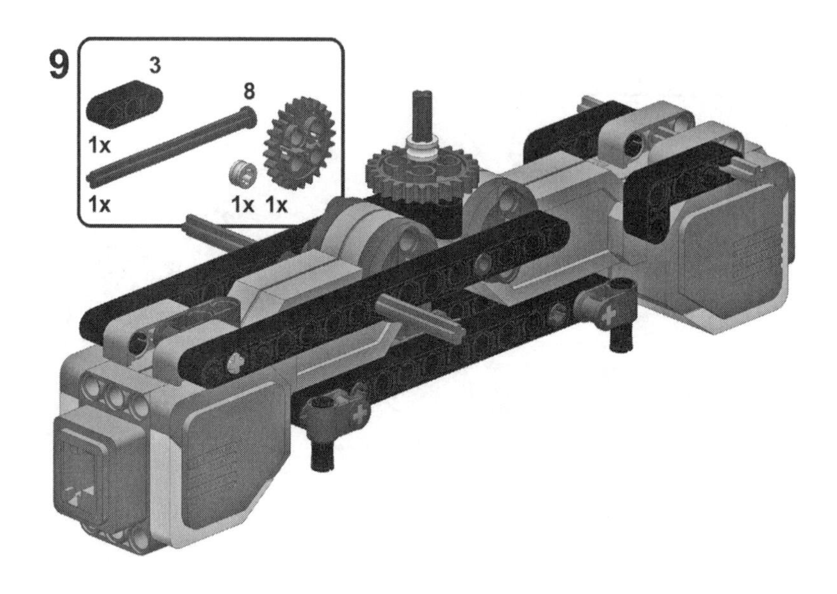
9
3
8
1x
1x
1x 1x

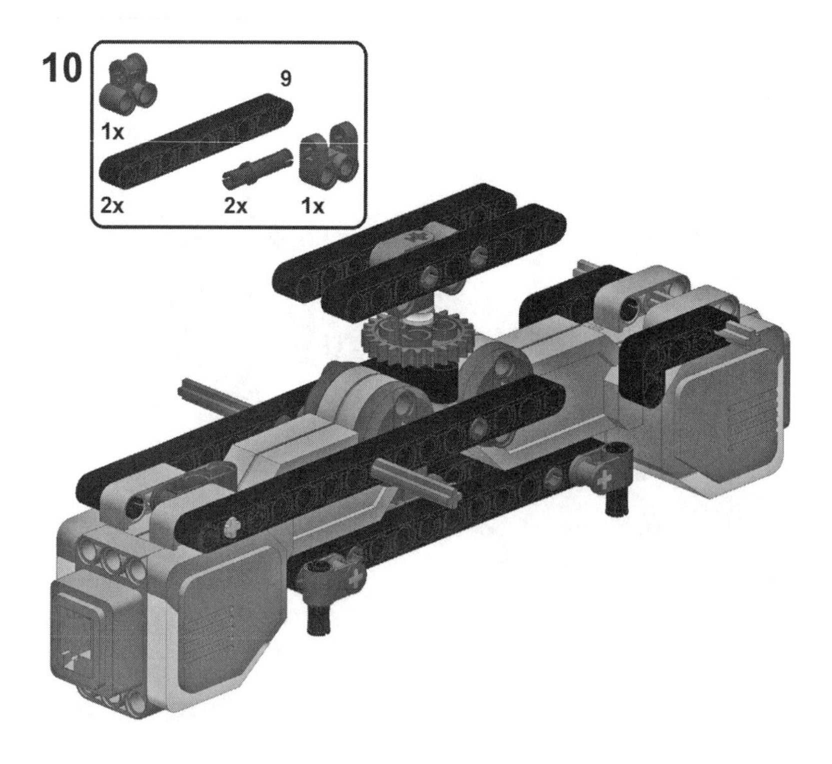
10
9
1x
2x
2x
1x

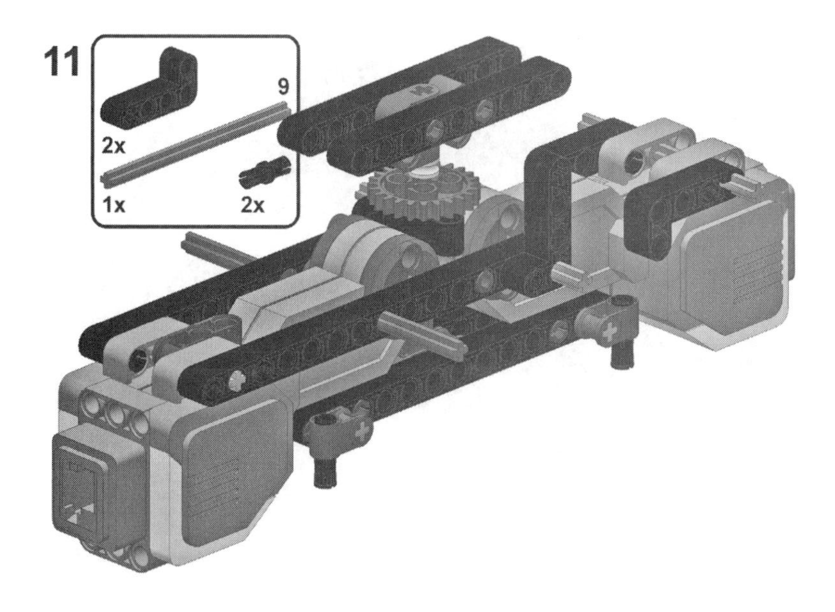
11
9
2x
1x
2x

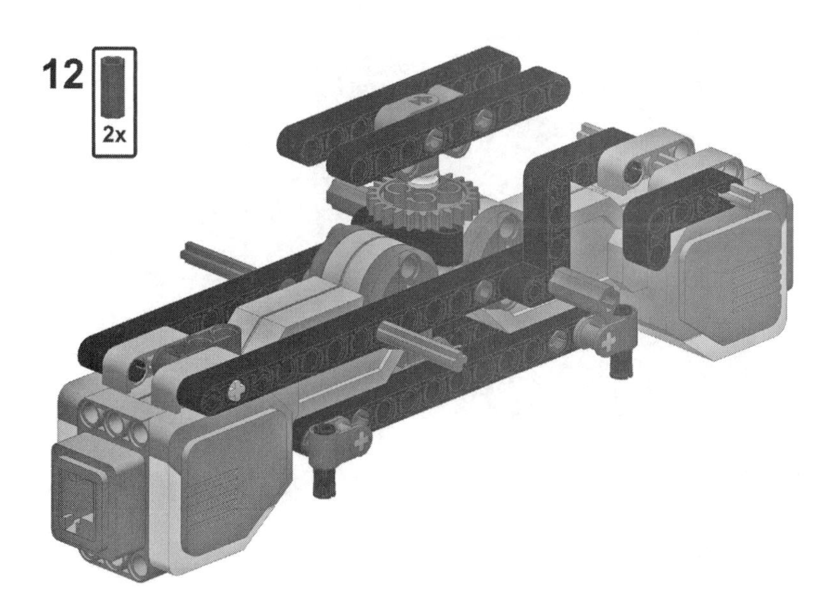
12
2x

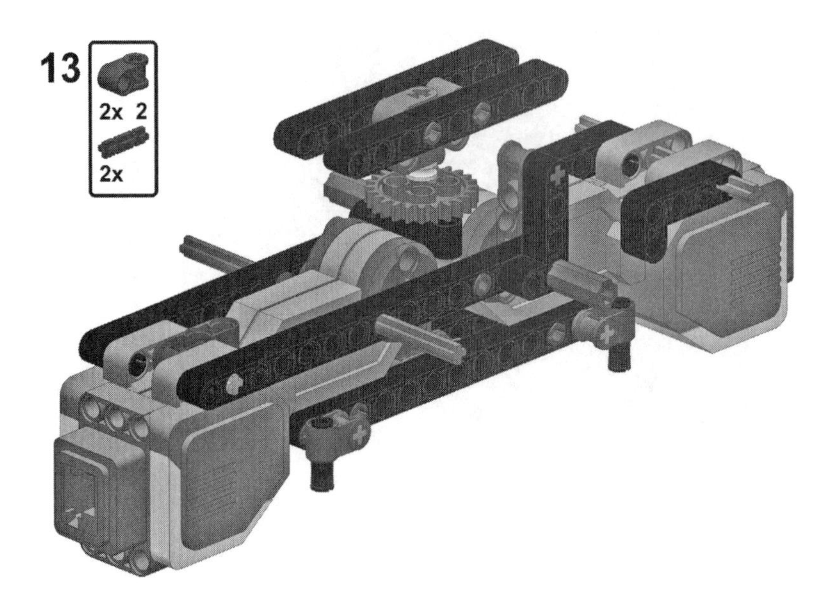
13
2x 2
2x

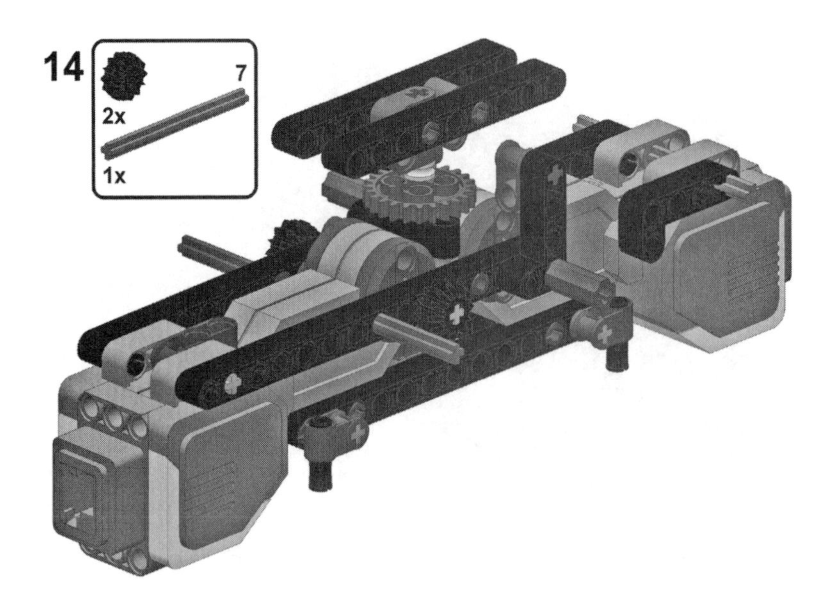
14
7
2x
1x

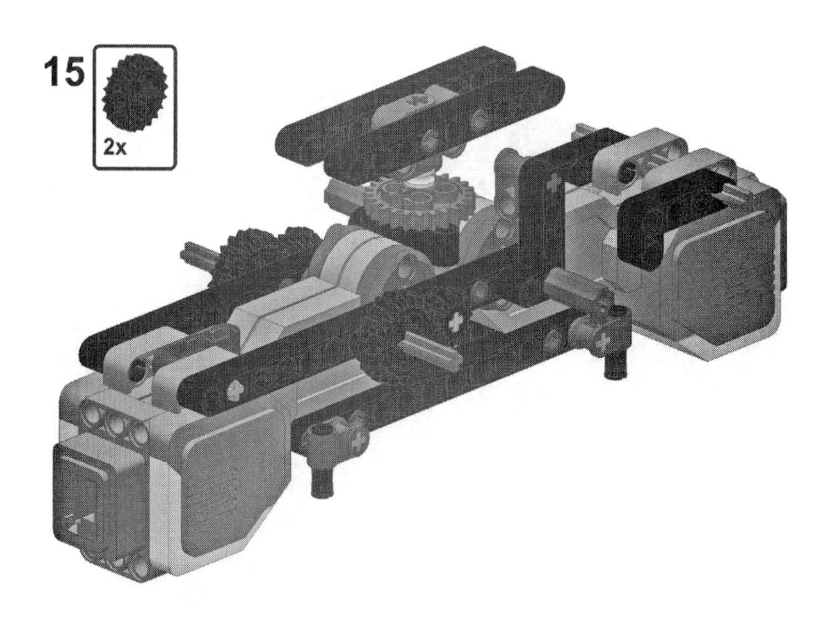
15
2x

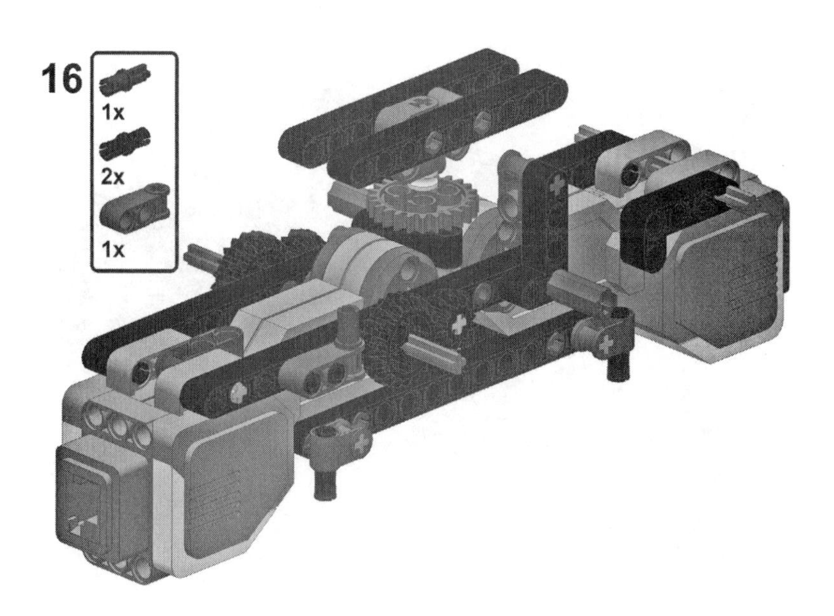
16
1x
2x
1x

17

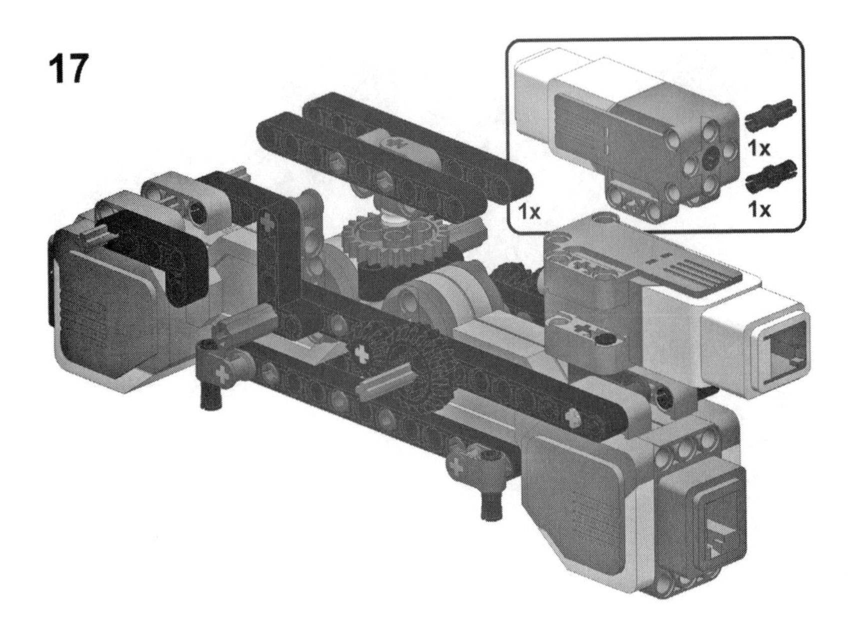

18

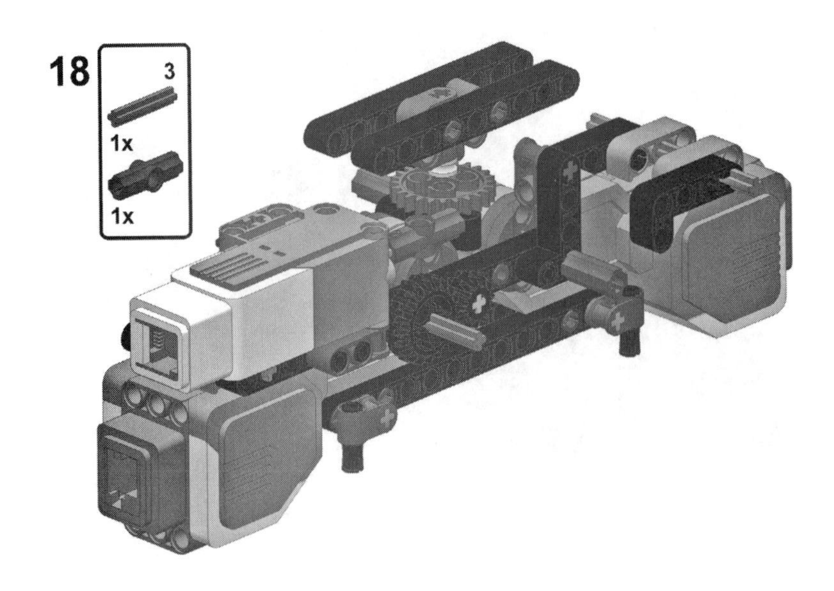

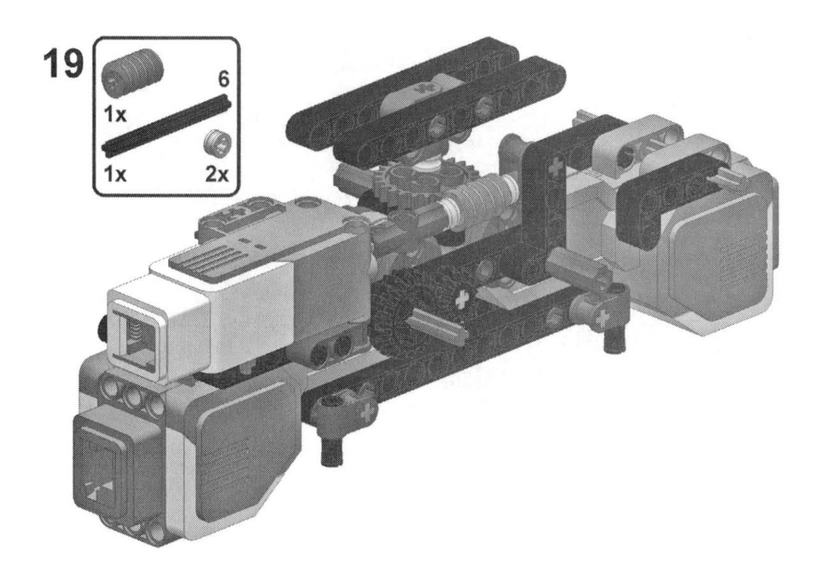
19
1x
6
1x
2x

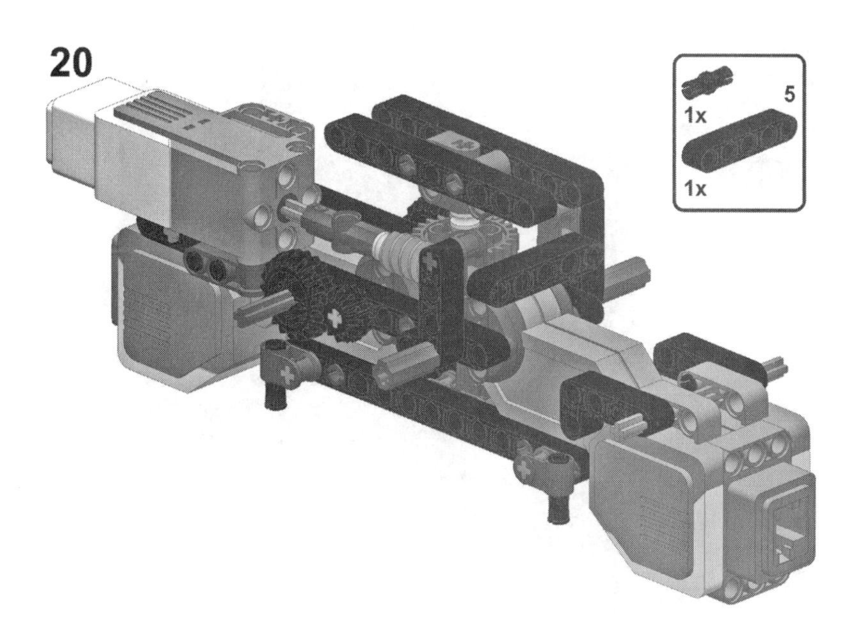
20
1x
5
1x

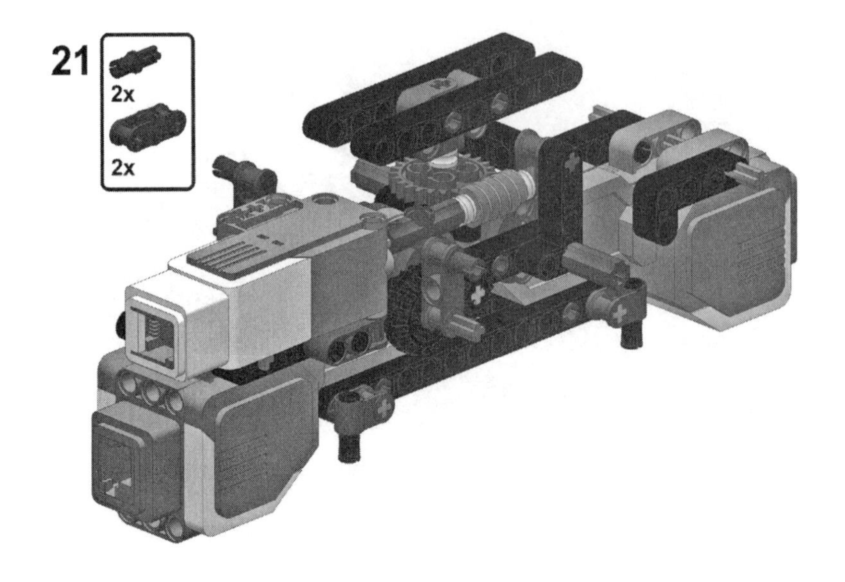
21
2x
2x

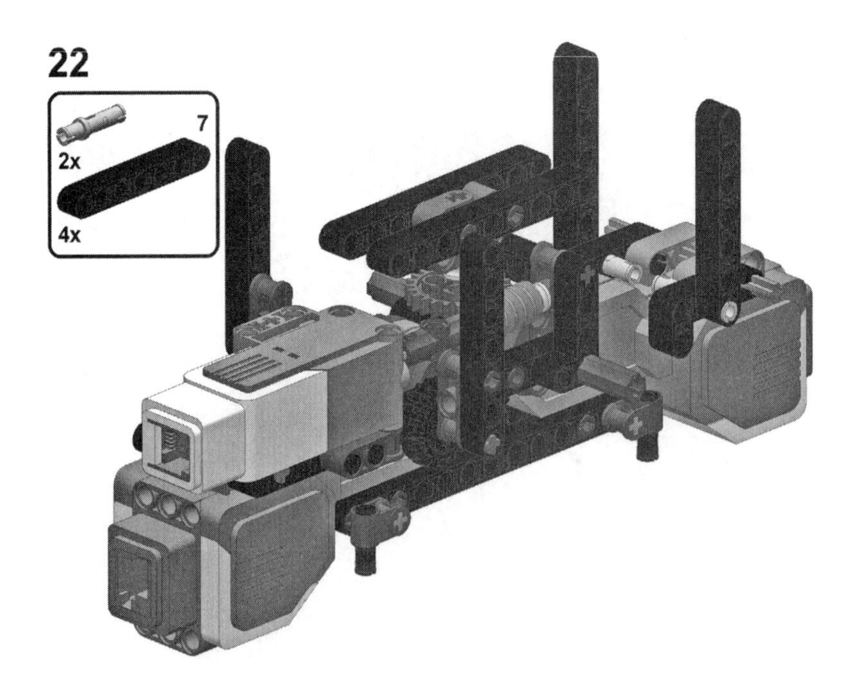
22
7
2x
4x

23

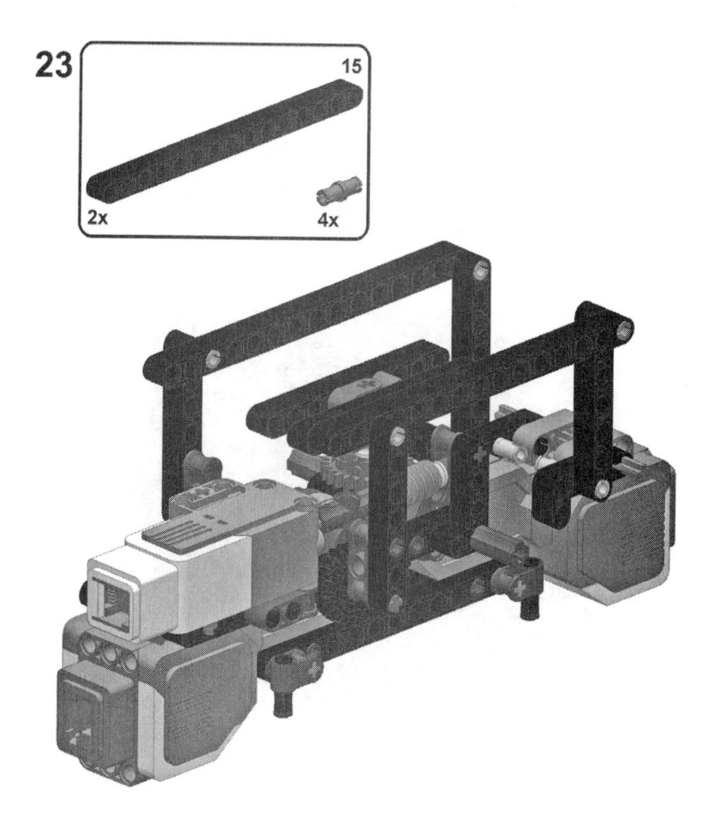

24

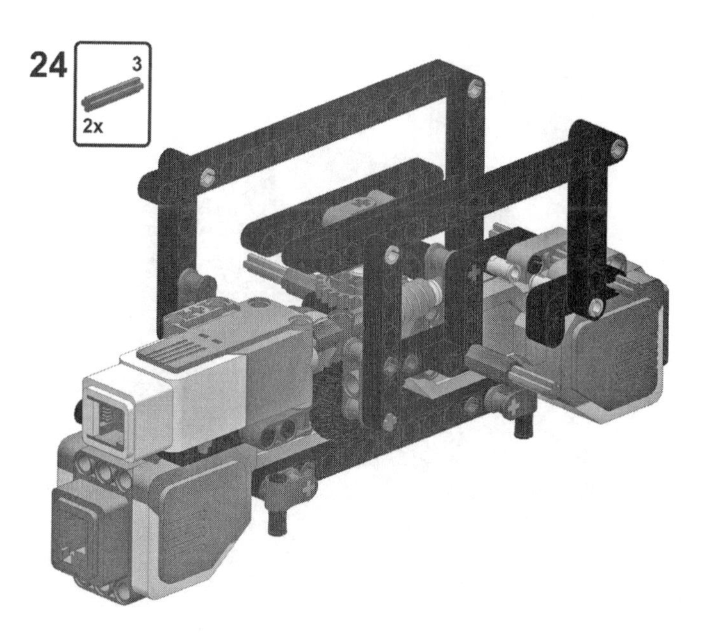

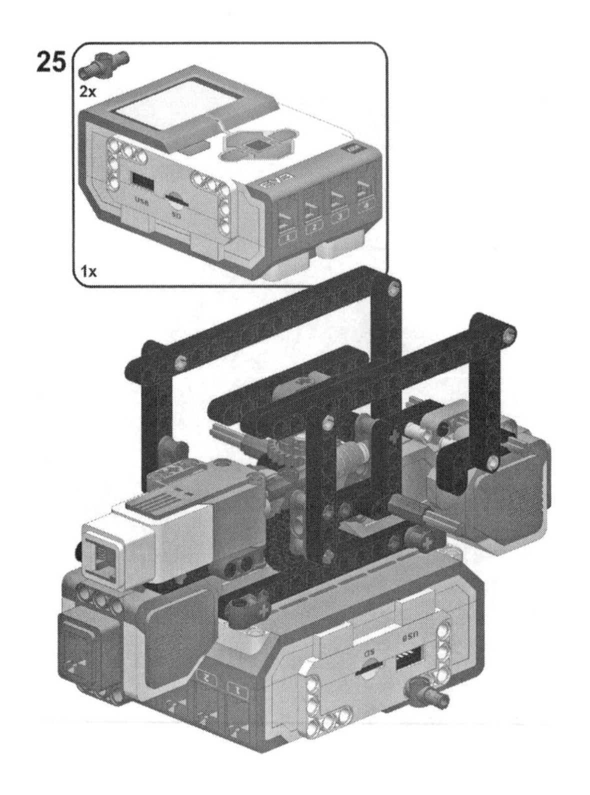
25
2x
1x

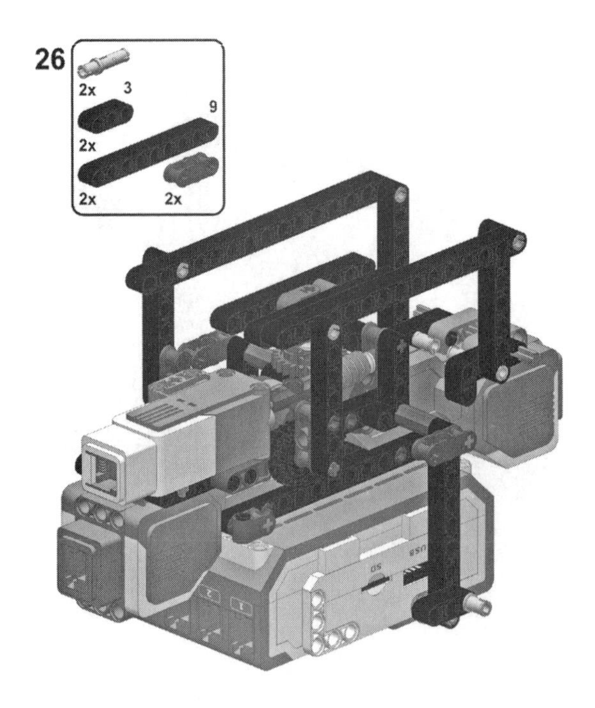
26
2x
3
9
2x
2x
2x

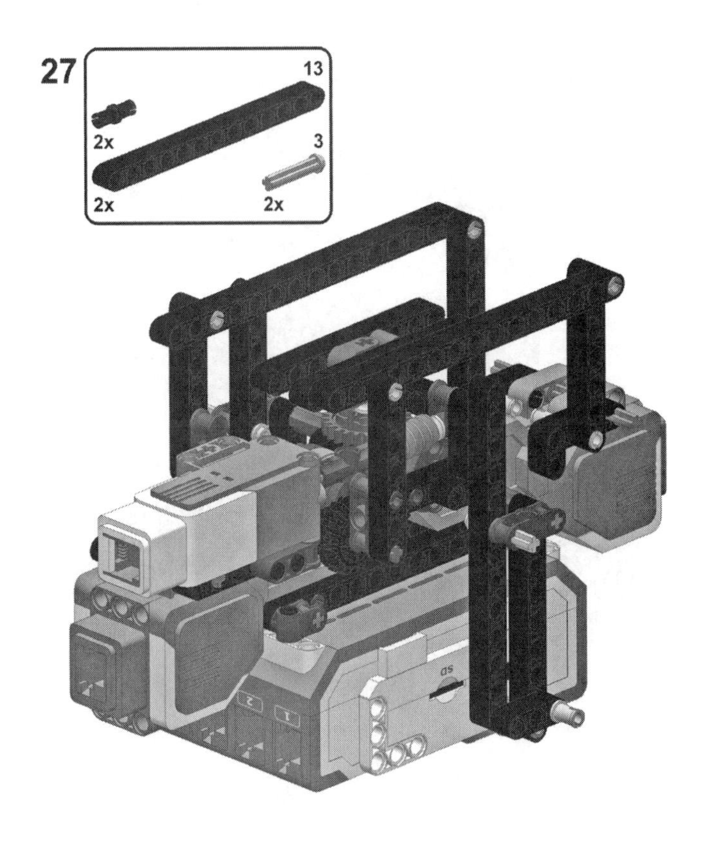
27
13
2x
3
2x
2x

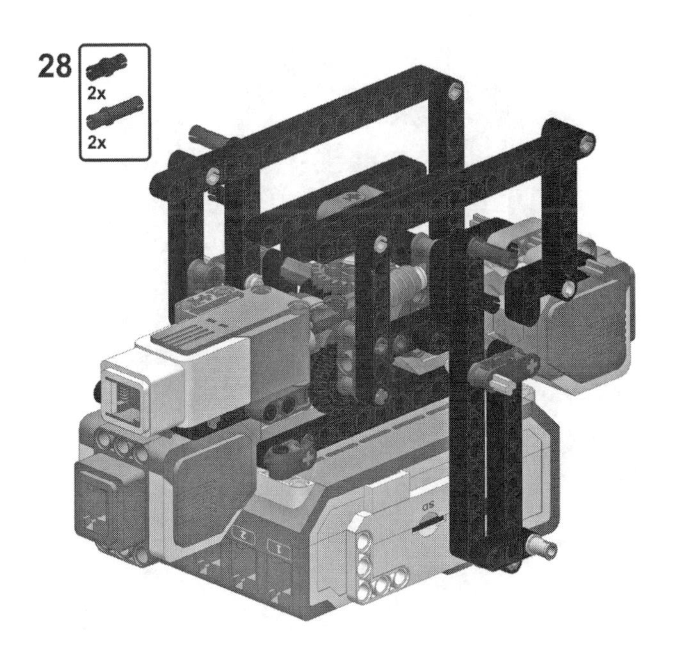
28
2x
2x

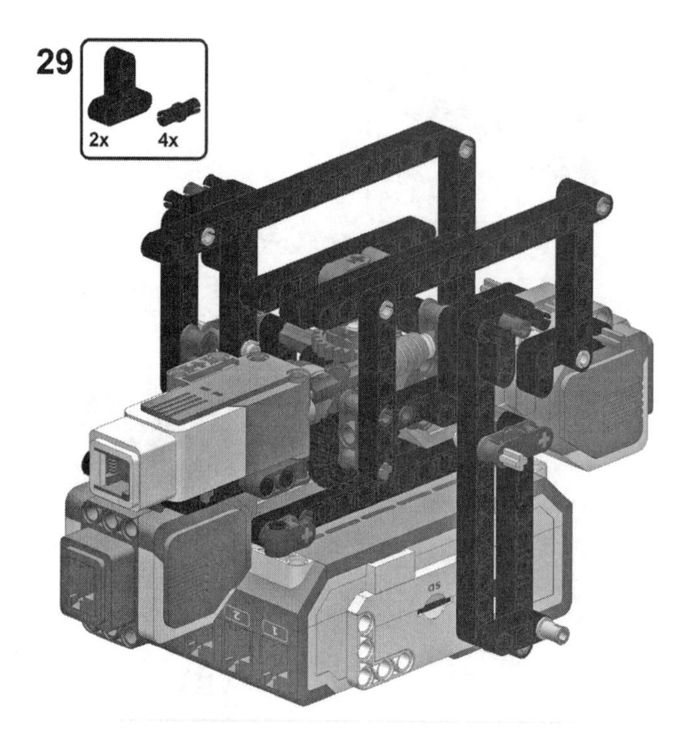
29
2x
4x

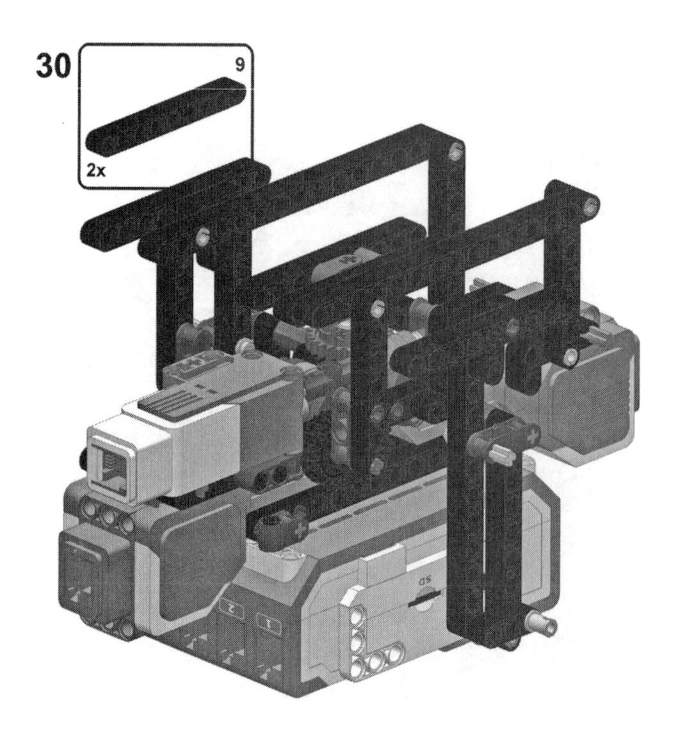
30
9
2x

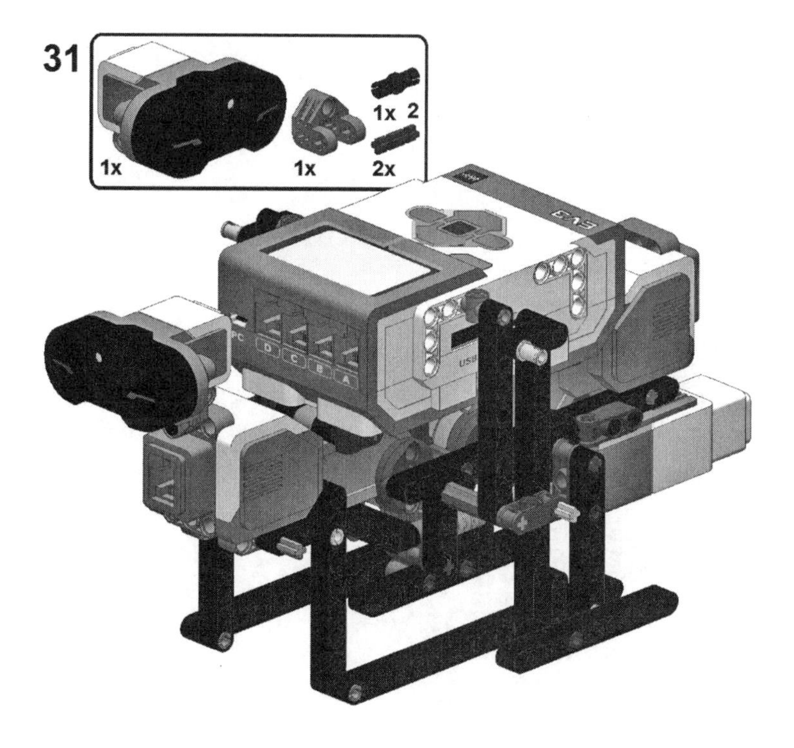

Cables

Now that we have a robot, it is time to connect the cables (see Figure 27-6). Connect a short cable from port A to the closest large motor. Connect a medium cable from port B to the opposite large motor. Finally, connect a medium or large cable from port C to the medium EV3 motor.

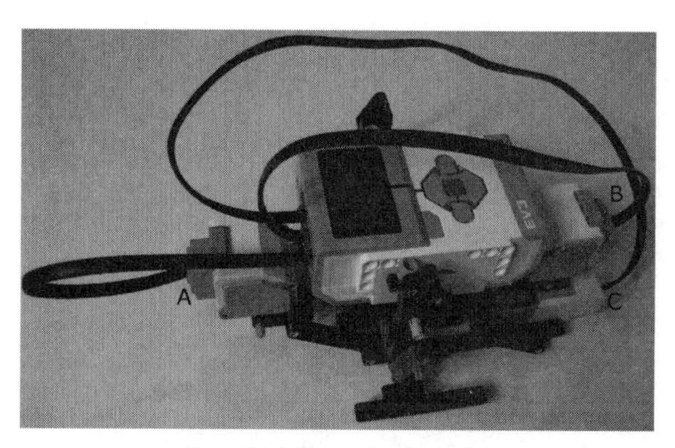

Figure 27-6: Connecting the Cables

NOTE: *You've probably seen walking robots that use one or two motors to perform a continuously cycling walking gait. Why does this robot use three? In short, so that it can walk precise distances. The lateral motor lifts the robot up and moves forward. If this ran continuously, it would only be able to move in gross increments of a set distance. By being able to interrupt a partial move, lower the base and pick up the legs, it allows the robot to move any fraction of a step, and therefore any distance.*

Walking Code

This is the moment we have been waiting for—to make this complex robot walk! We will code a short program to take the robot through its basic movements: walking forward and turns. As you will see, the code is very basic.

```
import lejos.hardware.motor.*;
import lejos.hardware.port.*;
import lejos.robotics.*;

public class Shambler {

  public static int STANDING = 145;
  public static int RETRACTED = 0;
  // movement gear ratio 20 to 12
  public static int BACKWARD = (int)(140 *
20.0/12.0);
  public static int FORWARD = 0;
  public static double RATIO = 24.0/1.0; // screw
gear ratio
  public static RegulatedMotor a_motor = null;
  public static RegulatedMotor b_motor = null;
  public static RegulatedMotor c_motor = null;

  public static void main(String[] args) {
    a_motor = new EV3LargeRegulatedMotor
(MotorPort.A);
    b_motor = new EV3LargeRegulatedMotor
(MotorPort.B);
    c_motor = new EV3MediumRegulatedMotor
(MotorPort.C);
```

```
    a_motor.setSpeed(70); // vertical lifter
    b_motor.setSpeed(110); // lateral movement
    c_motor.setSpeed(110); // rotation

    for(int i=0;i<4;i++) {
      stepForward(2);
      rotate(90);
    }
  }

  public static void stepForward(int steps) {
    for(int i=0;i<steps;i++) {
      a_motor.rotateTo(STANDING);
      b_motor.setSpeed(300);
      b_motor.rotateTo(BACKWARD);
      b_motor.setSpeed(110);
      a_motor.rotateTo(RETRACTED);
      b_motor.rotateTo(FORWARD);
    }
  }

  public static void rotate(int degrees) {
    c_motor.rotate((int)(RATIO * degrees));
    a_motor.rotateTo(STANDING);
    c_motor.setSpeed(1000);
    c_motor.rotateTo(0);
    c_motor.setSpeed(70);
    a_motor.rotateTo(RETRACTED);
  }
}
```

For the code to work, the robot's limbs must each start from a specific position (see Figure 27-7). Notice the vertical lifter is fully tucked in, the legs are fully forward, and the whole robot is resting on the rotating feet. Also, the rotating feet should be positioned so they are pointing in line with the rest of the robot. You can adjust this by using the Remote Control tool from the Tools menu on the EV3 (and plugging an IR sensor into any port).

Try it!

Set up a simple obstacle course for your robot on the floor, consisting of obstacles that your robot will navigate around. Next, figure out the moves it needs to make to get through the obstacle course and write them on a piece of paper (see Table 18-1 for an example). Finally, try altering the code in the main() method to make your robot move through the course. Keep in mind that each step your robot takes is about 7 centimeters.

1	Forward 21 cm (3 steps)
2	Rotate 45 degrees
3	Forward 35 cm (5 steps)
4	Rotate -180 degrees
5	Forward 70 cm (10 steps)

Table 27-1: Plotting moves through an obstacle course.

Try it!

With an IR sensor equipped to the front of the Shambler, try making it move around the room and change direction when it detects an obstacle in front of it.

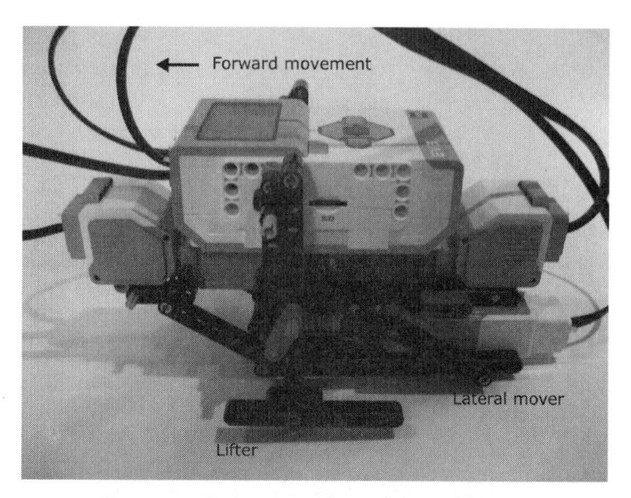

Figure 27-7: The starting position of the Shambler.

Now run the code and watch the robot move. As you can see, movement is much slower than a wheeled robot. Also, if you are using the robot on a hard floor, when it rotates it can sometimes over or under-rotate ever so slightly. This is because the central rotation foot is hard plastic, which allows the foot to slip slightly due to momentum once it starts rotating. In fact, you can spin the robot around like a top when the robot is in the default starting position. If the foot was rubberized, it would be more secure to the floor and produce more accurate rotations. To compensate, try using it on a rug.

CHAPTER 28

Adventures in Linux

TOPICS IN THIS CHAPTER

- SCP
- SSH
- Linux commands
- Vi
- Running Java code

In this chapter we'll examine the unique things we can do with Linux running on the EV3 brick. Because we are interacting with Linux over WiFi, we can use standard applications to connect to the Linux system and do things such as file management, executing Java code, and executing Linux commands.

Uploading Files with SCP

One of the most basic things you might need to do is to upload files to the brick that interact with your program. For example, you might find the need to upload a wav sound file, an image, or even map data to interact with one of your programs. You can do this natively with Linux using SCP, which stands for Secure Copy Protocol. The protocol allows you to browse files on a Linux system and perform basic file management.

If you are using Linux, chances are there are several SCP clients already installed on your system. For Windows, there are a number of SCP clients you can download for free. For this example we will use WinSCP.

1. Download and install WinSCP from http://winscp.net
2. Run WinSCP and you will see a login window (see Figure 28-1). Click New to start a new profile.

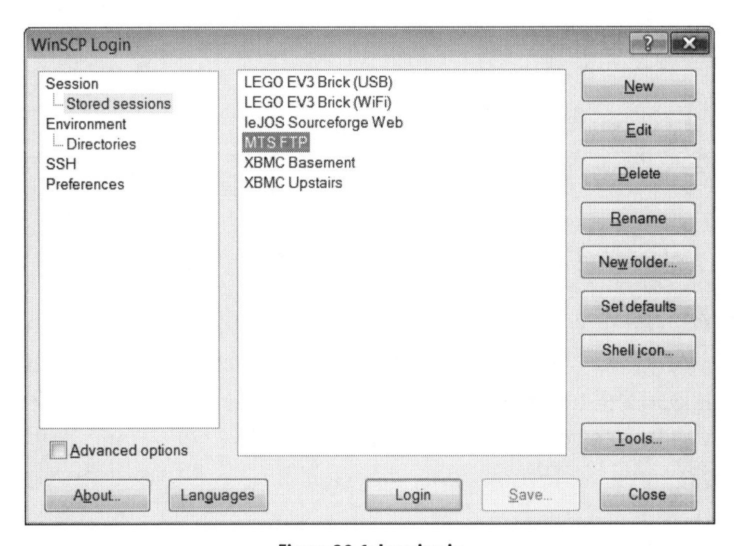

Figure 28-1: Logging in.

3. We can enter information on the next screen (see Figure 28-2). For host name, enter the 4-number IP address displayed on your EV3 when it boots up to the leJOS home screen. (If you are using Bluetooth or USB, use 10.0.1.1.) Port number remains 22. Username is "root", and there is no password. Be sure to change File protocol to SCP.

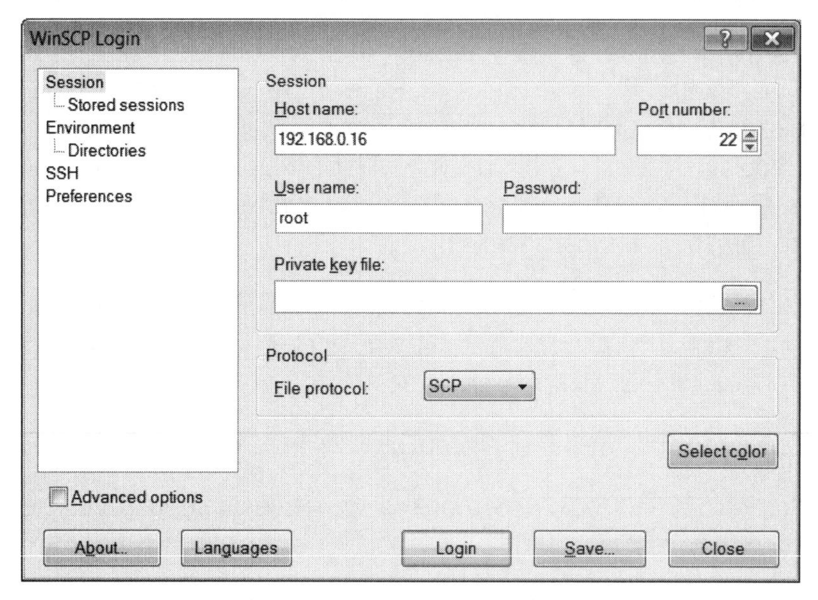

Figure 28-2: Entering connect information

4. Click Save... and enter a name for this, such as leJOS EV3

5. Now select the leJOS EV3 entry in the list, and select Login. The first time you login it will give a warning screen about the host key. Select Yes.

6. It will prompt you for password. Just click OK. Note: At this point an error or two might come up. You can safely ignore them by clicking OK.

That's it! You can now graphically browse through the EV3 file system (see Figure 28-3). Any text files you see can be double clicked to view them. If you want to upload files to the program directory, browse to /home/lejos/programs. Likely any data files your program might use will go here. You can also delete, copy, and rename files. When done, click Quit or hit F10 to disconnect.

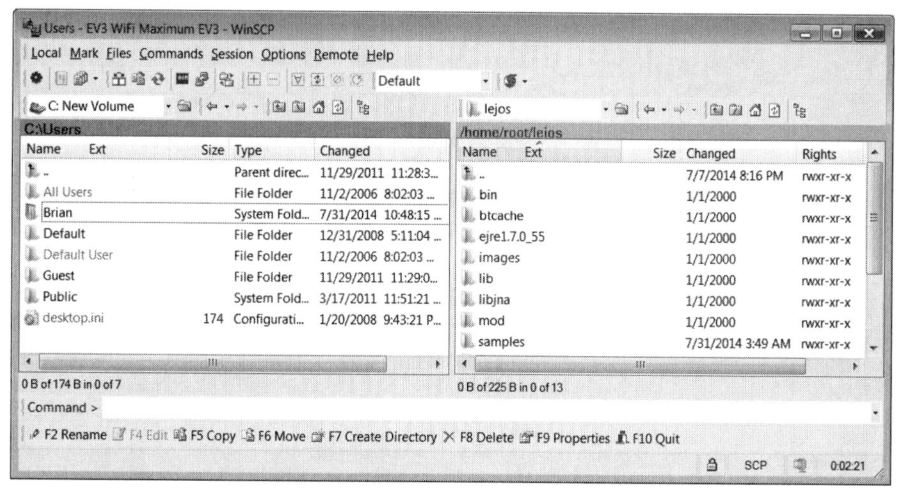

Figure 28-3: Browsing the EV3's Linux file system

Command Line with SSH

Using a terminal to edit files and run system tools is a big part of the Linux experience. In fact, with the EV3 there is no graphical GUI to interact with the OS, so all you have is the command line interface. But since there is no keyboard, and the LCD is quite small, you need to use a computer to remotely connect and issue commands.

Windows

1. Download and install PuTTY from www.putty.org
2. When you run PuTTY, the PuTTY Configuration screen greets you (see Figure 24-4). Enter the IP address of your EV3 brick (or 10.0.1.1 if using Bluetooth or USB), leave it as port 22, and select SSH. You can also click Save to save the session for later (you will need to give it a name such as LEGO EV3 WiFi).

> ***Website!***
>
> An alternate SCP client for windows is found on the same site as PuTTY, the SSH client used below. The tool's name is PSCP, found on the download page here:
>
> www.putty.org

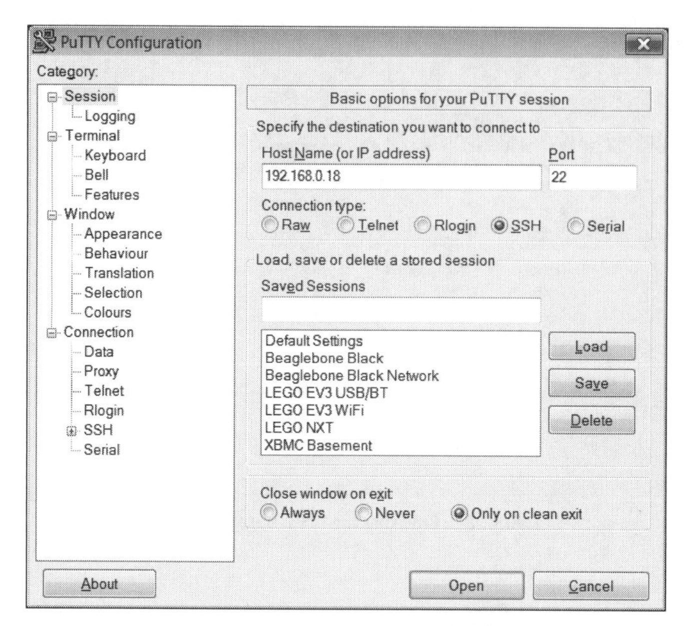

Figure 24-4: Entering the session information

3. Click Open. The first time you attempt to connect, PuTTY will ask you to save the host key. Answer yes.

4. Login as: root. When it asks for the password, hit enter.

You are now using the EV3 version of Linux at the command line (see Figure 28-5).

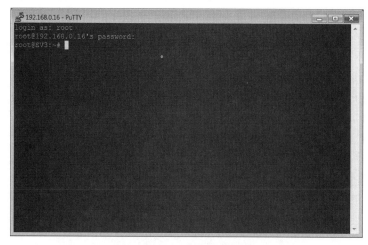

Figure 24-5: Logging into EV3 Linux

Try typing some of these commands:

```
hostname
date
ifconfig
lsusb
cd ..
ls
cd /home/root/lejos/samples
jrun -jar HelloWorld.jar
```

NOTE: *The last command to run a Java program will take a few seconds before it loads the JRE into memory. There is no friendly wait screen when launching from a terminal.*

Just for Fun!

The Linux operating system has several built in commands for doing a variety of common tasks, such as file management. The leJOS project uses BusyBox, which offers dozens of stripped down commands for embedded Linux systems like the one residing on your EV3. It includes common file management utilities such as ls, mkdir, cp, mv, and rm. It also includes tools such as vi (an editor), date, echo, tar, gzip/gunzip, uname, netstat and ping.

Editing a Text File

Editing a text file is also quite easy without a GUI tool such as WinSCP. Login to the EV3 as instructed above, then type:

```
cd /home/root/lejos
vi settings.properties
```

You should now see the contents of the leJOS properties file (see Figure 28-6).

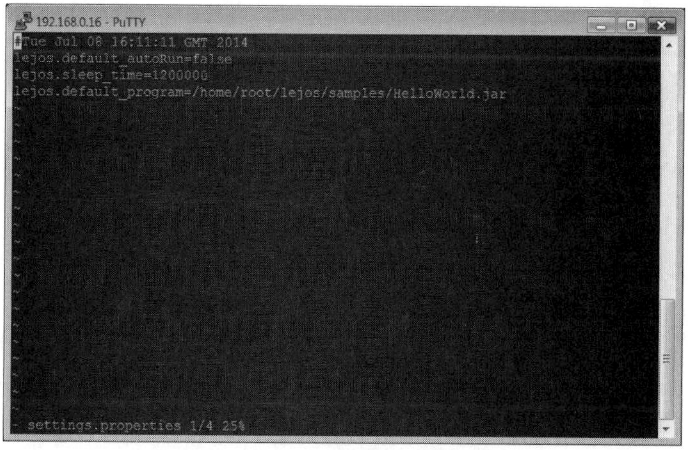

Figure 28-6: Editing in the Vi editor

One line specifies the default program to run when you select Run Default from the main menu. It might look something like this:

```
lejos.default_program=/home/lejos/programs/
RemoteControl.jar
```

To change this, hit Esc to turn on insert mode. Then move the cursor to where you want to begin editing and change the line to read the following:

```
lejos.default_program=/home/root/lejos/samples/
HelloWorld.jar
```

Now hit Esc again to exit insert mode and enter command mode. Now hit Z twice (shift z) in order to save and exit vi.

Now if you reboot the EV3 (or exit the menu and run it again as described in the Try It below) the default program becomes the HelloWorld program in the samples directory. From the leJOS EV3 menu, try running the default program to see this effect.

Try it!

You can shut down the EV3 menu in a number of ways. One way is to use the EV3 menu itself. Select the System, then choose Exit Menu. The second way is easier. Enter this command from SSH:

```
killall java
```

The LCD screen will go blank, but Linux is still running. You can restart the menu by typing the following:

```
cd /home/root/lejos/bin/utils
jrun -jar EV3Menu.jar EV3 1.2
```

The arguments above (EV3 and 1.2) are your hostname and version number of leJOS. The version number can be anything and will show up in the leJOS menu under version information. After a moment, you should see some start messages as the menu initializes.

To get your cursor back, press control-C. The menu will stop when you do this.

You can also shut down the brick by entering:

```
shutdown now -h
```

Give it a few moments for Linux to go through the shutdown sequence and the brick will turn off.

Uploading and Running Java Code

When you upload and run code on the EV3, it is actually the EV3 menu handling these transactions. In other words, the EV3 menu is always listening for events from your PC, running programs, and handling other remote commands. But did you know it's actually possible to interact with the EV3 brick without using the leJOS EV3 menu at all? Many advanced developers disable the menu entirely and upload and run code using SCP and SSH. This section will show you how you can accomplish that.

Normally you would use all text commands from a terminal in Linux. However, since most users are in Windows (and since Linux users are already familiar how to do this), I will describe the steps for Windows users using Eclipse. The following steps will allow you to upload a jar file to the EV3 and run it. You could compile your code using Java command line tools (the instructions for these are dead easy and available online), but we'll speed things up by using classes and jars that are all ready.

1. The first thing you need to know is where your Java jar files are kept. In Eclipse, open up a project you previously ran and right click on a class, then select Properties on the bottom of the list (see Figure 28-7). Note the location of your Java files.

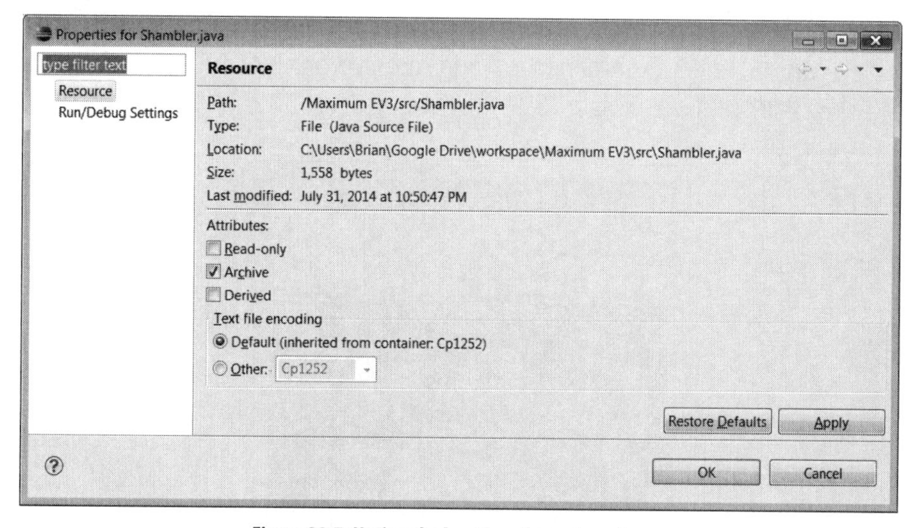

Figure 28-7: Noting the location of your Java classes

2. Now run WinSCP (or whichever SCP program you use). Follow the instructions above for how to connect to your EV3 brick.

3. On the left side of the WinSCP screen are your local files. Browse to the location of your Jar files as noted in step 1. There are in the directory above the src directory.

4. On the right side of WinSCP, browse to /home/lejos/programs (see Figure 28-8).

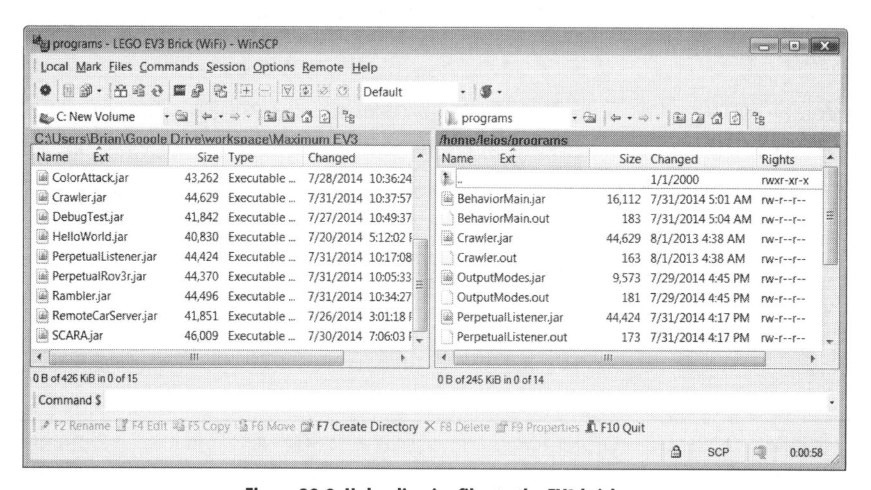

Figure 28-8: Uploading jar files to the EV3 brick

5. Now upload one of the jar files here by clicking a Jar file on the left, then click Copy (or press F5). After a moment the file will reside on the EV3.

6. In WinSCP, select Commands > Open Terminal or hit Ctrl-T. A Terminal window will open (see Figure 28-9).

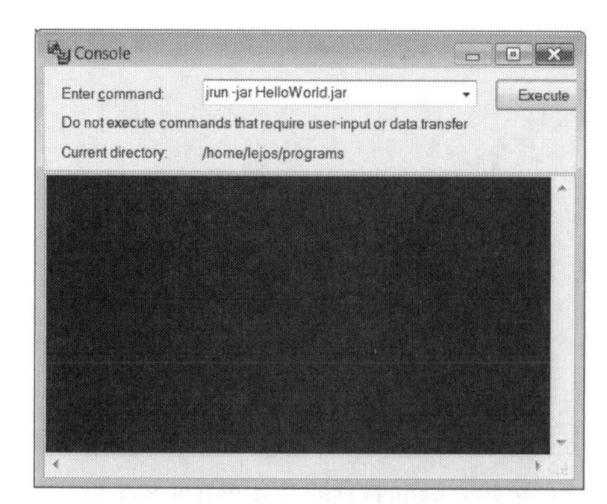

Figure 28-9: Entering commands in the console

7. Enter the following command and substitute "MyClass" with the name of the Jar file you uploaded to the EV3:

```
jrun -jar MyClass.jar
```

That's it! Hit enter or click the Execute button and your program will begin running on the EV3. To halt the program, enter killall java. To shut down the EV3 brick from a terminal, enter the following:

```
shutdown now -h
```

APPENDIX A: Robot Math

People seem to either love math or hate it. If you love robotics, you better learn to love math. This section covers some useful mathematics that can help you program your LEGO robots.

A.1 Circumference

The distance around the outside of a wheel can be important for LEGO creations, since this can be used to calculate distance traveled (see Figure A-1). Once you know the circumference, distance is measured by the number of wheel rotations multiplied by circumference. You can calculate circumference as follows:

Circumference = Pi * Diameter
The Pi value can be found in leJOS as Math.PI.

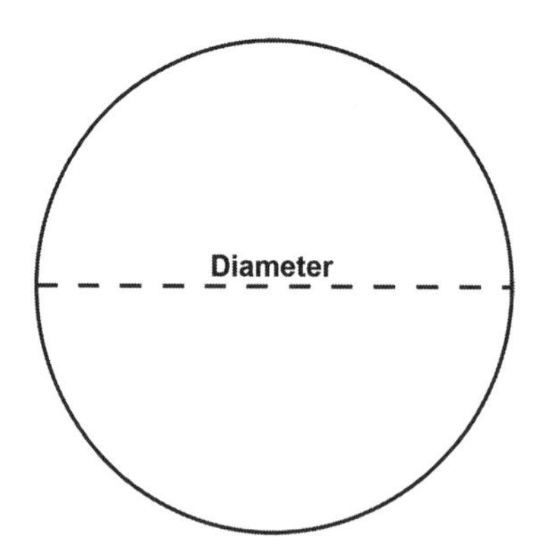

Figure A-1 Calculating circumference

A.2 Trigonometry

You are taught a lot of math in high-school, but the equations that keep coming back are the trigonometry equations. You remember: sine, cosine, and tangent. These equations go back to the time of Hipparchus, a Greek mathematician who was the first to use sine around 150 BC. Trigonometry is simply a branch of mathematics dealing with the relationships of the sides of triangles and angles.

Trigonometry is useful for calculating x,y coordinates for navigation, as well as calculating angles for arm movement in three dimensions. The Navigator classes in the leJOS API uses a lot of trigonometry to update coordinate values.

Angles can be measured in two ways. The first, most common form is degrees. A complete rotation in degrees is 360 degrees. However, engineers, scientists and mathematicians prefer using radians because the units are not arbitrarily arrived at. A complete rotation in radians is the value 2 * pi, or about 6.28 radians.

You can use whichever system you are more comfortable with. There are methods in the Math class for conversion between the two (see Math class in Chapter 20). In the coordinate system, zero degrees (or zero radians) always runs along the x axis, and positive rotation occurs counter clockwise (Figure A-2). Thus, when the robot is pointed north (along the Y axis) it is at 90 degrees.

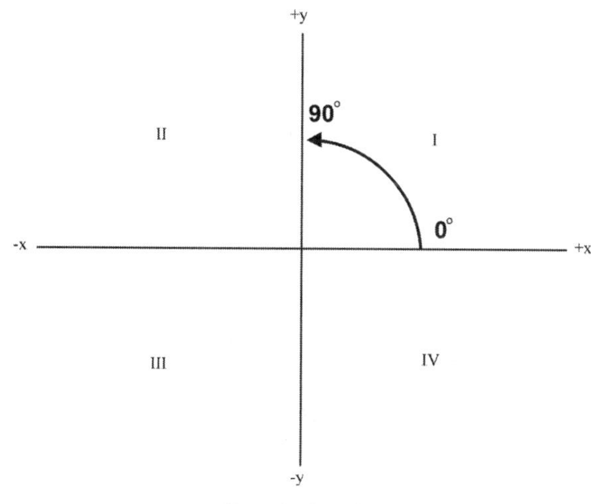

Figure A-2 Rotation

Let's examine the problem of navigation. Every time a robot moves a distance, the x and y coordinates will also change. For example, if the robot rotates 60 degrees to the left and travels twenty centimeters, both the x and y values will increase (Figure A-3). As you can see, this movement creates a triangle with a right angle. In trigonometry, when there is a right angled triangle with all angles known, and the length of one side is known, it is possible to calculate the lengths of all sides.

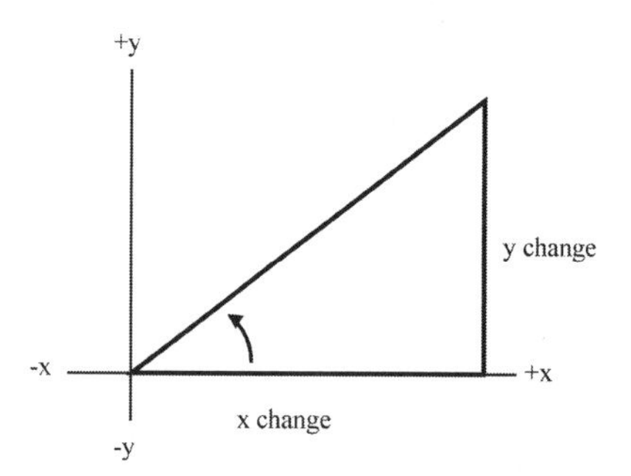

Figure A-3 Right angle trigonometry.

In trigonometry, the side of the triangle opposite the right angle is called the hypotenuse, the side opposite the angle in the calculation is the opposite, and the remaining side is the adjacent (Figure A-4). In order to calculate the new location we will need to calculate x (the adjacent) and y (the opposite). To do this, we will need to use calculus.

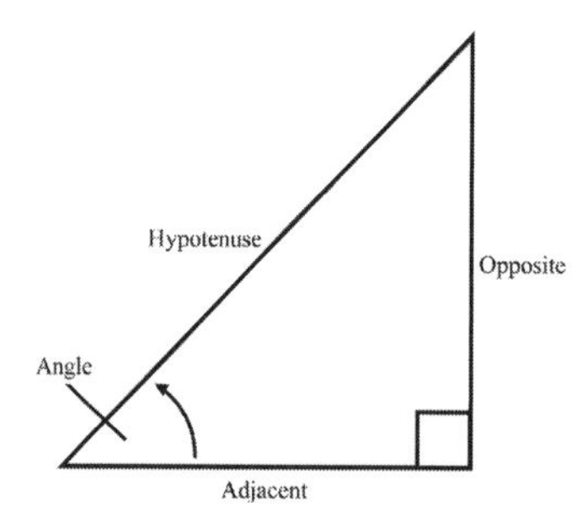

Figure A-4 The three sides of a right angle triangle.

In high school calculus you probably learned tangent, cosine, and sine (tan, cos, and sin). These functions are used to solve for the lengths of sides on a triangle. Many people use the mnemonic SOH, CAH, TOA (say it like a tribal chant) to remember the following equations:

Sin(angle) = Opposite/Hypotenuse
Cos(angle) = Adjacent/Hypotenuse
Tan(angle) = Opposite/Adjacent

We only need to know the opposite and adjacent, so only the first two equations are useful to us. Let's replace the technical terms with variables and rearrange the equations to make things simpler. The distance traveled by the robot, the hypotenuse, will be replaced by distance:

x = cos(angle) * distance
y = sin(angle) * distance

We can now use these equations to figure out the x and y coordinates after a robot has moved a distance across the floor. Let's imagine the robot has started at 0,0 and rotates positive 70 degrees (counter clockwise), then moves 25 centimeters (Figure A-5). In order to find new coordinates simply plug our values into the equations above to get:

x = cos(70) * 25
y = sin(70) * 25

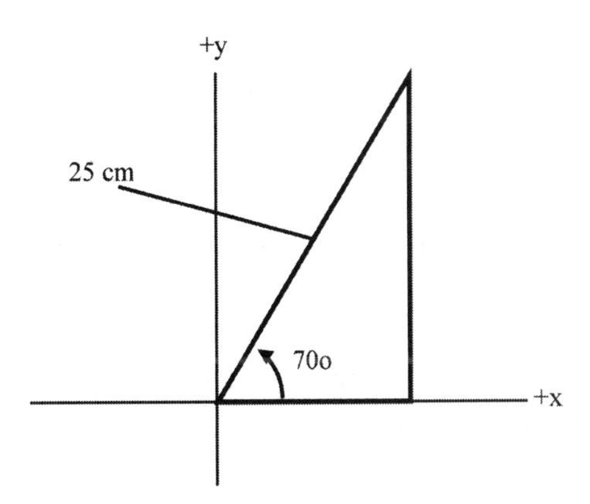

Figure A-5 Rotating 70 degrees and traveling 25 cm.

If you are using a calculator, make sure it is in degrees (DEG mode) and not radians (RAD mode). So the new x and y coordinates are 8.55 and 23.49. In Java, it is very easy to calculate these using Math class:

```
double x = Math.cos(Math.toRadians(70)) * 25;
double y = Math.sin(Math.toRadians(70)) * 25;
```

NOTE: *All methods in the java.lang.Math class use radians for angles. There are two methods available to convert back and forth between degrees to radians: Math.toRadians() and Math.toDegrees().*

A.3 Inverse Trigonometry Functions

In the preceding section we knew distances and angles, and used those to calculate x and y coordinates. However, sometimes we know the x and y coordinates and must calculate the angle. This is the problem faced when programming robot arms (see Chapter 19).

In this case we use inverse functions. Take a look at Figure A-4 again. If we know the length of each side of a triangle (x, y coordinates) then we can calculate the angles by rearrange the equations slightly:

angle = asin(Opposite/Hypotenuse)
angle = acos(Adjacent/Hypotenuse)
angle = atan(Opposite/Adjacent)

Asin, acos and atan are merely words for inverse sine, inverse cosine, and inverse tangent. In Figure A-5, pretend we know the x, y coordinates but don't know the angle indicated. We can use atan by entering the x and y values:

Angle = atan(y/x)
Angle = atan(23.49/8.55)
Angle = 69.99 (rounded to 70)

There is one problem with atan – it can't calculate angles greater than 90 degrees. However, using Math.atan2(y, x) you can calculate this accurately as atan2() can produce any angle between 0 and 360 degrees (or more accurately, between 0 and 2pi).

A.4 Law of Cosines

Sometimes you don't have a right angled triangle. In this case, as long as you know the lengths of all three sides of a triangle, you can still figure out the angles. In the arm project in chapter 19 we know the lengths of two parts of the arm (two sides) and we can calculate the distance between the ends of each arm. This means we can calculate the inner angle (see Figure A-6).

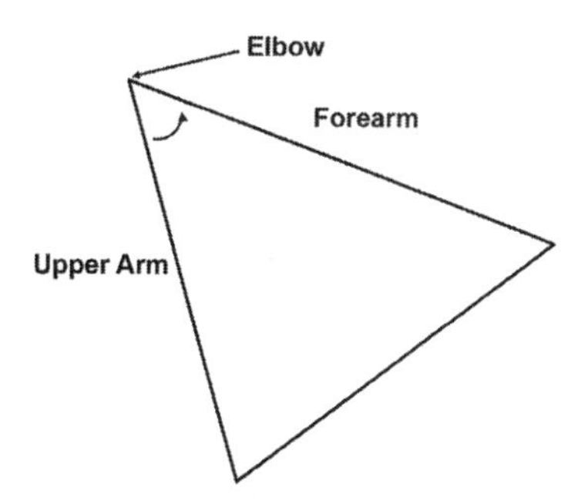

Figure A-6 Robot arm triangle.

The law of cosines applies to any triangle in which the lengths of all three sides are known (see Figure A-7). The law is as follows:

$c^2 = a^2 + b^2 - 2ab\cos A$

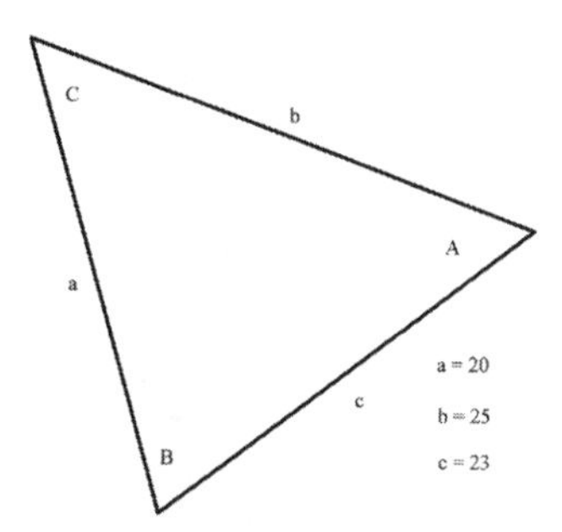

Figure A-7 Law of cosines.

In this form the equation isn't useful if we want to calculate angle A, so we rearrange as follows:

$$A = \text{acos}\left(\frac{a^2 + b^2 - c^2}{2ab}\right)$$

Now we merely plug in the values for a, b and c and we can determine the angle:

$$A = \text{acos}\left(\frac{20^2 + 25^2 - 23^2}{2 \text{ x } 20 \text{ x } 25}\right)$$

A = acos 0.496
A = 60.26 degrees

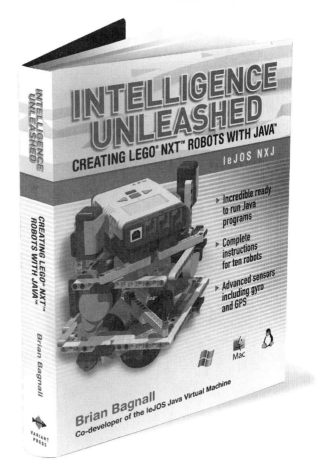